AQA

Biol

for GC

Keith Hirst
Mike Hiscock

Series Editor: Ke

www.heinemann.co.uk
✓ Free online support
✓ Useful weblinks
✓ 24 hour online ordering

01865 888058

Heinemann Educational Publishers
Halley Court, Jordan Hill, Oxford OX2 8EJ
Part of Harcourt Education

Heinemann is the registered trademark of Harcourt Education Limited

First published 2006

10 09 08 07 06
10 9 8 7 6 5 4 3 2 1

10-digit ISBN: 0 435 586041
13-digit ISBN: 978 0 435 586041

Edited by Patrick Bonham, Linda Moore and Anne Trevillion
Designed by Lorraine Inglis
Typeset by Ken Vail Graphic Design

Original illustrations © Harcourt Education Limited, 2006

Illustrated by Beehive Illustration (Martin Sanders, Mark Turner), Nick Hawken, NB Illustration (Ben Hasler, Ruth Thomlevold), Sylvie Poggio Artists Agency (Rory Walker).

Printed by CPI Bath

Cover photo: © Science Photo Library

Index compiled by Indexing Specialists (UK) Ltd

Picture research by Zooid Pictures Ltd

Acknowledgements
The authors and publisher would like to thank the following individuals and organisations for permission to reproduce photographs:

Keith/Custom Medical Stock Photo/Science Photo Library p iv R; Peter Gould/Harcourt Education p iv L; Corbis UK Ltd. p 2 TR, TL; Trl Ltd./Science Photo Library p 2 B; Bsip/Photolibrary.com p 3 TR; Richard Young/Rex Features/Rex Features p 3 TL; Burke/Triolo Productions/Foodpix/ Photolibrary.com p 3 B; Jack Sullivan/Alamy p 4 T; Mike Egerton/Empics p 4 B; Robert Harding Picture Library Ltd/Alamy p 6; Saturn Stills/Science Photo Library p 9; Liba Taylor/Corbis UK Ltd. p 14; Empics Sports Photo Agency/Empics p 20; Ian Hooton/Science Photo Library p 21; GettyImages/PhotoDisc p 22; Michael Donne/Science Photo Library p 23; Nick Sinclair/Science Photo Library p 25; Plainpicture/Photolibrary.com p 28; Zooid Pictures p 32 R; Cnri/Science Photo Library p 32 L; Ian Hooton/Science Photo Library p 34; Rex Features p 36; Tortel Sygma/ Corbis UK Ltd. p 42 T; Keren Su/Corbis UK Ltd. p 42 B; C. N. R. I./Phototake Inc/Photolibrary. com p 43; Niall Benvie/Photolibrary.com p 44 TR; Brandon D. Cole/Corbis UK Ltd. p 44 TL; Index Stock Imagery/Photolibrary.com p 44 B; Ifa-Bilderteam Gmbh/Photolibrary.com p 45; Doug Allan/Photolibrary.com p 46 L; Jonathan Blair/Corbis UK Ltd. p 46 R; David Tipling/Photolibrary. com p 47 TR; Frank Lane Picture Agency/Corbis UK Ltd. p 47 BR; Corbis p 47 L; Getty Images/ PhotoDisc p 48 TR; Bildhuset Ab/Photolibrary.com p 48 L; Phototake Inc/Photolibrary.com p 48 BR; Terry W. Eggers/Corbis UK Ltd. p 49 TR; Martin B. Withers; Frank Lane Picture Agency/Corbis UK Ltd. p 49 M; Ken Wilson; Papilio/Corbis UK Ltd. p 49 L; Botanica/Photolibrary.com p 49 BR; Nigel Cattlin/Alamy p 50; Densey Clyne/Photolibrary.com p 51; DR Jeremy Burgess/Science Photo Library p 53; David M Dennis/Photolibrary.com p 54; Holt Studios International Ltd/Alamy p 56; GettyImages/PhotoDisc p 57; David Hoffman Photo Library/Alamy p 58 L; Nick Cobbing/ David Hoffman Photo Library/Alamy p 58 R; Phototake Inc/Photolibrary.com p 59; Corbis p 60; David Fox/Photolibrary p 63; John Reader/Science Photo Library p 65; Getty Images/Photodisc p 66 L; NASA Goddard Space Flight Center (NASA-GSFC) p 66 R; Dobson agency/Rex Features p 67 L; Danny Lehman/Corbis UK Ltd. p 67 R; W.T. Sullivan III & Hansen Planetarium/Science Photo Library p 68 T; Photodisk/Photolink/Harcourt Education p 68 B; Getty Images/Photodisc/ pp 69 T & B; Peter Adams/Index Stock Imagery/Photolibrary.com p 71; GettyImages/PhotoDisc p 72; Corbis p 73; Simonpietri Christian Sygma/Corbis UK Ltd. p 74; Reuters/Corbis UK Ltd. p 74; Victor De Schwanberg/Science Photo Library p 75; Harcourt Education p 76; Alamy p 82 T; J. C. Revy/Science Photo Library p 82 M; Bsip/Science Photo Library p 82 B; Stonehill/Zefa/Corbis UK Ltd. p 83; Bill Longcore/Science Photo Library p 85 ; Herve Conge, Ism/Science Photo Library p 86 T; Andrew Syred/Science Photo Library p 86 BL; Biophoto Associates/Science Photo Library p 86 BR; Cnri/Science Photo Library p 87 T; Dr Gary Gaugler/Science Photo Library p 87 BL; Phototake Inc/Photolibrary p 87 BM; Science Photo Library p 87 BR; Andrew Lambert Photography/Science Photo Library p 88 T; Andrew Lambert Photography Science Photo/Science Photo Library p 88 M; Andrew McClenaghan/Science Photo Library p 88 B; Prof. P. Motta Dept. Of Anatomy University "La Sapienza", Rome/Science Photo Library p 89; Isifa Image Service s.r.o/ Alamy p 90 T; Cordelia Molloy/Science Photo Library p 90 B; Corbis/ p 92 ; Getty Images/20067/ p 94; Martyn F. Chillmaid/Science Photo Library p 95 T; Holt Studios/Frank Lane Picture Agency p 95 M; Nigel Cattlin/Frank Lane Picture Agency p 95 B; Jim Winkley/Ecoscene/Corbis UK Ltd. p 96 T; Photonica/Getty Images p 96 M; Peter Frischmuth/argus/Still Pictures p 96 B; Ron Giling/ Still Pictures p 97; Draper Tools Limited p 98 T; Malcolm Fielding, The BOC Group Plc /Science Photo Library p 18 BR; Worcester-Bosch Group p 98 BL; The Cover Story/Corbis UK Ltd. p 101 R; Jim Richardson/Corbis UK Ltd. p 101 L; Jack Fields/Corbis UK Ltd. p 104; Corbis p 108 L; H. Schmied/Zefa/Corbis UK Ltd. p 108 M; Digital Vision p 108 R; David T. Grewcock; Frank Lane Picture Agency/Corbis UK Ltd. p 109; Tudor Photography/Harcourt Education p 110 B; Picture Partners/Alamy p 110 T; Adrian Arbib/Corbis UK Ltd. p 111; Getty Images/29203 p 112 T; Getty Images/31104 p 112 B; WoodyStock/Alamy p 115; Holt Studios/Frank Lane Picture Agency p 116; Bettmann/Corbis UK Ltd. p 118; Cordelia Molloy/Science Photo Library p 119 T; Ashley Cooper/Corbis UK Ltd. p 119 B; Martyn F. Chillmaid/Science Photo Library p 124; Maximilian Stock Ltd/Science Photo Library p 126; Zooid Pictures p 48 T; Brand X Pictures p 128 B; David Nunuk/Science Photo Library p 130 T; Martyn F. Chillmaid p 130 BL; Tony Marshall/Empics p 130 BR; Steve Gschmeissner/Science Photo Library p 131; Getty Images/125087 p 132 T; Getty Images/27053 p 132 B; Mike Egerton/Empics p 133; Ted Kinsman/Science Photo Library p 134 T; Martyn F. Chillmaid/Science Photo Library p 134 B; Boulay/Bsip/Photolibrary p 136; Bettmann/ Corbis UK Ltd. p 137; Michael Donne/Science Photo Library p 138; Corbis p 139; Dung Vo Trung Sygma/Corbis UK Ltd. p 140; Lwa-Dann Tardif/Corbis UK Ltd. p 141; Andrew Syred/Science Photo Library p 142; Hattie Young/Science Photo Library p 148; Andrew Crowley/Daily Telegraph p 152 T; Phototake Inc./Alamy p 152 B; Ken Dumminger Sygma/Corbis UK Ltd. p 153; Martin Dohrn/ Science Photo Library p 162 T; Eye Of Science/Science Photo Library p 162 M; Mauritius Die Bildagentur Gmbh /Photolibrary p 162 B; Chris Fredriksson/Alamy p 163 L; Ken Wagner/Phototake Inc/Alamy p 163 R; Jason Burns/Dr Ryder/Phototake Inc./Alamy p 167; C. Lyttle/Zefa/Corbis UK Ltd. p 169; Phototake Inc/Alamy p 170; Nigel Cattlin/Holt Studios International Ltd/Alamy p 172; Troy and Mary Parlee/Alamy p 174 plainpicture/Wolff, M./Alamy p 175 T; Bill Bachman/Alamy p 175 B; Sears Wiebkin/zefa//Corbis UK Ltd. p 178 T; Dan Chung / Reuters/Action Images p 178 B; Samuel Ashfield/Science Photo Library p 180; Associated Press/Empics p 182; Nordicphotos/ Alamy p 183; Helen King//Corbis UK Ltd. p 184; Chris Priest/Science Photo Library p 191; Harcourt Education p 192; Tina Manley/Alamy p 193; Corbis/ p 194 Ta; p 194 Tb; Comstock Images/ p 194 Tc; John A. Rizzo/ p 194 Td; AKG – Images p 194 M; Qcumber/Alamy p 194 B; Bettmann/Corbis UK Ltd. p 195; Maximilian Weinzierl/Alamy p 196 TL; Dennis Kunkel/Phototake Inc./Alamy p 196 TR; Samuel Ashfield/Science Photo Library p 197; Martyn F. Chillmaid p 198; Getty Images/19216/ p 200 L; Nigel Blythe/Cephas Picture Library p 200 R; Cephas Picture Library/Alamy p 201 T; Ian Shaw/Cephas Picture Library p 201 B; Eye Ubiquitous/Hutchison p 202; Phototake Inc./Alamy p 203; John Durham /Science Photo Library p 206 T; Phototake Inc./Alamy p 206 B; Marlow Foods Ltd p 208 T; Marlow Foods Ltd p 208 B; Owen Franken//Corbis UK Ltd. p 212 T; Jim Holmes/Panos Pictures p 212 B; National BioDiesel Board p 213; Solar Engineering Services p 215 T; Solar Engineering Services p 215 B; Digital Vision/ p 218 T; Nigel Cattlin/Frank Lane Picture Agency p 218 B; David Wall/Alamy p 220 TL; Ricardo Funari / BrazilPhotos/Alamy p 220 TM; Richard Wareham Fotografie/Alamy p 220 TR; Nic Hamilton/Alamy p 220 B; Martin Bond/Science Photo Library p 223.

The authors and publisher would like to thank the following individuals and organisations for permission to reproduce copyright material:

British Heart Foundation Coronary heart disease statistics, p 13 M, p 15 T, p17 M; Smokenders Program, p 24 B; Pittet D. et al *Effectiveness of a hospital-wide programme to improve compliance with hand hygiene* The Lancet 356 (9238), p 37 B; Blackburn T. and Hawkins B. *Bergmann's rule and the mammal fauna of northern North America* Ecography 27 (6) Blackwell Publishing, p 46 T; NASA Earth Observatory, p 72 T; *Gardening Which?*, p95; Food and Agriculture Organisation of the United Nations, Fisheries Technical Paper T407 – Integrated agriculture-aquaculture A primer, p 114 T, M; © 2001, The Washington Post, p 157; Assessment and Qualifications Alliance (AQA), p 160, p224; Eurotransplant, p 192. All Crown Copyright material is reproduced with the permission of HMSO and the Queen's Printer for Scotland.

Every effort has been made to contact copyright holders of material reproduced in this book. Any omissions will be rectified in subsequent printings if notice is given to the publishers.

Tel: 01865 888058 www.heinemann.co.uk

How to use this book

This book has been designed to cover the new AQA GCSE Science curriculum in an exciting and engaging manner and is divided into four units: B1a, B1b, B2 and B3.

The book starts with two double page spreads focusing on 'How science works', which show you how scientists investigate scientific issues, including those in our everyday lives.

Each unit in the book is broken down into separate sections, for example unit B1a consists of sections 1 and 2. Each section is introduced by a double page introductory spread which raises questions about what is covered in the section, acts as an introduction to the section, and includes a box encouraging you to think about what you are going to learn in the section.

The introductory spread is followed by double page content spreads which cover what you need to learn, but which also cover 'How science works' and the procedures you need to be familiar with to enable you to produce your internally assessed work: the Practical Skills Assessment and the Investigative Skills Assignment.

Each section includes at least one 'ideas, evidence and issues' spread which either focuses on interpreting data and evidence or on evaluating the role of science in society and the issues that affect all our lives.

Throughout the content pages there are in-text questions to test your understanding of what you have just learnt and to further your appreciation of how science can be used and what the issues are surrounding the development of science and technology.

At the end of each unit there are two double page spreads of questions to test your knowledge and understanding of the module. They will also prepare you for the kinds of questions you will meet in exams.

The words displayed in bold in the text also appear in the glossary at the end of the book, together with a definition.

Contents

B3

How science works

▲ Not all scientists work in laboratories.

Why study science?

Even if you are not going to become a scientist, it is important that you know how scientists work. Science affects almost every aspect of our lives – sometimes in very obvious ways, like the development of new technology or new drugs that can be used to treat us, and sometimes in less obvious ways, like additives in our food that we may be unaware of.

We need to know what scientists are up to, so that they cannot 'pull the wool over our eyes' or 'blind us with science'. We need to be able to understand the way in which scientific experiments are carried out and the way in which information is collected. We need to be able to tell the difference between facts and opinions and to judge whether the information, or the people providing it, are biased in any way. Then we can make our own judgements based on an understanding of the facts.

Scientists frequently use a number of technical terms, and often these have meanings that are slightly different from the meanings in everyday speech. These terms have been printed in **bold** and are explained in the text and in the glossary at the end of the book.

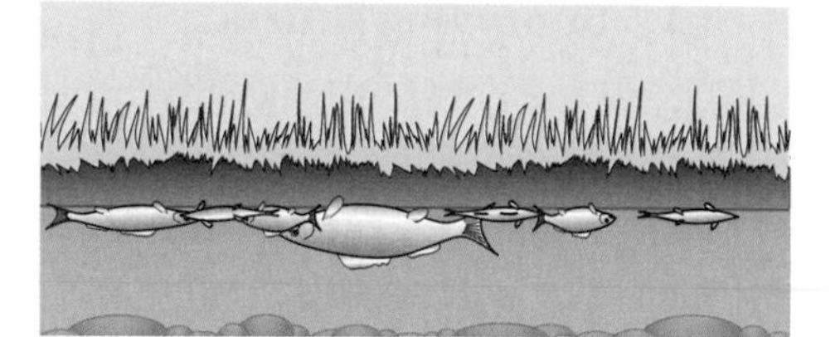

▲ Observing the fish dying in this river started a local investigation.

General principles

Many scientific investigations start with observations. These need to be made carefully and should be unbiased. They are often the basis for investigations or classification of things.

One of the skills that you need to learn is the ability to distinguish between scientific fact and opinion.

Local residents noticed that large numbers of fish seemed to be dying in the river. The river flows past several factories but then flows through farmland. They suspected a detergent factory was the cause of the pollution.

▲ This factory discharges waste into the river.

River pollution blamed on detergents factory

Local residents have become alarmed at the number of fish dying in the River Sabrina, and have blamed the detergents factory of Lees & Co. Local farmer Fred Guest, who has lived by the river for 30 years, said, 'I've never seen anything like it. I know it's caused by Lees because they use chemicals that contain cyanide, and I've seen them discharging waste into the river.'
A spokesperson at Lees & Co. said, 'It's nothing to do with us. The cause of the problem is all the nitrates that local farmers are putting on the land in fertilisers. They get washed down into the river. There was never any problem until recently – they used to use natural manure.'

Question

a *Are the statements made by*
(i) Farmer Fred Guest, and
(ii) the spokesperson for Lees & Co.
giving facts or opinions? Give a reason for your answer.

As a result of protests, a team of scientists was called in to carry out an investigation. They decided to sample the water in the river at different places.

Question

b *What do you think that the scientists should be measuring in the water?*

▲ Map showing section of the river and four sampling points.

Designing an investigation

When you are designing an investigation you need to plan carefully to make sure that you are measuring the correct **variables** at suitable **intervals** (length of time between measurements) and over a suitable **range**. You need to be ready to measure values between the upper and lower ones that will occur, for example daytime temperature.

The **independent variable** is the one that you deliberately alter. The **dependent variable** is the one that changes as a result of this. Other variables may also affect the outcome, and these need to be controlled or monitored. These are called **control variables**.

You also need to think about whether you need to repeat any of the measurements to improve **reliability**. A value becomes more reliable the more times it is measured, for example taking a reading five times and working out an average.

It is important your results and conclusions are **valid**. Valid results are ones that answer the original question. A valid result can be matched by other scientists following your method. They should get the same result, which validates your result.

The scientists sampled the water at four different points.

The scientists also needed to choose their measuring instruments carefully. In any investigation, it is important to make measurements using the most suitable instruments. Different instruments measure to different levels of **precision**. A tape measure measures to the nearest 1 centimetre and a ruler to the nearest 1 millimetre. You also need to use instruments effectively to get **accurate** results. An accurate reading is one that is nearest to the true result.

Questions

c *For each letter on the map, A, B, C and D, say why you think that the scientists chose this particular point.*

d *What variables do you think that the scientists should have controlled or monitored to make sure that it was a fair test?*

Looking at the evidence

Presenting the data

The scientists presented their **data** in a table.

The table only shows the average of the measurements. It is often important to see all of the original data in order to see whether there was a wide variation and whether there were any **anomalous** results.

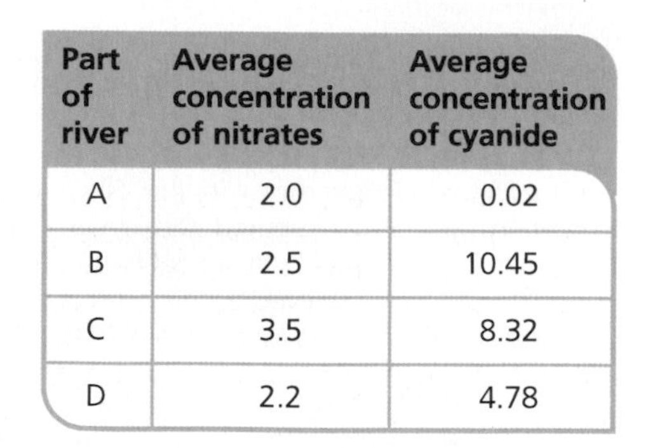

Part of river	Average concentration of nitrates	Average concentration of cyanide
A	2.0	0.02
B	2.5	10.45
C	3.5	8.32
D	2.2	4.78

Questions

a *In this investigation, what was (i) the independent variable, and (ii) the dependent variable?*

b *What is meant by the term 'anomalous results'?*

c *What should be done with anomalous results before calculating the average?*

d *What other important information is missing from the table of results?*

Displaying the data

Often it is useful to display the data in a graphical form. There are several ways of doing this, and you need to know which is the best one to use. If you choose the wrong one, it can be very unhelpful and at worst misleading.

Three of the most common methods are shown below.

A **pie chart** is often used when you want to show the percentage or fractional contribution of several parts to the whole.

A **bar chart** is often used when you are dealing with variables that are **discrete**, **ordered** or **categoric.**

A **line graph** is often used when you are dealing with variables that are **continuous**.

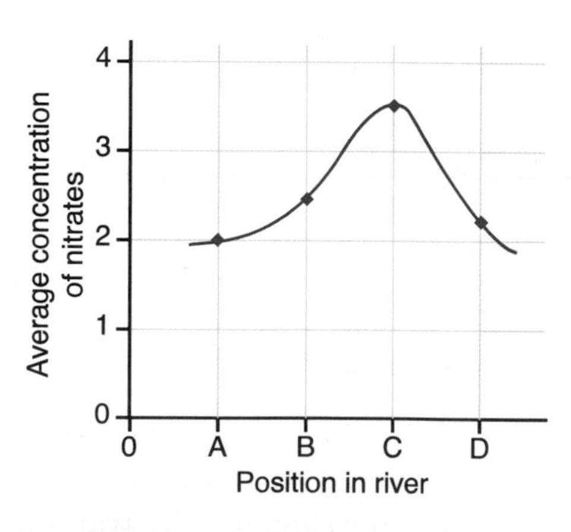

Question

e *Which of the three methods of displaying data in graphical form do you think the scientists should have used in this case? Explain your answer.*

A **discrete variable** is a variable whose value is restricted to whole numbers.

An **ordered variable** is a variable that can be ranked, e.g. first, second, third, etc.

A **categoric variable** is a variable that can be described by labels, for example by colour or size.

A **continuous variable** is one that can have any numerical value, for example time, speed or length.

Analysing results and drawing conclusions

The scientists issued the following statement:

```
Although both pollutants are present in above average quantities,
we cannot yet determine which, if either of them, is responsible
for killing the fish. This would require more data.
```

If it were clear that cyanide or nitrates caused the fish to die there would be a **causal** link between one variable causing a change in the second.

Science and society

The managers of the detergent factory looked at the data and made the following statement: 'This has proved that it is not our factory that has caused the problem: it is the nitrates in the water. There is a significant increase in nitrate levels just after the river has passed the farmland.'

Farmer Fred Guest's response to this was: 'Rubbish! The increase in cyanide is far greater than the increase in nitrates. It is the factory's pollution that is killing the fish.'

Sometimes the conclusions that people come to may be influenced by other things and not just the scientific evidence. For example, for many years tobacco companies were accused of misleading the public about the effects of smoking. They might have wanted to do this so that they did not lose sales of their products.

Question

i *Can you think of any reasons why the company operating the detergent factory might have been biased in coming to their conclusion?*

Your work as a scientist

These are the sorts of problems and questions that you will have to deal with in your science course. In your practical work you will need to think carefully about all of these points:

- observing
- designing an investigation
- making measurements
- presenting data
- looking for patterns and relationships
- coming to conclusions
- considering the relationship between science and society.

Finally, remember that sometimes it is difficult to collect enough evidence to be able to answer a question properly. There are also questions that cannot be answered by science alone, but need moral or social judgements to be made.

Question

f *Which of the following additional data do you think would be most useful to the scientists? Give a reason for your answer.*

- *A chemical analysis of the dead fish*
- *Sampling at different points in the river*
- *More samples at the same points in the river*
- *More samples over a longer period of time*

Questions

g *Do you think that either of these claims is reasonable? Explain your answer.*

h *Suggest one way in which it might be possible to prove that one of the claims is correct.*

B1a 1 Body conditions and health

The human body can perform amazing feats.

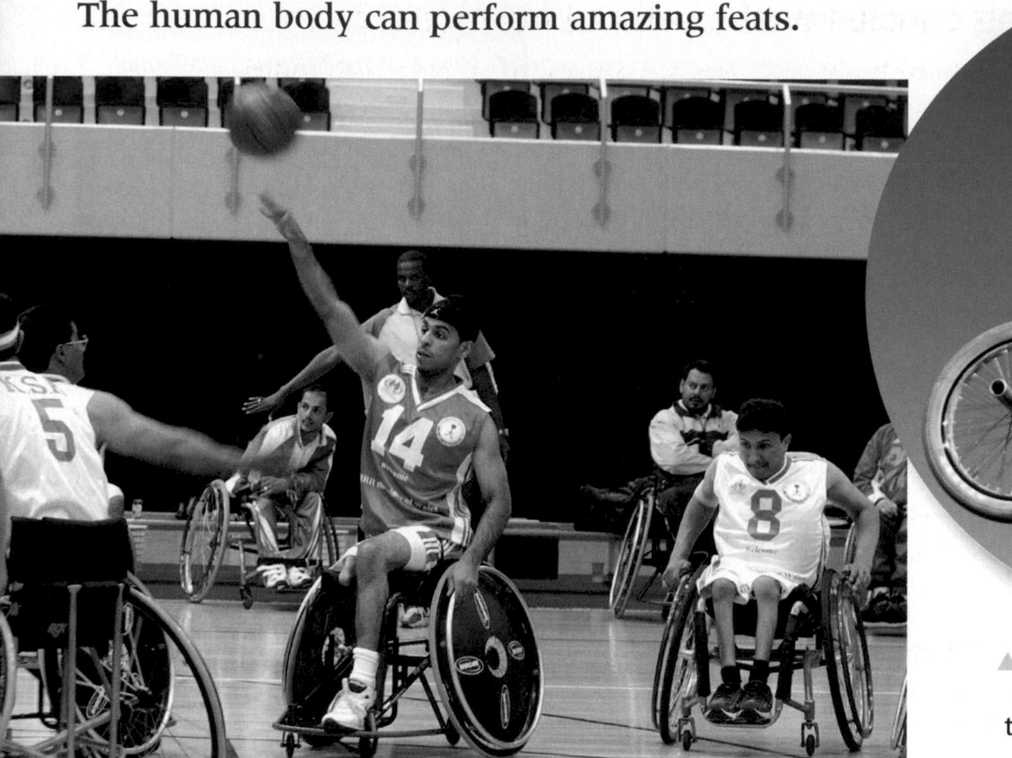

These players are very skilful. The actions of their arms and upper bodies are perfectly coordinated.

To do this, dozens of muscles in the rider's body must contract in exactly the right sequence. These muscles are coordinated by the nervous system.

Passing on information

Injuries have damaged the spinal cords of these athletes. The spinal cord acts like a telephone cable – it carries information. Damage to the spinal cord, which connects the legs to the brain, means that information in the legs can't reach the brain – the person has no feeling in his legs. It also means that information can't get from the brain to the leg muscles, so the legs are paralysed.

Our nervous system allows us to collect information about the world around us. Some people are born with **sense organs** that do not work correctly. The other senses may become much keener to compensate. For example, many blind people are able to judge the direction and distance of a sound source much better than a sighted person.

The dog is this blind man's eyes.

Where would we be without hormones?

Many of our body processes are controlled by chemicals called **hormones.** These are produced by organs called **glands**. The hormones pass from glands into the bloodstream, which transports them around the body. Each hormone affects one or more organs, known as the target organs.

How different would our lives be if scientists had not discovered how hormones worked?

Without insulin treatment, people with type 1 diabetes would die soon after birth.

Patrick Steptoe – the pioneer of IVF. Without his work thousands of 'test-tube' babies would not have been born.

Making decisions

Every day you make decisions that affect your health. For example, choosing what to eat and drink is a decision you make several times each day. These decisions may be influenced by what you read in newspapers and magazines and see on TV.

Good food, bad food

A recent survey monitored more than 200 TV adverts and found that 95% of the food products advertised during children's programmes contained high levels of sugar, fat or salt. Junk foods are also advertised using interactive websites and toys or games. This advertising encourages young children to eat more and more unhealthy foods.

A parents' campaign group is trying to get a ban on food advertising aimed at children. They say, 'A car can't run on bad petrol, and our kids can't run on bad food'. Recently PepsiCo, the company that makes products such as Pepsi and Walkers crisps, announced it is to restrict advertising to children.

Many people say they need more advice on how to find out if a food is healthy. Information on food labels helps people to compare foods and make healthy choices. However, a recent survey of 70 food products found that the information on some labels was inaccurate. One 'kid's pizza' contained 47% more sugar than stated on the label.

Think about what you will find out in this section

How does the nervous system help us to respond to external and internal changes?	What are the pros and cons of using hormones to control human fertility?
How can we evaluate the effects of food and exercise on health?	Can we trust the information in food advertisements and on labels?
How is the female menstrual cycle controlled?	

B1a 1.1 Keeping a balance

When we exercise

During a marathon, runners top up with 'sports drinks' several times. Why do they need to do this?

The graph compares the rate of heat production and the body temperature of a marathon runner during a race with those of the same athlete at rest.

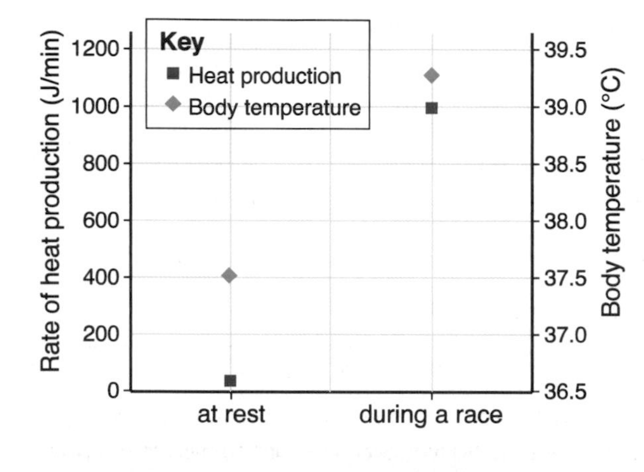

Question

a *(i) By how much does body temperature rise during a marathon?*
(ii) Calculate the percentage increase in heat production by a marathon runner during a race.

Refuelling during a marathon.

The marathon runner's body is cooled by sweating. For every 2.5 MJ of heat produced by the muscles, about 1 litre of sweat is evaporated. A marathon runner loses about 5 litres of sweat during a race.

Athletes drink sports drinks to replace the fluid they lose as sweat when they are running. If you have ever tasted sweat you will know that it is a bit salty. The saltiness is due to ions – salt consists of sodium ions and chloride ions.

Replacing salt and sugar

When we sweat we lose a lot of water, but not quite as many **ions**. This leaves us with more ions in our blood than normal. If the balance of ions and water changes in our bodies, cells do not work so well. Sports drinks help to replace both the water and the ions.

Sports drinks also contain glucose. This helps to top up the athlete's blood sugar levels during the marathon. To work properly, body cells need a constant supply of glucose for their energy needs. This glucose is supplied by the blood.

Sales of sports drinks are rocketing.

Question

b *Why does the athlete's blood sugar level fall during the race?*

Posters for a sports drink say that it is the 'water designed for exercise'. The eye-catching television advert for this drink shows an athlete made of water running, doing cartwheels and back flips, diving into a large pool and swimming away. A voice says, 'Imagine water redesigned for exercise and for better hydration than water alone.'

The drink contains 2 g carbohydrate and 35 mg sodium per 100 ml – and provides 10 calories. The ingredients are: water, glucose syrup, citric acid, acidity regulators, flavouring, sweeteners and vitamins.

Questions

c *Are the makers of this sports drink justified in calling it 'water redesigned for exercise'? Explain the reasons for your answer.*

d *Explain how this drink might help someone who is playing sport.*

Why is our body temperature kept at 37 °C?

Enzymes control most of the chemical reactions in our bodies. Enzymes work best at a particular temperature. When we are healthy our internal body temperature is 37 °C. This is the temperature at which the enzymes found in human cells work best.

The graph shows how temperature affects the rate of a reaction involving an enzyme.

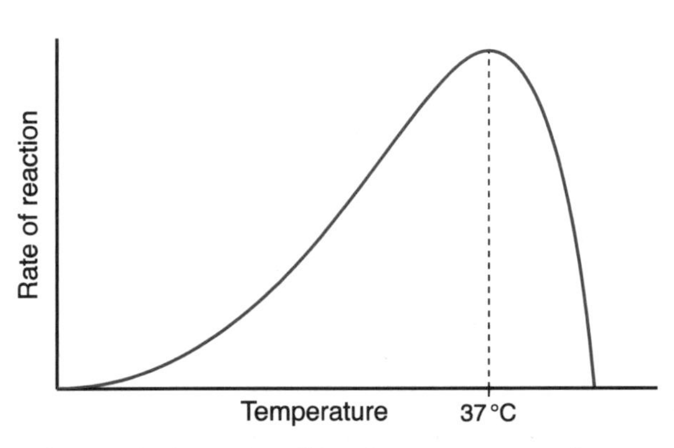

▲ This graph shows the effect of temperature on the rate of an enzyme-controlled reaction. If our bodies were cooler than 37 °C, the chemical reactions in our cells would be much slower. If we heat enzymes above 45 °C their structure changes and they stop working.

Balancing the water budget

To stay healthy the body needs to balance the gain and loss of both water and ions. Besides losing water when we sweat, we lose water in the air we breathe out. This is why a mirror becomes misty if we breathe on it.

The kidneys control the balance of water and ions in the body. They do this by producing fluid called urine. Urine contains the excess salts and water that the body does not need. It also contains other waste materials.

The amount of water entering the body should balance the amount of water leaving the body.

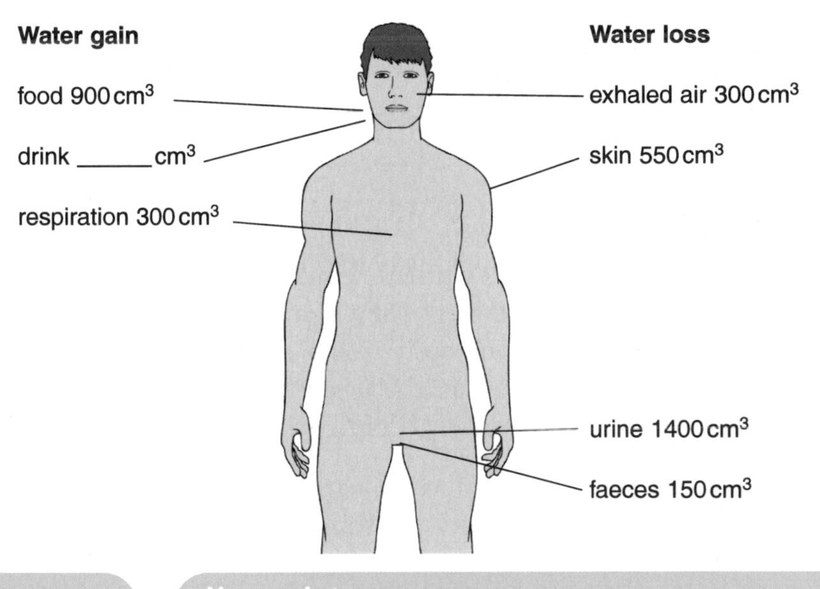

Question

e *(i) How much water did this person drink in one day?*
(ii) What proportion of water loss was via the skin?
(iii) A person suffering from severe dehydration continues to produce urine, making the body even more dehydrated. Suggest why the body continues to produce urine under these conditions.

Key point

- It is important to keep the internal conditions of the body under control. These include:
  - water content
  - blood sugar levels
  - temperature.

Alarm systems

Detectors and sensors

Most houses have fire alarms that contain a smoke detector. Detectors are sensitive to changes in the environment – in this case smoke. Our bodies have detectors to alert us to internal and external changes. These sense organs have specialised cells called **receptors**. A change in the environment that can be detected by a receptor cell is called a stimulus.

▲ Smoke detectors in the alarm keep you safe.

The human body has receptors that are sensitive to:

- light
- sound
- touch and pressure
- chemicals
- changes in temperature
- changes in position.

The receptors that are sensitive to changes in position are found in the ears. These receptors help us to keep our balance.

Question

a *Where in the human body are the other types of receptor found?*

The nervous system

The brain and the spinal cord consist of millions of **neurones**, grouped into fibres called **nerves**. They carry information to the brain. Nerves from parts of the body below the head enter or leave the spinal cord, rather than the brain itself. The spinal cord carries information to or from the brain.

The central nervous system collects information from body receptors, makes sense of it and then sends messages back via the nerves to the organs that need to respond. Information passes along neurones as electrical **impulses**. For example, if temperature receptors in your fingers tell the central nervous system that they have touched something extremely hot, a message is sent to the muscles in your hand. The muscles respond by contracting, moving your hand away from the heat. A quick, automatic response like this is called a **reflex action**. We do not need to learn how to do it, or think about doing it.

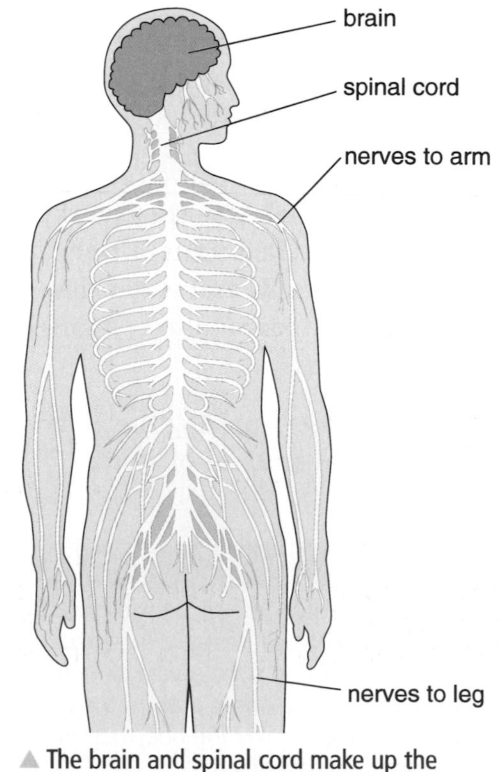

▲ The brain and spinal cord make up the **central nervous system**.

Reflex actions

Three types of neurone are involved in this type of reflex action. A **sensory neurone** carries impulses from the receptor to the central nervous system. A **relay neurone** then carries impulses between the sensory neurone and a **motor neurone**. The motor neurone carries impulses to the organ that will allow the body to respond to the stimulus. This organ is called an **effector**.

Information is passed from one neurone to another at a junction called a **synapse**. When an electrical impulse reaches the end of one neurone, it releases a chemical transmitter. This chemical passes across the synapse to the next neurone and causes an impulse to be sent along it.

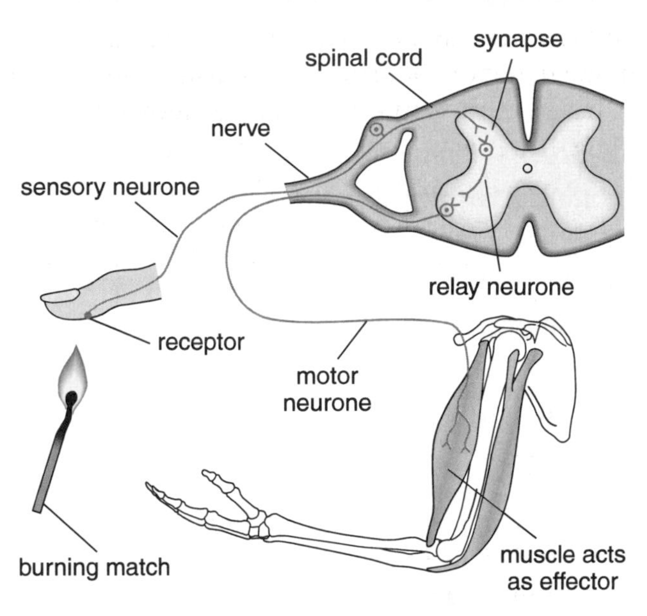

▲ How your nervous system stops you from getting burned.

Question

b *(i) What type of receptor is involved in this reflex action?*
(ii) In which organ is the receptor found?

The effector that moves our hand away from a hot object is a muscle. Not all our reflexes depend on muscle movement. Other reflexes help the body to work properly. When you smell food that you like, your mouth will automatically begin to water. In this case it is not a muscle that brings about the reaction, but the salivary glands in the mouth. This is the mouth-watering reflex.

Question

c *What is the effector in the mouth-watering reflex?*

Pupil in bright light

Pupil in dim light

The width of your pupils depends on light intensity. The iris of your eye controls the width of the pupil. The width can vary between 1.5 and 8 mm. The pupil gets narrower in bright light. This is a reflex action.

Question

d *Design an experiment to measure the width of the pupil under different light intensities. Remember you must not touch the surface of the eye. Explain what you will do to make sure your experiment produces reliable results.*

Key points

- The central nervous system coordinates the body's reactions to stimuli. Receptors detect stimuli and send information along neurones to the brain or spinal cord, which makes effectors react.
- Information from receptors is taken to the central nervous system by sensory neurones. Information is carried from the central nervous system to effectors by motor neurones. Relay neurones carry information within the central nervous system.

Controlling pregnancy

Changing methods of birth control

Women began taking substances to prevent pregnancy 4000 years ago, when Chinese women drank mercury. Methods of preventing pregnancy are called contraceptives. In the centuries that followed:

- Greek women drank diluted copper ore.
- Italian women drank tea made from willow leaves and mules' hooves.
- African women drank gunpowder mixed with camel foam.
- Canadian Indian women drank alcohol brewed with dried beaver testicles.

Until the twentieth century, contraceptive methods were often based on 'old wives' tales' rather than science.

A scientific approach to designing contraception began in 1937, when scientists discovered that a hormone could prevent the release of eggs in female rabbits. By the 1950s scientists had worked out how hormones controlled human fertility. They were able to use this knowledge to produce birth control pills.

Question

a Which one of the old wives' tales could possibly be linked to hormones? Give the reason for your answer.

The menstrual cycle

Every month, an egg develops inside a female ovary. At the same time, the lining of the womb becomes thicker, ready to receive a growing embryo. If the egg is not fertilised, the womb lining breaks down, causing bleeding. The monthly cycle of changes that take place in the ovaries and womb is called the **menstrual cycle**. The menstrual cycle is controlled by several hormones. The action of the hormones involved is summarised in the diagram.

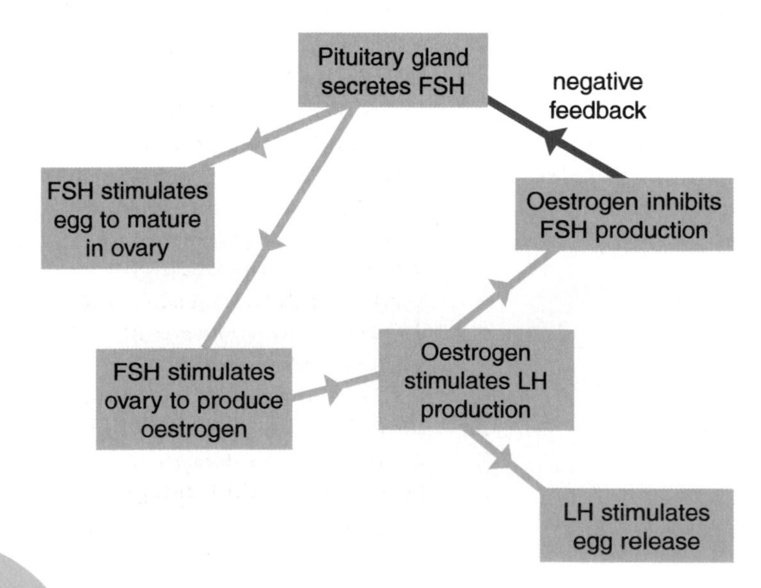

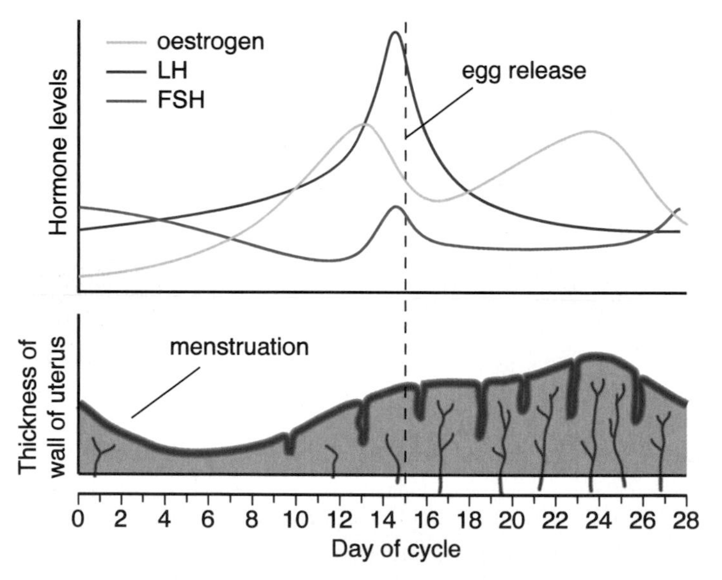

Questions

b During the menstrual cycle, which hormone is (i) the first to be secreted; (ii) the last to be secreted?

c What is the relationship between oestrogen concentration in the blood and the thickness of the lining of the uterus?

d What effects does oestrogen have on the pituitary gland?

e Which hormone do you think causes egg release?

The pill

A woman can take the **contraceptive pill** to stop her from becoming pregnant. The pill contains a hormone that has the same effect on the pituitary gland as oestrogen. It stops the pituitary gland making FSH. This means that no eggs will mature in the ovaries.

Benefits and problems

The pill decreases the chance of getting cancer of the womb by 50% and of cancer of the ovaries by 40%. However, there is an increased risk that women will develop blood clots. The table shows some data about this risk. There are two types of contraceptive pill available – 'old type' and 'new type'.

Situation	Risk in cases per 100 000
women not on the pill	5
women taking 'old type' pill	15
women taking 'new type' pill	25
women who smoke	100
pregnant women	60

Question

f *Imagine you are a doctor. Using the data in the table, what would you say to a woman who wanted to go on the pill and was worried about side effects?*

Fertility drugs

If a woman's own level of FSH is too low, her ovaries will not release eggs and she cannot become pregnant. Infertility can be treated by injecting FSH into the blood. FSH acts as a fertility drug by stimulating the ovaries to produce mature eggs.

Unfortunately, the treatment does not always work. Or sometimes it may cause more than one egg to be released. This can result in twins, triplets, quadruplets or even more!

▲ A daily task for many women who don't want to get pregnant.

Key points

- Hormones are chemicals which control many processes in the body such as the menstrual cycle.
- In the menstrual cycle:
  - FSH stimulates eggs to mature in the ovary
  - LH causes egg release
  - oestrogen causes the lining of the womb to increase in thickness.
- Artificially produced hormones (contraceptives) can be given orally to women to help control fertility.
- FSH is also called the fertility drug because it can be given to women to stimulate egg release and help them get pregnant.

Evaluating the benefits

Interfering with nature?

Sixty-six-year-old Adriana Iliescu became the world's oldest mother when she gave birth to a daughter following fertility treatment.

Ms Iliescu is a retired professor who lives alone. She said she had delayed having a child so she could concentrate on her academic career. Ms Iliescu became pregnant through IVF, using donated sperm and eggs, and this was her third attempt at having at baby.

Question

a *Should Ms Iliescu have been given fertility treatment? Give arguments for and against.*

Test-tube babies

Many women are infertile because of blocked oviducts. This means that eggs cannot travel down to the uterus. Nor can sperm travel upwards to meet an egg. This type of infertility can be treated by using ***in vitro*** **fertilisation** (IVF). This literally means fertilisation in a test tube, which led to the term 'test-tube baby'.

The first stage is to obtain eggs from the woman. She is given injections of LH. Eggs are then collected just before they are released.

The eggs are mixed with sperm from the father outside the body and the fertilised eggs are allowed to divide to form embryos. At the stage when the embryos are still just a ball of cells, they are inserted into the woman's womb.

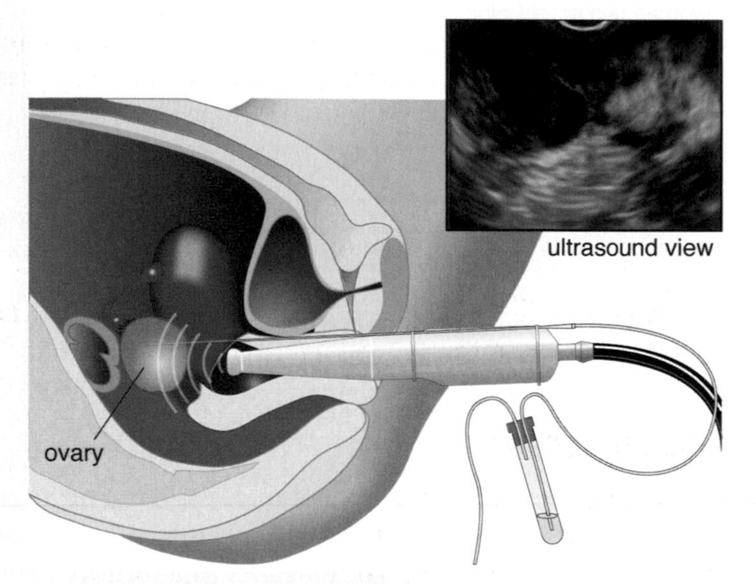

▲ Using ultrasound to view the ovary, the doctor inserts the needle through the wall of the vagina into the ovary and removes the eggs for use in IVF.

Questions

b *(i) Explain why LH is used in IVF treatment.*
(ii) Three embryos are usually inserted into the mother, even if she only wants one child. Suggest why.

c *(i) Explain why FSH can be used as a fertility drug.*
(ii) A drug called clomiphine is often used instead of FSH. This drug blocks the effect of oestrogen on the pituitary gland. Explain how clomiphine works as a fertility drug.
(iii) Explain why hormones similar in their effect to oestrogen can be used as contraceptive drugs.

What are the statistics?

The average success rate for IVF treatment using fresh eggs in the UK is shown right.

Women aged	Success rate (%)
under 35	27.6
35–37	22.3
38–39	18.3
40–42	10

Question

d *How is the success of IVF treatment affected by age? Suggest an explanation for this.*

The typical cost of a cycle of IVF treatment is approximately £3000. On top of this, the couple will have to pay for the costs of consultation, drugs and tests. The single biggest risk from IVF treatment is multiple births, and particularly triplet births. Many women decide to abort one of these triplets. Multiple births carry potential health risks for both the mother and the unborn child:

1 Multiple birth babies are more likely to be premature and the babies below normal birth mass.

2 The risk of death before birth or within the first week is more than four times greater for twins and almost seven times greater for triplets than for single births.

3 The incidence of cerebral palsy – a form of brain damage – is approximately five times higher for twins and approximately 18 times higher for triplets than for single births.

Questions

e *Currently 0.5% of IVF births are triplets – down from almost 4% in the early 1990s. 20% of IVF births are twins. Use the information above to explain this trend.*

f *Imagine that you are a doctor. What advice would you give to a couple who were considering IVF treatment? Use the information above in your answer.*

Thirsty work

The table gives the composition of some sports drinks. Osmolarity is a measure of the effect of glucose and ions on cells. Blood plasma and body cells both have an osmolarity of about 290 units. Solutions with osmolarity higher than 290 will cause water to leave cells. Solutions with an osmolarity of less than 290 units will cause cells to swell.

Sports drink	Carbohydrate (g/l)	Sodium ions (mmol/l)	Chloride ions (mmol/l)	Osmolarity units
Coca Cola	105	3	1	650
Dioralyte	16	60	60	240
Gatorade	62	23	14	349
Isostar	73	24	12	296
Lucozade	180	0	0	658
Lucozade Sport	64	23	14	280

Question

g *Which drink would be: (i) best for a marathon runner; (ii) worst for a marathon runner? Explain the reasons for your answers.*

Key points

- There are benefits and problems arising from the use of hormones to control fertility which need to be considered carefully before use.
- We can evaluate the claims of manufacturers for their sports drinks by considering what the body needs and whether the drink supplies it.

Healthy eating

Changing lifestyles

This article shows how worried medical experts are about the effect of poor diet and lack of exercise on young people. Experts are warning that many children are overweight and therefore have more chance of developing serious health problems in later life, including heart disease, diabetes and high blood pressure. The risk of health problems increases the more overweight a person becomes.

Each year a national survey is carried out to measure the health of adults and children. The survey involves an interview with a health professional and a visit by a nurse. Over 16 000 adults and 4000 children were involved in the 2004 survey. The graphs show the trend in the number of adults who are very overweight. People are described as being **obese** when they are so overweight that their health is seriously damaged.

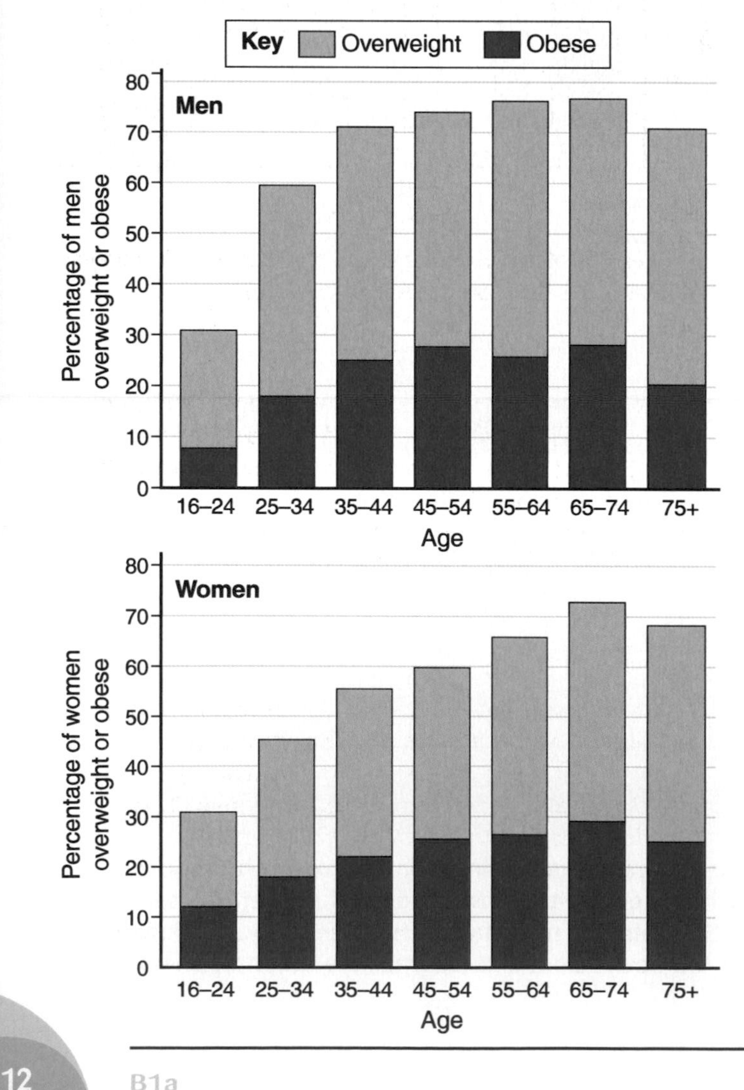

BRITISH CHILDREN TOP LEAGUE FOR UNHEALTHY LIVING

From our health correspondent.

The largest study of youth health reveals that British children live on sugary snacks and almost no fruit and vegetables.

The World Health Organisation report, based on surveys of more than 160000 children in 35 countries, found that the dietary habits of Britain's young were among the worst.

Doctors recommend at least five portions of vegetables daily for a healthy diet. More than two-thirds of children aged 11–15 admitted that they did not eat even a single portion of vegetables a day. A third of 11-year-olds drank at least one sugary drink a day, as well as eating sweets and chocolate every day. Snacks and sugary drinks are high energy foods.

Poor diet and increasingly inactive lifestyles are blamed for a massive increase in the number of people who are overweight.

Questions

- **a** *Use the information in the newspaper article to explain why poor diet and inactive lifestyles are blamed for the increase in obesity.*
- **b** *In the 25–34 age group: (i) what percentage of men are obese; (ii) what percentage of women are overweight?*
- **c** *Suggest why obesity increases with age.*

You are what you eat

The amount and type of food that you eat has a major effect on your health. Food provides the energy you need to stay alive and be active. It also provides the proteins, vitamins and minerals your body uses to grow and to replace damaged cells and tissues. By eating a varied diet you are more likely to get everything you need to keep your body healthy. Deficiency diseases can occur when the body doesn't get enough of a certain vitamin or mineral. These diseases are avoided by eating the right kinds of food. For example, you need vitamin C to develop resistance to disease. Eating fresh fruit provides your body with the vitamin C it needs. The diagram on page 13 shows the types and proportions of different foods that make up a healthy balanced diet.

The energy balance

You need to eat enough food each day to provide the energy your body needs. A diet may provide all the vitamins and minerals a person needs, but it can still be unbalanced by providing too much or too little energy. If you take in more energy than you use, your body stores the extra energy as fat and you put on weight. The key to maintaining a healthy weight is to balance your energy intake with your energy output.

High and low metabolisms

Your body converts the energy in food into a form it can use by **respiration**. The rate at which reactions are carried out in the cells of your body is called the **metabolic rate**. Your metabolic rate is affected by:

- the amount of exercise you do
- the proportion of muscle to fat in your body
- the genes you have inherited.

Lifestyle trends

Too much food and a lack of exercise can make a person put on weight. The less exercise you take and the warmer the climate, the less food you need. It is much easier now than ever before for people in the developed world to become overweight. In the developed world, social and technological changes mean that many people have very inactive lifestyles. Sitting in front of the television or computer rather than playing sports, and travelling by car rather than walking for short distances, mean that people need to eat less. But at the same time, high-calorie foods, such as snacks and processed meals, are more common than ever.

People who are active and exercise regularly are usually fitter and healthier than people who take little exercise. Your metabolic rate increases during exercise and remains high for some time afterwards, so fitter people generally have a healthy weight.

The graph shows the results of a survey of over 6000 young people from all over the UK. Each person was asked to complete a questionnaire about the amount of exercise they carried out during a typical week.

▲ These are the proportions of different foods that make up a healthy **balanced diet**. Does this match your diet?

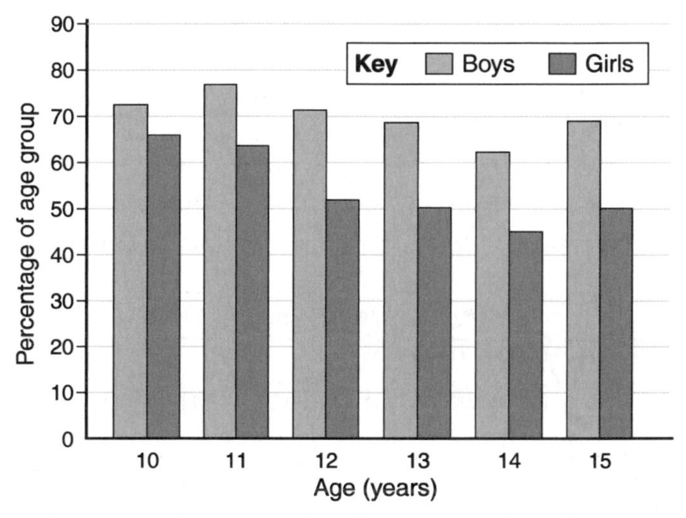

▲ Percentage of young people taking part in at least 60 minutes of physical activity a day (national survey 2002).

Questions

d *Medical experts recommend that young people carry out at least 60 minutes of moderate exercise every day. What percentage of 15-year-old boys achieve this recommendation?*

e *Suggest a reason for the change in activity levels among girls as they get older.*

Key points

- A balanced diet supplies the correct amount of energy and nutrients for a healthy body.
- The metabolic rate is the rate at which the chemical reactions in the body happen. This depends on how much you exercise, the proportion of muscle to fat in the body and inherited factors.
- People who exercise regularly are usually fitter than those who don't.

B1a 1.6 Too much or too little

Eating too little

In some developing countries people are suffering from health problems linked to a lack of food. People become **malnourished** if their diet does not provide the energy, protein, vitamins and minerals their bodies need. The photograph shows the effects of famine in Africa, which has led to millions of people going without food and having to rely on international aid. As well as making people lose too much weight and feel extremely weak, a lack of food also causes:

- **deficiency** diseases
- reduced resistance to disease
- irregular periods in women.

▲ As well as dying from hunger, people are also more likely to become infected with disease as a result of malnutrition.

Eating too much

In the developed world eating too much food is a cause of malnutrition. Over-eating and taking too little exercise are leading to high levels of obesity and diseases linked to excess weight. The food you eat and the lifestyle you choose to lead will have a long-term effect on your health. Overweight people who do little exercise and have poor eating habits are more likely to develop long-term health problems such as arthritis, diabetes, high blood pressure and heart disease.

Arthritis and being overweight

Your joints move easily because the end of each bone is covered with a layer of **cartilage**. The cartilage may get worn away with use, allowing the bones to scrape against each other. This is called **arthritis** and it makes the joints feel stiff and painful.

Overweight people have more weight pressing down on their knee and hip joints. As a result, the cartilage in these joints is more likely to wear away, causing arthritis.

FAT EPIDEMIC WILL CUT LIFE EXPECTANCY

The childhood obesity epidemic caused by poor nutrition and lack of exercise is creating a health crisis. The average life expectancy is expected to drop for the first time in more than a century.

The chairman of the Food Standards Agency said obesity was a 'ticking time bomb' and was one of the most serious issues facing the nation.

Diabetes and controlling blood glucose

Diabetes is an illness where the body cannot control the amount of glucose in the blood. One form of diabetes, called type II diabetes, usually develops in people over the age of 40. The risk of developing diabetes increases as body mass increases. Diabetes is three times more common in people who have gained an extra 10 kg in mass. Type II diabetes was recently diagnosed in children for the first time. All reported cases so far have been in overweight children.

The graph shows the trends in the prevalence of diabetes and obesity.

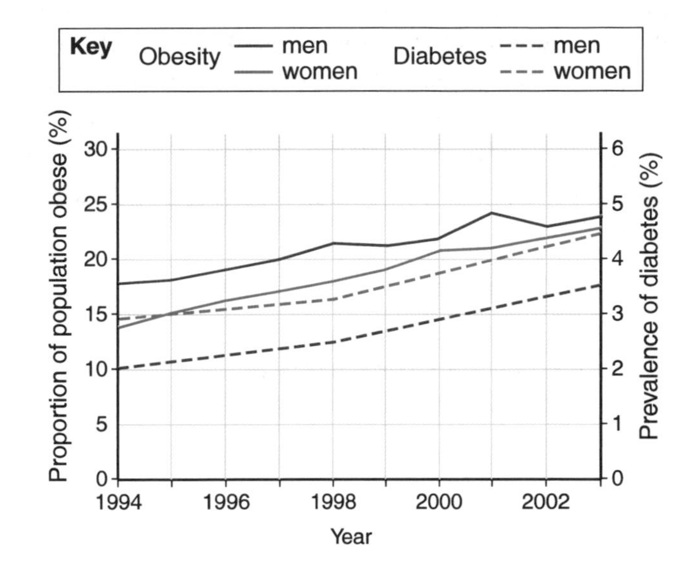

Questions

a *Between 1994 and 2003 what was the percentage change of women: (i) with diabetes; (ii) who were obese?*

b *Describe the evidence that shows that diabetes and obesity are linked.*

Slimming programmes

People who are overweight can lose weight gradually by following a sensible slimming programme. This involves eating less energy-rich food and taking more exercise. The information on food labels can help you choose foods that are lower in fat and energy. Normal slimming programmes are unlikely to make someone lose too much weight. But a small number of people become so worried about being fat that they start to eat too little to meet the body's basic energy needs. Their weight can become dangerously low. There are health risks for people who lose too much weight too quickly.

The graph shows the changes in body mass of two young women who are trying to maintain a healthy weight. Bes controls her weight by avoiding foods high in fats and sugars, and taking regular exercise. Jo is always switching from one diet to another and often goes for days eating very little.

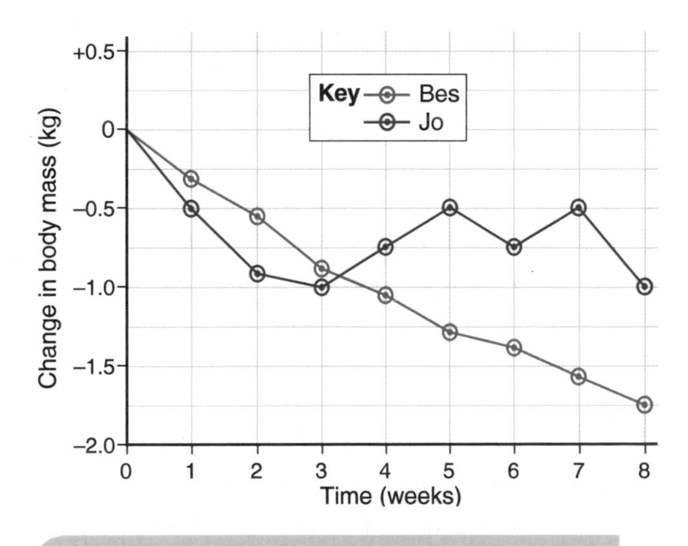

Questions

c *How much weight did (i) Bes and (ii) Jo lose over the 8 weeks?*

d *Calculate the mean mass lost per week by the two women.*

e *Why is Jo's crash dieting referred to as 'yo-yo' dieting?*

f *Suggest reasons why Bes's methods to control her weight are healthier than Jo's.*

Key points

- In the developed world, eating too much and exercising too little are leading to obesity and related diseases including arthritis and diabetes.
- In the developing world, famine leads to infection due to reduced resistance to disease, and to irregular periods in women.
- Data can be used to evaluate slimming programmes.

Keeping your heart healthy

Heart disease

It is important to have a healthy heart. A poor diet and lack of exercise can lead to heart disease. If you are aware of the risks, you can try to avoid them.

Your heart is a powerful muscular pump working throughout your life to get blood to every tissue in your body. Heart muscles also need a blood supply to keep working. Blood flows to the heart muscles along coronary arteries. In a healthy heart, the walls in these arteries are smooth and blood flows easily. When fat deposits build up within the walls of the arteries their diameter is reduced. Blood flow is more difficult and the heart muscle receives less oxygen. This weakens the heart. If an artery becomes completely blocked, blood carrying oxygen doesn't get to some of the heart muscle. This causes a heart attack.

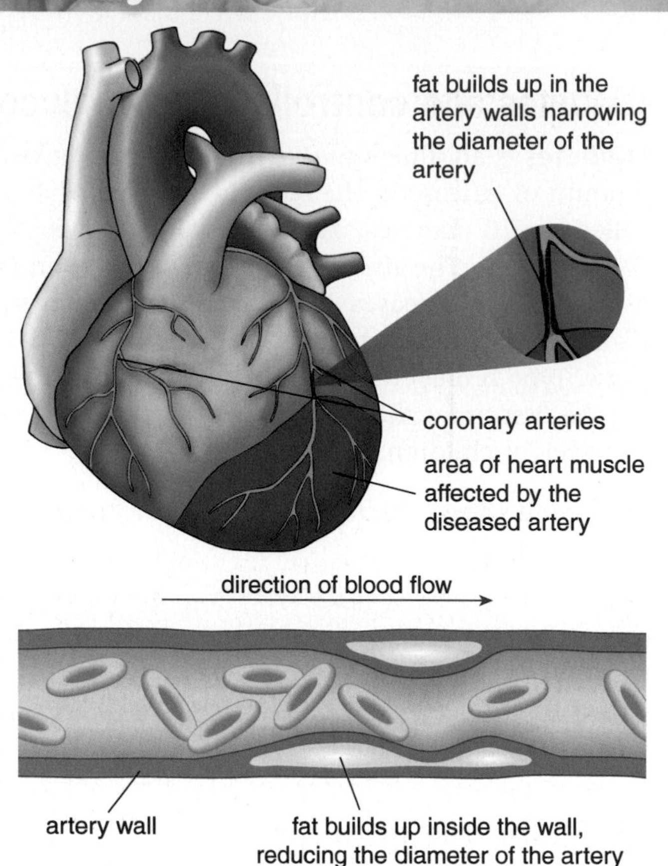
fat builds up in the artery walls narrowing the diameter of the artery
coronary arteries
area of heart muscle affected by the diseased artery
direction of blood flow
artery wall
fat builds up inside the wall, reducing the diameter of the artery

▲ Artery in a diseased heart.

Question

a *Explain how the heart muscle in the shaded area on the diagram will be affected by the narrowing of the artery.*

Cholesterol

Cholesterol is a fatty substance that is mainly made in your liver. Your liver makes cholesterol from the saturated fats in your food. Cholesterol plays a vital role in how every cell works. Too much cholesterol in the blood can increase your risk of getting heart disease.

Cholesterol is transported in the bloodstream attached to proteins. The combination of protein and cholesterol is called **lipoprotein**. There are two types of lipoprotein:

- low-density lipoproteins (**LDLs**), which carry cholesterol from your liver to your cells
- high-density lipoproteins (**HDLs**), which carry the extra cholesterol that your cells don't need back to your liver.

Your risk of heart disease is increased if you have a high level of LDL (called 'bad' cholesterol). A high level of LDL causes fat to build up on the artery walls, causing heart disease. HDL helps to prevent cholesterol building up in arteries. HDLs are 'good' cholesterol. A high level of HDLs and a low level of LDLs is good for heart health.

The amount of cholesterol your liver produces depends on a combination of diet and inherited characteristics. The amount and type of fat you eat can change the amount of cholesterol in your blood. Saturated fats, found in animal fat, increase blood cholesterol levels. Unsaturated fats (called monounsaturated and polyunsaturated fats), found in fish and vegetable oils, may help to reduce blood cholesterol levels. Processed foods often have a high proportion of fat. Consuming too many processed foods can contribute to high cholesterol levels.

Question

b *Explain why a high level of HDLs and a low level of LDLs is good for heart health.*

Different populations around the world have diets containing different amounts of cholesterol. This graph shows the dietary cholesterol levels and the incidence of heart disease in different countries.

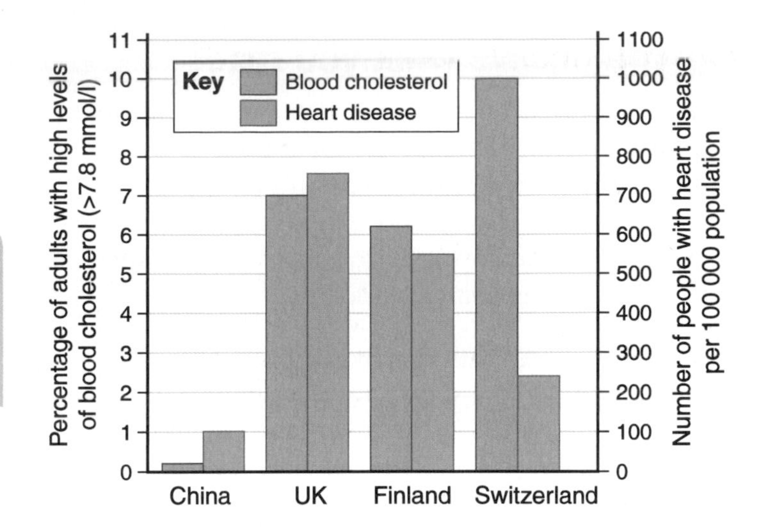

Questions

- ***c*** *Describe the evidence that shows that high levels of cholesterol increase the risk of heart disease.*
- ***d*** *Describe the evidence that shows that other factors as well as cholesterol levels may lead to heart disease.*

Fighting high blood pressure

People with **high blood pressure** have a greater risk of developing heart disease. Blood pressure is measured in the arteries. A certain pressure is needed to keep our blood flowing. Blood pressure in the arteries changes as your heart beats and then relaxes. The highest pressure occurs at the moment the heart contracts and forces blood into the arteries. The lowest pressure occurs as the heart relaxes between beats.

The bar chart shows how blood pressure varies with age. The data was obtained using people with similar weight and taking readings under similar conditions, for example while resting.

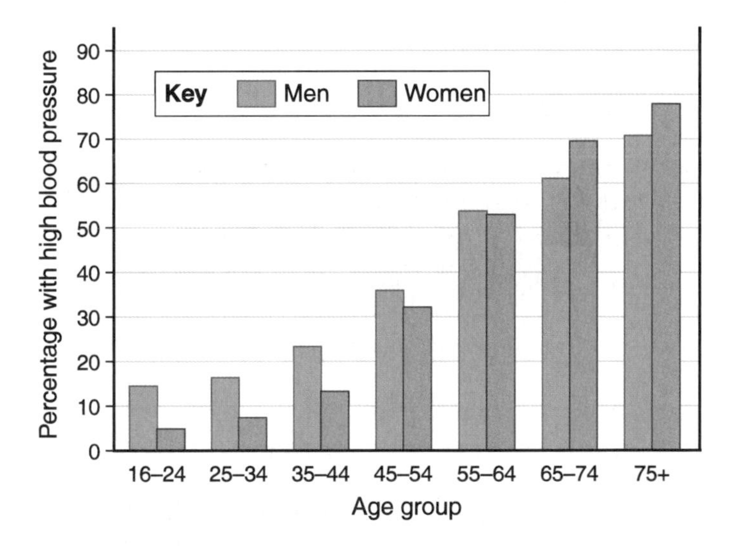

Questions

- ***e*** *In which age group does the percentage of males with high blood pressure become higher than in females?*
- ***f*** *In which age group does the percentage with high blood pressure exceed 50%?*
- ***g*** *This survey looks at the effect of age on blood pressure. Give two features of the people used in the survey that should be kept as similar as possible. Explain why these features should be similar.*

Too much salt

Too much salt in your diet can lead to increased blood pressure. Health experts recommend that adults consume less than 6 g of salt a day. Currently, the average daily intake in the UK is far too high, with the vast majority of adults consuming more than 9 g of salt each day. About 75% of the salt people consume comes from processed foods such as canned soups, take-aways and ready-made meals.

Key points

- High cholesterol levels in the blood increase the risk of heart disease and blocked arteries.
- There are two types of cholesterol. Low-density lipoproteins (LDLs) are bad and can cause heart disease. High-density lipoproteins (HDLs) are important for a healthy heart.
- Saturated fats increase bad cholesterol. Monounsaturated and polyunsaturated fats reduce levels of bad cholesterol.

Do you believe what you read?

The latest trends in slimming diets and slimming products appear regularly on TV and in magazines, claiming that they can help people lose weight effortlessly. Some adverts use 'evidence' and scientific jargon to try to convince people that a diet really works.

The advert below shows how one company tries to get people to buy their slimming pills as a way of losing weight. The advert shows 'proof' to try to convince you that this product works. Many companies use adverts like this. Most of the 'evidence' is based on reports from just one or two people showing dramatic weight loss in a short time. Having an understanding of how evidence is used to provide reliable results enables you to question what you read and to make informed decisions.

Medical experts say that there is only one way to lose weight – eat healthy foods and keep energy intake below energy expenditure, and aim to lose a sensible amount of weight.

FLAB2FITNESS
These pills really worked for me
FLAB2FITNESS

Question

a *Adverts showing photographs of a person before and after losing weight by using a slimming product used to be common. This type of advertising has now been banned. Suggest reasons why the evidence in this kind of advert should be regarded as unreliable.*

Slimming drugs warning

SLIMMING DRUGS WARNING

A warning against 'miracle' slimming drugs has been issued after a survey by trading standards officials.

Officials investigated a variety of slimming products offered for sale over the internet. They found that three-quarters of the products tested made false claims. Most companies could not provide reliable evidence to back up their weight-loss claims. These are some of the claims that companies are making about their products:

- tablets that enable the body to burn fat before food is digested
- pills that allow people to lose weight without dieting or exercising
- a product that burns fat while people are asleep.

Health experts warn that if a product or a diet programme sounds too good to be true, it probably isn't good for you and isn't true.

A scientific evaluation

Meal replacements are widely used as a way of losing weight. Meal replacements, such as Slim•Fast, are energy-reduced products that contain added vitamins and minerals.

Three hundred people were interviewed to take part in a study to compare a meal-replacement diet with a conventional low-calorie diet. Sixty-six people were chosen to take part in the study. All these had a similar level of health and fitness. Those chosen to take part were divided at random into two groups. One group received the meal replacement and the other group, the control group, received a conventional low-calorie diet. This method of comparison is called a randomised control trial.

Weight change was measured after 3 months and after 6 months. The results are shown in the table.

Time interval	Mean weight loss (kg)	
	Meal replacement group	Control group
after 3 months	6.0 ± 4.2	6.6 ± 3.4
after 6 months	9.0 ± 6.9	9.2 ± 5.1

The scientists concluded that there was no significant difference between the two groups. The meal replacement is as effective as conventional low-calorie dieting in achieving weight loss.

Questions

b *Suggest why the data is presented using figures such as 6.0 ± 4.2.*

c *Do you agree with the scientists' conclusions from the survey? Support your answer using information from the survey.*

Key points

- Data can be used to evaluate claims about slimming products and slimming programmes.
- It is necessary to distinguish between opinion based on valid and reliable evidence and opinion based on non-scientific ideas.

Drugs and disease

Drugs are chemicals that affect the processes in our bodies. Most medical drugs are beneficial in treating illness if they are used in the right way. But misusing recreational or medical drugs can harm our bodies.

Alcohol

Many drugs are extracted from natural substances. They have been used by people around the world as part of different cultures for medicines and recreation for thousands of years. Alcohol has been fermented from fruit and grain since at least ancient Greek times.

Alcohol is the oldest known drug. It is a drug that affects the nervous system. People drink it because it can cause an excited mood and lack of inhibitions. But even small amounts slow down the body's reactions. This is why it is illegal to drive a car after drinking a certain amount of alcohol. Larger amounts of alcohol can lead to a loss of self-control. People behave in ways that they wouldn't when sober. Drinking more alcohol than your body can process can make you lose consciousness and even fall into a coma.

MAY DAY CELEBRATIONS END IN TRAGEDY

Hundreds of students followed a centuries-old tradition and jumped off Magdalen Bridge at dawn on May Day into the River Cherwell at Oxford. But most of them were drunk after all-night parties and they ignored police warnings that the river was too low this year. The result was at least ten students with serious injuries to spines, legs and ankles.

▲ Would the students do this if they were sober?

Worrying reports

You are probably protected against measles, mumps and rubella. The MMR **vaccination** is given to young babies to provide life-long protection against these three diseases. Deciding whether to have a young child vaccinated can be a very worrying decision for parents when they read reports saying that the MMR vaccination may harm their child. The following reports appeared in recent newspapers.

WE WISH WE'D HAD THE MMR

Cases of measles are soaring as parents reject the MMR vaccine amid fears that it may be harmful.

One parent whose daughter, Clara, almost died from the complications of measles said,

'Even now it hasn't sunk in that my daughter almost died of measles. As she lay in intensive care the doctors said it was touch and go. I thought measles was just a harmless childhood illness. I'd had it as a girl with no problems. Now it could rob me of my daughter.

All my other children had the MMR jab. But when Clara was due for her jab she had a bad cold so I delayed the injection. That delay almost cost me my daughter.'

WHY I WOULDN'T GIVE MY BABY THE MMR JAB

No one said being a parent is easy. The most difficult problem I have faced so far is the question of MMR – the measles, mumps and rubella vaccine.

Over the past six months it has become the most hotly debated subject among the mothers I know. Before my daughter was born I ignored this issue. Now she is one year old and I have to make a decision. But what shall I do?

When I see my daughter running around I feel sick at the thought that I could do her harm. All the information about the dangers of MMR has left me very confused. Only one of my friends is getting her child vaccinated with MMR. The information provided for parents doesn't convince us that the vaccine is safe. I can't take the risk of letting my child have the vaccine.

Who to believe?

The safety of the MMR vaccination has often been reported on TV and in newspapers. Following these reports, many parents chose not to have their child vaccinated using MMR. The main issues about the safety of MMR are:

- parents have the responsibility to decide
- nearly all the evidence gathered from several countries showed that MMR was safe
- the research of one doctor raised concerns about health risks
- some reports presented information carefully, but some reports emphasised parents' worries and fears
- recent evidence shows that the research findings about health risks were not reliable.

Think about what you will find out in this section

What are the dangers of using common drugs such as alcohol and tobacco?	Is there enough evidence to suggest a link between taking cannabis and developing mental illness?
The advantages and disadvantages of being vaccinated against diseases.	How scientists ensure that new medicines are safe to use.
How scientists ensure the reliability of their evidence.	How the treatment of disease has changed.
How the credibility of science suffers as a result of any bias in research.	

What's in a drug?

Why do people use drugs?

Some drugs are so dangerous that they are illegal in Britain. An example is highly addictive heroin, which leads to severe health problems in many users, including mental health problems. But it isn't only illegal drugs that are dangerous. Science has shown that tobacco and alcohol cause thousands of deaths in Britain every year. People continue to use these drugs from habit, or to change their mood. Even though smoking and drinking are legal for adults, people need to know what these drugs do to their bodies in order to understand the risks they are taking.

The NHS has to spend far more money on treating the effects of legal drugs than illegal drugs, because far more people use them. Many people would argue that these drugs should be made illegal. Many cities are now banning smoking in public places. Some councils think that a much higher tax should be put on alcohol to discourage binge drinking.

▲ Heroin addicts have to inject themselves frequently or they suffer severe withdrawal symptoms.

More than a quarter of 15- and 16-year-old European girls admitted binge drinking says survey of 100 000 students

Nearly 2000 English children aged 14 and under are admitted to hospital each year because of drunkenness, a rise of 13.5% over 6 years

38% of teenagers in Britain – 42% of boys and 35% of girls – admit trying illegal drugs

How do people get addicted to drugs?

Drugs are chemicals that change the way the body works. Most medical drugs have been developed by scientists to help cure or relieve the symptoms of illness. They are prescribed by doctors or sold in pharmacies. Recreational drugs, such as alcohol and nicotine (a drug in tobacco), are used to change the way a person feels and even thinks.

All drugs change the chemical processes in our bodies. When your body gets used to the change, it may become dependent on the drug. People become **addicted** to drugs when their whole body chemistry changes. Some people using sleeping tablets or tranquillisers as medicines can get addicted to them too.

Hard drugs like heroin and cocaine are very addictive. If an addict stops taking the drug, their body loses its natural balance and they suffer **withdrawal symptoms**. This can make them very ill. Even people drinking a lot of coffee can suffer very mild withdrawal symptoms if they stop drinking it.

Questions

a *(i) Explain what is meant by addiction.*
(ii) Many smokers are now trying to give up the habit. How does stopping smoking affect them?

b *Explain why drugs can alter the way we behave.*

c *What are the possible effects of drug damage: (i) to the lungs; (ii) to the liver?*

Breathalysers are used to deter people from drinking and driving.

Alcohol

Drinking too much alcohol in the long term can lead to alcohol addiction, known as alcoholism. Many alcoholics die young because of damage to their liver or brain. It is important that people understand the risk of long-term health problems, so they can make an informed decision about how much and how often to drink. The table shows estimates of the cost of drinking alcohol to the UK.

Cost group	Cost (£ millions)
Health care costs	1500
Workplace and wider economy costs	
• lost output due to absenteeism	1800
• lost output due to reduced employment	2100
• lost output due to premature death	2500
Crime costs	
• alcohol-related offences	1750
• property/health and victim services	2500
• costs in anticipation of crime (alarms, etc.)	1500
• emotional impact costs for victims of crime	4600
Drink driving	
• Criminal Justice System costs	77
• cost of drink-driving casualties	430

Estimated overall costs of alcohol misuse.

Question

d *(i) The population of the UK is approximately 60 000 000. How much do the effects of drinking alcohol cost per head of population?*
(ii) The costs are divided into four groups. Which is the highest group of costs? Suggest a reason for this.

The tables give data about trends in drinking alcohol in Britain. There is one unit of alcohol in half a pint of beer or one glass of wine.

Weekly alcohol consumption by age	1988	1992	1996	2000	2004
More than 14 units					
16–24	15	17	22	33	33
25–44	14	14	16	19	19
45–64	9	11	13	14	14
65 and over	4	5	7	7	7
More than 35 units					
16–24	3	4	5	9	10
25–44	2	2	2	3	3
45–64	1	1	2	2	2
65 and over	0	0	1	1	1

Females: trend in percentage of females drinking more than 14 and more than 35 units of alcohol per week.

Weekly alcohol consumption by age	1988	1992	1996	2000	2004
More than 21 units					
16–24	31	32	35	41	37
25–44	34	31	30	30	29
45–64	24	25	26	28	28
65 and over	13	15	18	17	15
More than 50 units					
16–24	10	9	10	14	12
25–44	9	8	6	7	8
45–64	6	6	5	7	8
65 and over	2	2	3	3	3

Males: trend in percentage of males drinking more than 21 and more than 50 units of alcohol per week.

Questions

e *How is alcohol consumption related to age?*
f *How did alcohol consumption change between 1988 and 2004?*

Key points

- Drugs change the chemical processes in the body so people may become addicted to them and suffer withdrawal symptoms.
- People use legal and illegal drugs recreationally. The impact of legal drugs on health is much greater than that of illegal drugs as more people use them.
- Alcohol slows down the reactions of the nervous system, helping people to relax, but too much can lead to unconsciousness or coma and liver and brain damage.

Smoking kills

Tobacco

Tobacco smoke is a mixture of hundreds of different chemicals. Many of these chemicals harm the human body. Some of these are known as **carcinogens**. They stimulate cells to divide rapidly, which can lead to cancer of the mouth or lungs. Tobacco affects the lungs in other ways as well.

STOP
SMOKING
for free help and advice call
0800 328 8537

Evaluating ways to stop smoking

The NHS is very keen to reduce the number of smokers. It provides money for each local health authority to fund helplines.

There are two common methods of stopping smoking:

- *'Cold turkey'*. This is stopping without any kind of aid. Withdrawal symptoms are very severe in the initial few days, but they fade away within the first two or three weeks. Most people give up smoking using this method.
- *'Nicotine replacement therapy'* (NRT). NRT is clinically proven to be twice as effective as the 'cold turkey' method. NRT eases withdrawal symptoms while the smoker gets used to not smoking and the dose is gradually reduced. NRT methods include gum, skin patches and lozenges.

Most 'stop smoking' courses involve counselling and NRT. Even so these methods have a low success rate.

Question

a *(i) Suggest one reason why NRT is twice as effective as 'cold turkey'.*
(ii) Suggest one reason why most people giving up smoking do so by 'cold turkey' rather than by NRT.

The table gives the results of an evaluation of one 'stop smoking' course.

	% who did not stop smoking (number = 46)	% who stopped smoking for a time (number = 107)	% who stopped smoking completely (number = 93)
Total	18.7	43.5	37.8
Gender			
Male	18.0	44.0	38.0
Female	15.0	45.9	39.1
Age			
18–39	19.5	45.5	35.1
40–49	15.9	46.3	37.8
50+	12.5	41.6	45.8
Income			
less than £20 000	31.3	46.8	32.9
£20 000–£30 000	6.9	55.2	37.9
£30 000–£40 000	11.4	45.7	42.9
£40 000–£50 000	20.4	34.7	44.9
more than £50 000	13.3	49.3	37.3

Questions

b *How many people took part in the course?*

c *What is the relationship between failure to stop smoking and age?*

d *What is the relationship between stopping smoking completely and income? Suggest one explanation for this relationship.*

e *One person running the course said, 'The data proves that the course was more successful in stopping women smoking than men.' Was he correct to say this?*

The man who saved lives

Sir Richard Doll began his study of lung cancer in 1948. Deaths from lung cancer had risen 50-fold in the previous few years. Many scientists blamed pollution, while others, including Sir Richard, thought the tar being used on Britain's new roads was responsible.

Sir Richard visited 2000 people with suspected lung cancer. He found that nearly all these were heavy smokers. He came to the conclusion that there was probably a link between smoking and lung cancer. But most cancer workers did not accept his findings. They said that smoking could not be the cause of lung cancer because they knew that some non-smokers develop lung cancer. People didn't realise that lung cancer could have several causes. Smoking is only one cause of lung cancer.

To convince them, Sir Richard decided to look at people's smoking habits and see whether that could predict who would contract lung cancer. He chose doctors for his sample. Within two and a half years, he found 37 doctors who had died from lung cancer. All were smokers and a high proportion smoked heavily. Sir Richard had found that smoking was the most important cause. The association between cancer and lung cancer was now very clear.

▲ Sir Richard Doll, the scientist who discovered the link between smoking and cancer. From *Cancer World*, 31 December 2004.

Question

f *(i) Explain, as fully as you can, why doctors were reluctant to accept the conclusions of the first survey.*
(ii) Explain why the second survey gave powerful evidence of the link between lung cancer and smoking.

Underweight babies

One of the most poisonous substances in tobacco smoke is **carbon monoxide**. This stops red blood cells from binding with oxygen, so a smoker's blood carries less oxygen around the body.

The graph shows the relationship between birth mass and the number of cigarettes smoked by the mother.

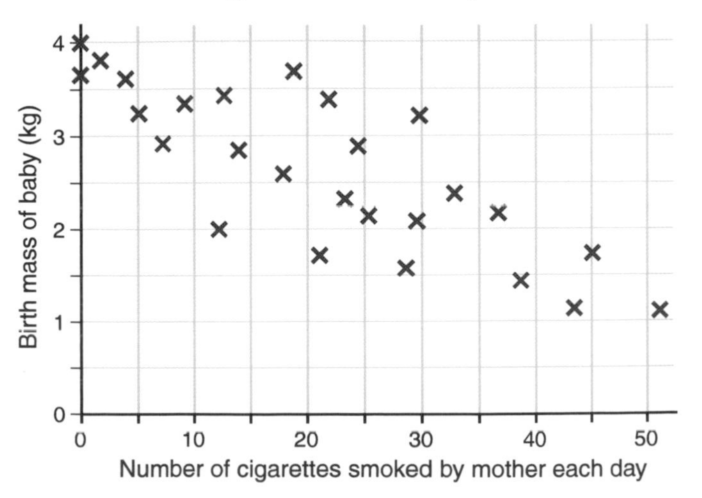

Questions

g *(i) From this data, what is the relationship between birth mass and the number of cigarettes smoked by the mother?*
(ii) How valid and reliable is this data?

h *Give one factor, other than smoking, which could also affect the birth mass of the babies. Explain how this factor could be related to the smoking habits of the mother.*

Key points

- Nicotine is the addictive substance in tobacco smoke. Tobacco also contains carcinogens and carbon monoxide. Carbon monoxide in the blood reduces the oxygen in the blood so that pregnant women who smoke may give birth to underweight babies.
- There are a number of different ways people can be helped to give up smoking.

Developing new drugs

Developing new medicines

The treatment of disease is always being improved by the development of new drugs. Before a new drug can be used, it is put through several tests and has to pass each stage. The first stage is to test the drug in laboratories to find out if it is toxic. People who are being treated for a medical condition may then be asked if they would like to be part of a **clinical trial**. This is a research project to find out if a new drug works better than an existing one or if it has any side effects. A drug can only be used after it has been thoroughly tested and trialled to make sure that it is safe.

An unsafe drug

Thalidomide is a drug that was first used in Britain in 1958. It was thought to be a safe drug because it had been tested and trialled. However, the drug was not approved for use in the US because medical experts there thought there was insufficient proof of the drug's safety.

Thalidomide was given to women in the first few months of pregnancy for use as a sleeping pill and to overcome the morning sickness women feel at the start of pregnancy. Many women who took the drug gave birth to babies with limbs that weren't properly formed. The drug was banned worldwide in 1961 after it was discovered that it caused tragic birth defects. The total number of babies damaged by thalidomide throughout the world was about 10 000.

The newspaper article shows that thalidomide is being used again to treat leprosy. This is a disease which affects the skin and nerves in the hands and feet. In severe cases of leprosy the skin dies, causing hands and feet to become deformed.

In 1964 a critically ill leprosy patient was in such pain that he hadn't slept for weeks. He was given thalidomide and slept soundly for 20 hours. After receiving more thalidomide his pain disappeared. Six other patients were then treated and showed similar results. Following randomised trials on 173 leprosy patients, 92% were relieved of their symptoms. A follow-up study by the World Health Organization involving 4552 leprosy patients showed improvement in 99% of patients.

THALIDOMIDE – should we risk another tragedy?

Thalidomide, the drug banned 30 years ago, has been given approval for use again. It will be given to leprosy sufferers but with strict restrictions to minimise any risks.

Question

a *Before a female leprosy patient can be given thalidomide she must first be tested for pregnancy and then use birth control while taking the drug. Explain why these restrictions are necessary.*

Testing a new drug

The Heart Protection Study was carried out to test the effectiveness of a drug called simvostatin. This drug is taken by people who are at risk of having a heart attack.

To assess the long-term effects of taking statins, a study was carried out by a team of heart specialists at a leading UK hospital. The study involved 20 536 patients with heart disease, aged between 40 and 80. The health of these patients was monitored closely over a 5-year period.

A total of 10 269 of the patients took a simvostatin tablet daily, whilst 10 267 received a placebo every day. The placebo was a tablet that had no effect on cholesterol levels. Patients were randomly placed into the 'statin' group or the 'placebo' group and were not told which tablets they were receiving during the trial.

Heart drug could save lives

The Heart Protection Study provided proof that thousands of lives could be saved each year by using drugs called **statins** to lower the amount of cholesterol in the blood. This eight-year research project was carried out by the British Heart Foundation and the Medical Research Council.

At the end of the study the researchers concluded that taking statins over 5 years would prevent major circulation problems occurring in:

- 100 of every 1000 patients who have already had a heart attack
- 80 of every 1000 patients with heart disease
- 70 of every 1000 patients with diabetes.

The main conclusion of the study was that simvostatin is safe and reduces the risk of people having a heart attack or a stroke.

▲ Statins are so valuable in preventing heart disease and are so safe that they can now be bought without a doctor's prescription.

Questions

b *This type of study is called a randomised controlled trial.*
(i) What feature of the study was randomised?
(ii) How was a control used in the study?

c *A heart disease patient has a 1% chance of having a heart attack. How would this risk change after treatment with statins?*

d *It is important that studies to assess drugs are highly reliable. What features of this study show that the findings are reliable?*

Key points

- New drugs are tested to see if they are toxic and then trialled in clinical trials with people to check if they are effective and whether they have any side effects.
- Thalidomide is an example of a new drug that was tested and trialled and thought to be safe. It was given to pregnant women, resulting in terrible defects and the banning of the drug.
- Statins are a new drug being developed and tested which lower the cholesterol level in the blood and so treat and prevent heart disease.

A harmless joint

Many people smoke cannabis as a recreational drug – it helps them to 'chill out'. Most of these people think that cannabis is harmless. But the mother of the young man described in this article would not agree with them. Many scientists believe that cannabis can cause psychological problems. But other scientists are uncertain whether cannabis actually causes these problems.

The active ingredient in cannabis is a chemical called THC. Enthusiasts have bred cannabis plants which contain 20 times more THC than the wild variety. People who smoke this cannabis are much more likely to develop psychological problems than those who smoke the wild variety.

My son sat with me on a hospital bench outside the hospital canteen. Suddenly, he looked up and said,

'Oh, mother, you don't know how terrible it is to be Hitler.'

'You're not Hitler,' I said. 'Your voices are only your own thoughts.' He looked up.

'You really believe that?'

'I do,' I said.

He was in better form than he had been. At this moment he was not complaining that the nurses were plotting to kill him. He had told me that cannabis was the most dangerous of the many drugs he had taken, because it was the cannabis which had triggered the paranoia, and it was the drug he feared most. He died of heroin poisoning in a dealer's flat in 2000.

Does cannabis cause mental illness?

Patients with mental illness may lose contact with reality or suffer from delusions. Here are some results of research into cannabis and mental illness.

New Zealand scientists followed 1000 people born in 1977 for the next 25 years. They interviewed people about their use of cannabis at the ages of 18, 21 and 25. The questions were about their mental health. The researchers took into account factors such as family history, current mental disorders, and illicit substance abuse.

The scientists' findings were:

- Mental illness was more common amongst cannabis users.
- People with mental illness did not have a greater wish to smoke cannabis.
- Cannabis probably increased the chances of developing mental illness by causing chemical changes to the brain.
- There was an increase in the rate of mental illness symptoms after the start of regular use of cannabis.

▲ Many people think that smoking a joint is harmless – but is it?

Scientists studied 45 000 Swedish male conscripts (men called up for army service). This was 97% of the male population aged 18–20 at that time. They followed these men for the next 15 years. They found the men who smoked cannabis heavily at the age of 18 were six times more likely to develop schizophrenia in later life than those who did not smoke cannabis.

Question

a *(i) What type of scientific research was this?*
(ii) What were the control variables in the investigation?
(iii) What did the investigation show about the link between cannabis and psychosis?

Question

b *Why do many scientists think that this study gives powerful evidence for a link between cannabis use and schizophrenia?*

A study in 2004 looked at 600 same-sex twins, one of whom was dependent on cannabis and one of whom was not. Scientists found that the twin who was dependent on cannabis was three times more likely to think about suicide than their co-twin who was not cannabis-dependent.

Question

c *Why is data from twin studies regarded as powerful evidence of a link between cannabis use and mental illness?*

Many people think that cannabis should be legalised. Here are some of their views.

Cannabis is a safer drug than alcohol. It leads to fewer deaths, both direct and indirect, suppresses the violent tendencies that alcohol releases, and has fewer long-term effects on health.

Four million people in the UK have used cannabis in the last month. Society is not disintegrating as a result. Are you saying that each and every one of these 4 million is a desperate loser, a waster, a drop-out? Most of them will be people you know, holding down good jobs, good lives, good families.

Let's look at the empirical facts – cannabis use exists, and is not doing the UK any harm. (Nor is Holland sinking into a pit of cannabis-inspired debauchery, and they decriminalised it nearly 25 years ago!)

The 'evil dealer' is another myth. Most people buy from friends. Cut out the dealers that do exist by decriminalising cannabis and encouraging people to grow their own.

Question

d *Some people say that there is enough evidence to ban the use of cannabis. Others say that its use should be legalised. Discuss this in groups then report back to the whole class.*

Key points

- It is important to evaluate the claims made about the effect of cannabis on health, and to look at the link between cannabis and addiction to hard drugs before legalising it.
- The link between smoking tobacco and lung cancer only gradually became accepted.

Defence against disease

Pathogens

Bacteria and **viruses** are types of **microorganism**. Microorganisms that cause disease are called pathogens. Many diseases are caused by viruses and bacteria getting into the body. Once bacteria or viruses are inside the body they may reproduce very quickly. Some bacteria release poisons called **toxins** which make you feel ill.

Viruses can only reproduce inside living cells. When a virus gets into a body cell it uses it to make thousands of new viruses. The new viruses burst out of the cell ready to invade other body cells. This damages or even destroys the cell.

The table shows some common diseases caused by viruses and bacteria.

Diseases caused by bacteria	Diseases caused by viruses
tonsillitis	colds
whooping cough	flu (influenza)
typhoid	measles
tuberculosis	mumps

Question

a *When people cough and sneeze they release tiny droplets of moisture into the air. Pathogens can stick to the droplets and get into your body when you breathe in. Name two diseases from the table which can be spread in this way.*

Spreading disease

Bacteria and viruses can pass from one person and infect another. This is how some diseases spread and affect many people. You can become infected by pathogens in the air you breathe, the food you eat and liquids you drink, and by touching someone. By making sure that your environment is clean you lessen the chance that you will become infected.

Cells to fight pathogens

Your body has different ways of protecting itself against pathogens. White blood cells are specialised cells that defend your body against pathogens.

There are several different types of white blood cell. Some **ingest** (take into the cell) any pathogens that they come across in your body. Once the pathogen is inside, the white blood cell releases enzymes to digest and destroy it.

Other white blood cells release chemicals called **antibodies** which destroy pathogens. Antibodies can only destroy a particular bacteria or virus, so white blood cells learn to make many different types. For example, when a flu virus enters the body antibodies are made which destroy the flu virus. After the virus has been destroyed, flu antibodies remain in the blood and act quickly if the same pathogen enters in the future. White blood cells also produce **antitoxins**. These are chemicals that prevent the toxins made by pathogens from poisoning your body.

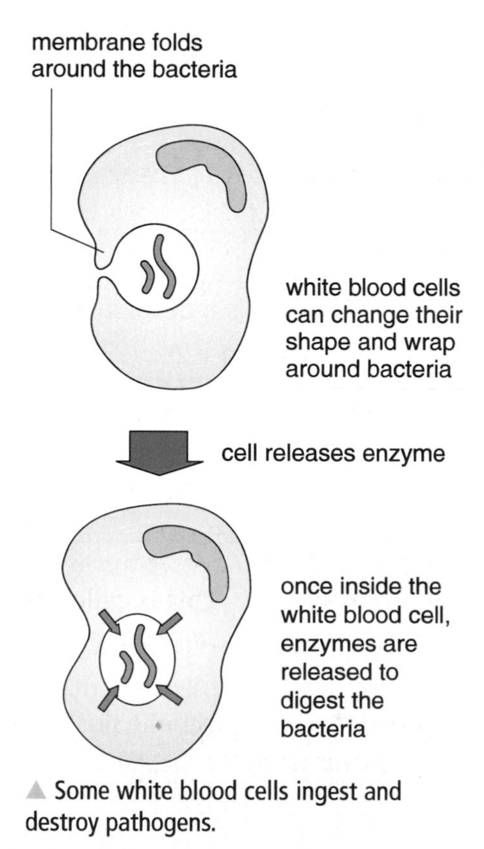
membrane folds around the bacteria
white blood cells can change their shape and wrap around bacteria
cell releases enzyme
once inside the white blood cell, enzymes are released to digest the bacteria

▲ Some white blood cells ingest and destroy pathogens.

Question

b *Describe the ways that white blood cells prevent pathogens causing disease.*

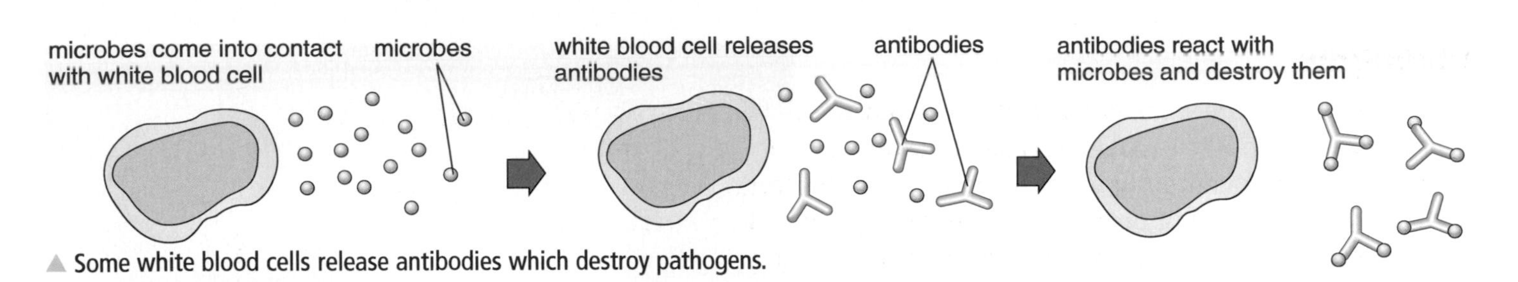

▲ Some white blood cells release antibodies which destroy pathogens.

Life-long protection

Once your white blood cells have destroyed a type of pathogen you are unlikely to develop the same disease again. This is because your white blood cells will recognise the pathogen the next time it invades your body and produce the right antibodies very quickly to kill the pathogen before it can affect you. This makes you **immune** to the disease.

Question

c *Some pathogens, such as the virus that causes the common cold, keep changing (mutating). Other viruses, such as the virus that causes chicken pox, tend not to change. Use this information to explain why people keep catching colds but usually only get chicken pox once.*

The graph shows what happens when someone is infected by a particular pathogen. The graph also shows what happens when the person is infected a second time by the same pathogen.

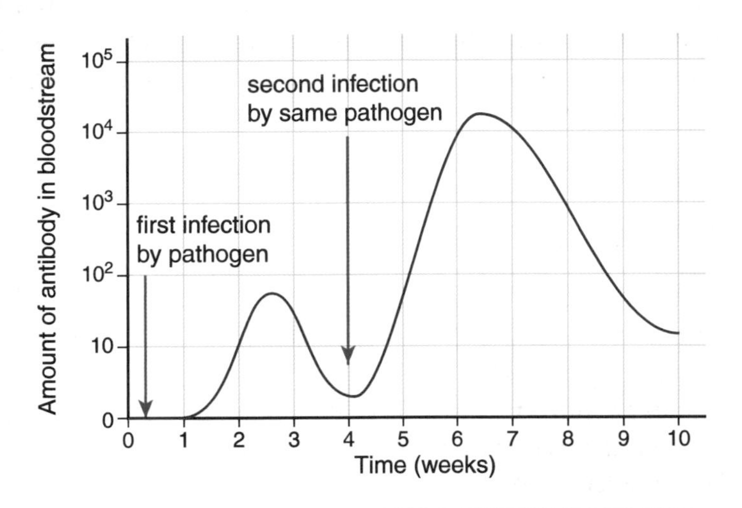

Questions

d *How long did it take to start producing antibodies: (i) after the first infection; (ii) after the second infection?*

e *Explain why antibodies were produced more quickly after the second infection.*

f *Suggest why the person did not become ill after the second infection.*

A global threat

Flu is a viral disease that affects many people every year. Most people recover within 1–2 weeks, but flu can cause serious illness and death, especially in very young children and old people. When an outbreak of flu affects thousands of people in a country it is called a flu **epidemic**. Sometimes flu spreads very rapidly around the world, affecting people in many countries. This is called **pandemic** flu. These alarming headlines show the concern about a possible pandemic outbreak of flu.

BIRD FLU BIGGER THREAT THAN TERRORISM

Is your country ready for bird flu?

New flu strain could kill millions around the world

Key points

- Microorganisms such as bacteria and viruses which cause disease are called pathogens. They produce toxins which make us feel ill.
- White blood cells protect the body against pathogens by ingesting them, producing antibodies to destroy them or producing antitoxins which counteract the toxins produced by pathogens.

Treating and preventing disease

Feeling ill

If all your defence systems fail and a pathogen gets inside your body, it starts to reproduce rapidly. You will eventually start to feel ill and show the **symptoms** of disease. The symptoms are the effects that the disease has on your body, such as a high temperature and headaches.

Some medicines relieve the symptoms of disease. For example, people take **painkillers** to ease aches and pains. Painkillers do not kill the pathogens that cause the disease. They just make you feel better while your body fights the pathogens.

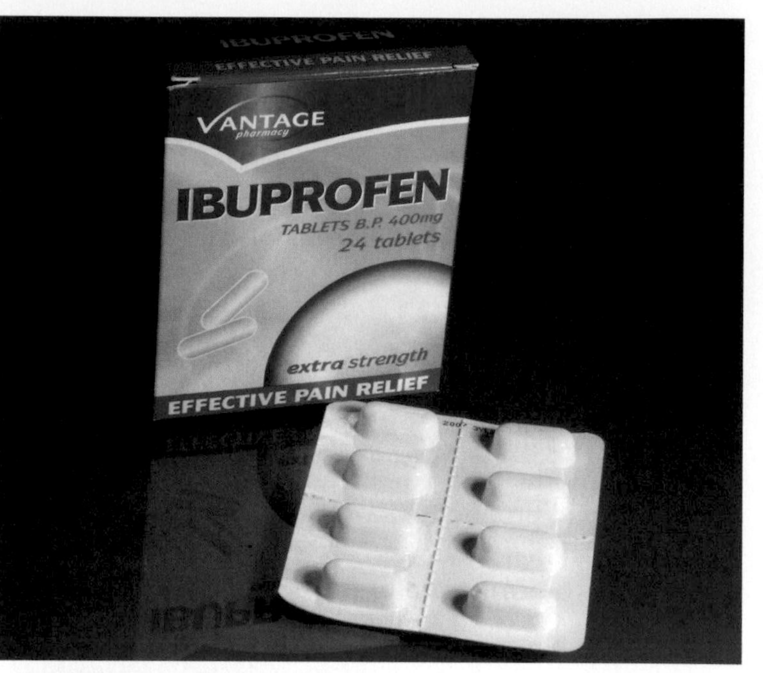

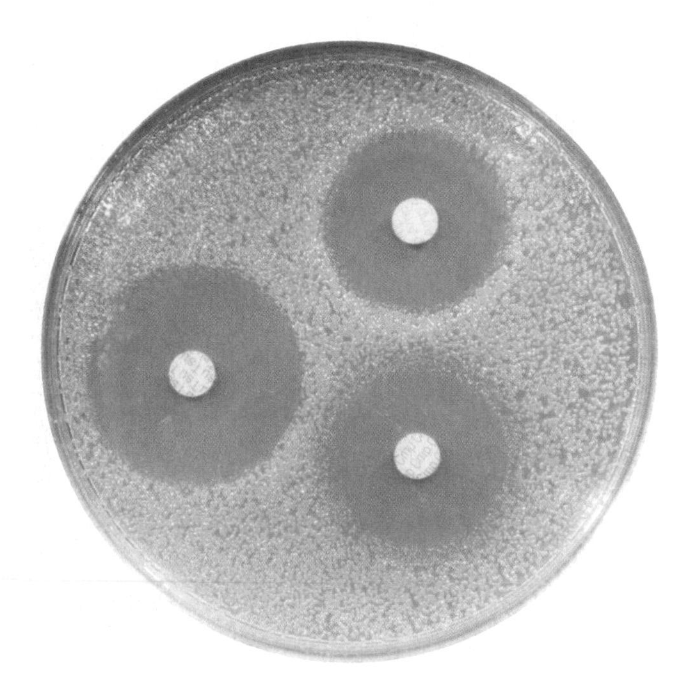

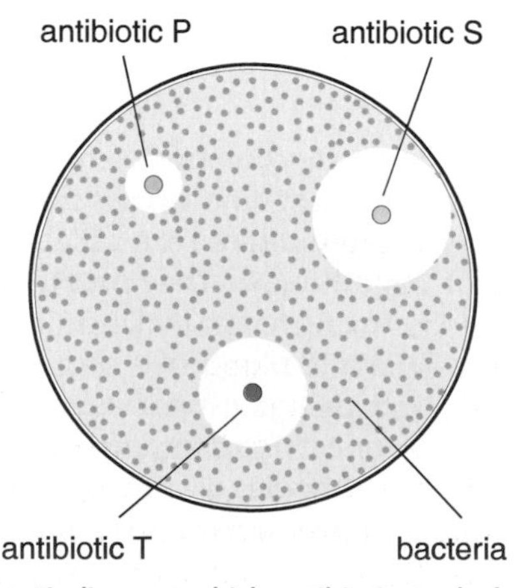

▲ Finding out which antibiotic works best.

Killing bacteria

Antibiotics are medicines that help to cure diseases caused by bacteria. You take antibiotics to kill bacteria that get inside your body. Doctors use many different antibiotics to treat people. **Penicillin** was the first antibiotic to be discovered.

Antibiotics can't kill viruses. Because viruses live and reproduce inside body cells, it is difficult to develop medicines that kill viruses without damaging body cells and tissues.

The effect of antibiotics on bacteria can be measured in the laboratory. This is done by using small discs of paper containing antibiotic. The discs are placed in a dish containing bacteria growing on a gel. The photograph shows the effect of the antibiotic. The clear zone that forms around the disc is where bacteria have been killed. A more effective antibiotic will leave a wider clear zone.

The diagram shows the results of testing different antibiotics.

Questions

a *Which antibiotic kills the most bacteria? What evidence supports your answer?*

b *To which antibiotic do the bacteria show resistance? Use evidence from the diagram to explain your answer.*

Resistance to antibiotics

Many types of bacteria have developed **resistance** to antibiotics (they are no longer killed by antibiotics). When an antibiotic is used, the non-resistant bacteria are killed but a small number of resistant bacteria remain. The resistant bacteria survive and reproduce. Continued use of the antibiotic causes the number of resistant bacteria to increase. This is an example of natural selection. To prevent more and more types of bacteria becoming resistant, it is important to avoid overusing antibiotics. This is why antibiotics are not used to treat non-serious infections like a sore throat. Doctors should only prescribe an antibiotic to treat a serious disease. By avoiding overusing antibiotics, you increase the likelihood that they will work when you really do need them.

Question

c *Explain how natural selection could lead to an increase in the number of antibiotic-resistant bacteria when patients fail to complete their treatment.*

Changing viruses

Flu viruses are always changing (mutating) to produce new strains. Pandemic flu occurs when a new flu virus is produced that is very different from previous strains. Because the new strain is so different people will have no immunity to it. This allows the new strain to cause more serious illness and to spread quickly from person to person.

A number of Asian countries have recently been affected by bird flu. The virus that causes bird flu has also infected a small number of people. Scientists are worried that the virus could eventually combine with a human flu virus. This could produce a new virus that could cause a deadly pandemic.

Question

d *Explain why the bird flu virus combined with a human flu virus would spread very rapidly and cause a pandemic.*

Key points

- Painkillers relieve the symptoms of diseases but do not kill the pathogens.
- Antibiotics such as penicillin can cure bacterial diseases by killing the bacteria. They do not work on viruses.
- Many bacteria have become resistant to antibiotics due to natural selection.
- Due to an increased understanding of antibiotics and immunity, the way we treat disease has changed.

Protecting against disease

A quick jab

It is not always necessary to suffer from a disease once before you become immune to it. When you were a young child you were probably **immunised** to protect you from very harmful diseases, such as whooping cough, measles and polio. **Immunisation** usually involves injecting or swallowing a **vaccine** containing small amounts of a dead or weak form of the pathogen. Because the pathogen is weak or inactive, the vaccine does not make you ill but your white blood cells still produce antibodies to destroy the pathogen. This makes you immune to future infection by the pathogen. Your white blood cells will quickly recognise the pathogen if it gets into your body and respond by producing antibodies. The pathogen does not get a chance to reproduce enough to make you ill.

Question

a *Explain why being given a vaccine protects you against a disease but does not cause you to develop the disease.*

The level of antibody in the blood after some vaccinations does not get high enough to give protection. In this case a second, or booster, injection of vaccine a few weeks later is needed. The graph shows the level of antibodies in a person's blood following a first and second injection of a vaccine.

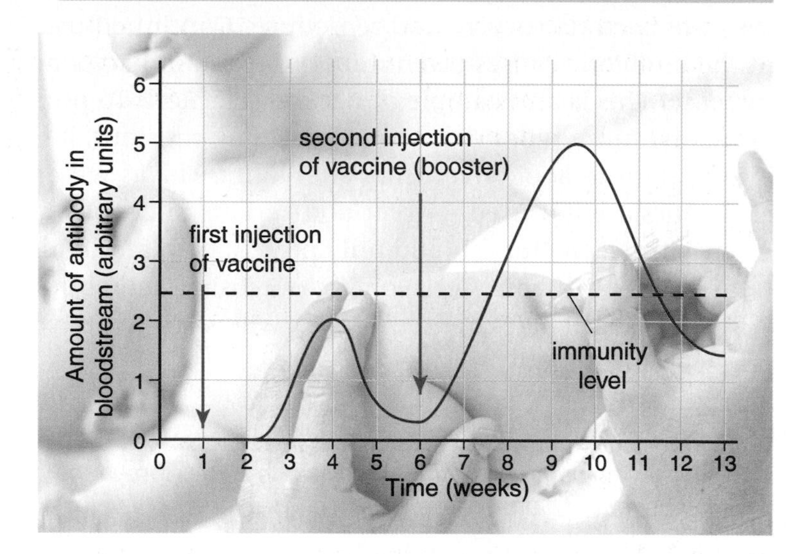

Questions

b *What was the difference in arbitrary units in the level of antibody between the first and second injection?*

c *Explain why the person became immune after the second injection but not the first.*

Immunisation programmes

Immunisation provides protection against several diseases that used to be very common in children. An example is the use of MMR vaccine – a combined vaccine that makes your body develop **immunity** to measles, mumps and rubella. Each of these diseases is caused by a virus that is easily spread from someone with the disease to someone who is not immune.

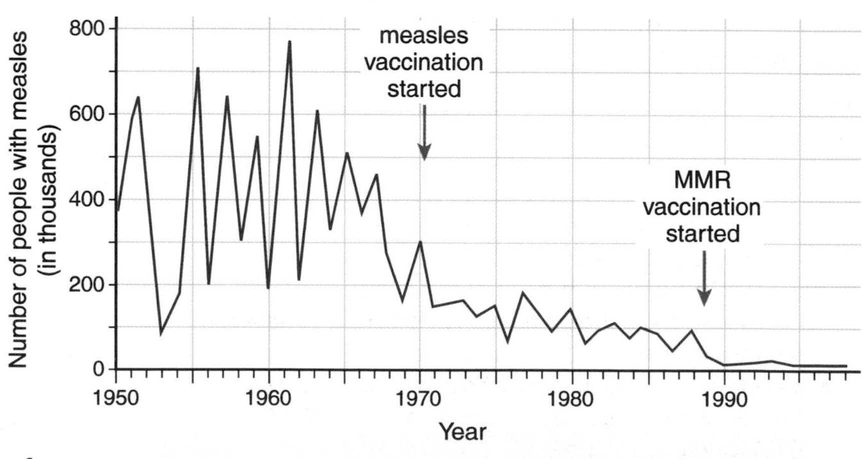

Vaccines such as MMR have saved millions of children from illness and even death. Before a measles vaccine was available, an average of 250 000 children developed measles and 85 children died every year. The graph shows the effectiveness of the immunisation programme against measles.

Question

d *(i) What was the maximum number of cases of measles in any one year before a vaccine against the disease was introduced?*
(ii) What was the maximum number of cases of measles in any one year after the introduction of the measles vaccine?

Concern about vaccines

Children who are not vaccinated are much more likely to develop serious illnesses. Whooping cough is a disease that can cause long bouts of coughing and choking, making it hard to breathe. The disease can be very serious and can kill babies under 1 year old. More than half the babies under 1 year old with whooping cough need to be admitted to hospital and many need intensive care.

In the 1970s parents were concerned about possible side effects of the whooping cough vaccine, and fewer children were vaccinated against whooping cough. As a result major outbreaks of whooping cough occurred, with thousands of children being taken into hospital.

Recent concerns about side effects of the MMR vaccine have led to a decrease in the number of babies receiving the vaccine. In a measles outbreak in 2002, 18 out of 20 patients had not received the MMR jab.

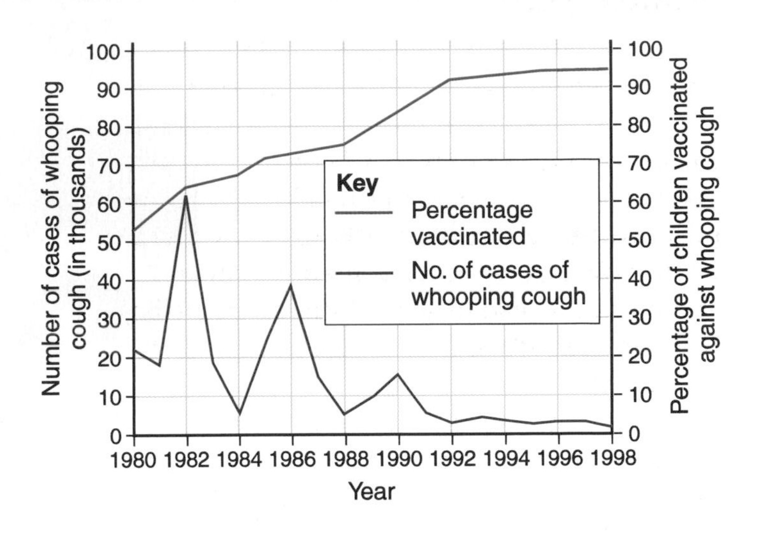

Question

e *(i) Explain why two major outbreaks of whooping cough occurred in 1982 and 1986.*
(ii) Describe the relationship between the percentage of children being vaccinated and the number of whooping cough cases since 1990.

Protecting against flu

Elderly people could become seriously ill if they caught flu. Doctors recommend that the elderly are vaccinated against flu each winter. The bar chart shows the number of people in the UK aged 65 and over and the percentage of those who were vaccinated against flu.

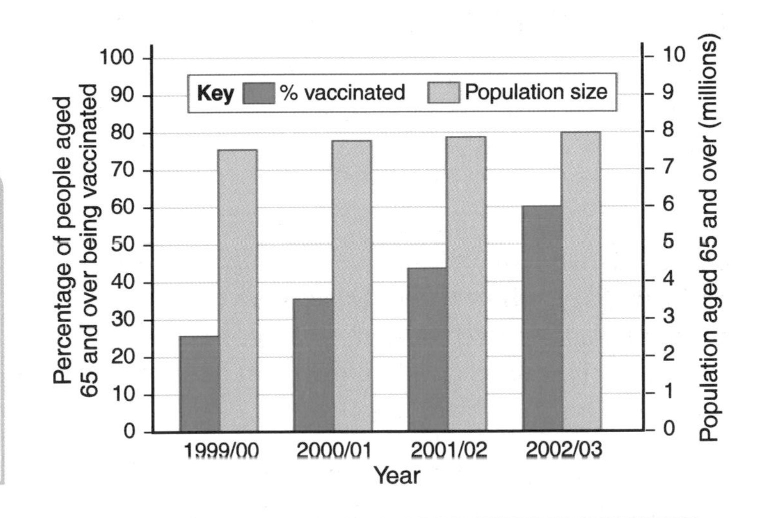

Questions

f *(i) Explain why doctors advise people to be vaccinated against flu every year.*
(ii) What evidence in the bar chart shows that the campaign to encourage the elderly to get vaccinated is working?

g *Calculate the total number of people aged 65 and over who were vaccinated against flu in (i) 1999/2000 and (ii) 2002/03.*

Key points

- People can be immunised using dead or inactive forms of a pathogen in a vaccination. This stimulates the white blood cells to produce antibodies and makes the body immune.
- MMR is an example of a vaccine. Some vaccinations may appear to have side effects and so it is necessary to weigh up the advantages and disadvantages of being vaccinated against a particular disease.

MRSA

MRSA stands for methicillin-resistant *Staphylococcus aureus*. *Staphylococcus aureus* (*S. aureus* for short) is a common type of bacterium often found on the skin of healthy people causing no harm at all. Sometimes *S. aureus* invades the skin and gets into the bloodstream causing serious infections. These serious infections need to be treated with antibiotics. Serious infections are more likely to occur in people who are already unwell.

HOSPITAL SUPERBUG KILLS BABY

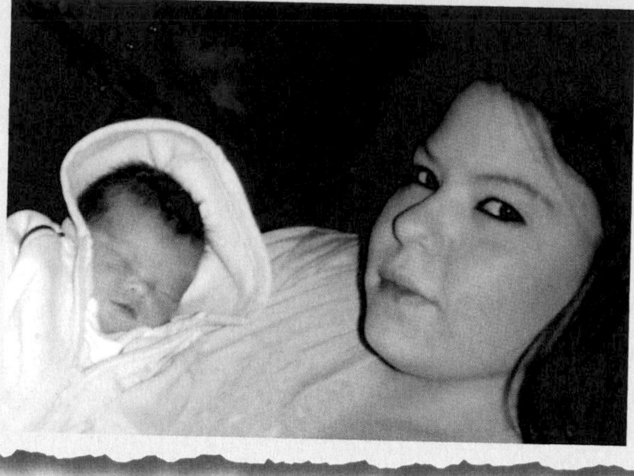

A one-day-old baby boy was killed by the hospital superbug MRSA.

Baby Luke was only 36 hours old when he died.

Luke was born showing no signs of bad health.

Antibiotic resistance

Methicillin is a powerful antibiotic drug. Antibiotics have been used to kill bacteria very successfully for many years. Bacteria that are not killed by antibiotics continue to multiply. This is why you are always advised to complete a course of antibiotics, even if you start to feel better. If you do not complete a course of antibiotics, it is likely that some bacteria will survive. The bacteria that survive will be more resistant to the antibiotic.

Questions

- ***a*** *Explain why patients are advised to complete a course of antibiotics.*
- ***b*** *Explain why the overuse of antibiotics has led to the development of so-called 'superbugs'.*

Antibiotics can still be used against MRSA. The infection may require a higher dose over a longer period of time to kill the more resistant bacteria. Alternatively a different antibiotic can be used to which the bacteria have less resistance.

Questions

- ***c*** *What was the percentage increase in antibiotic-resistant types of* S. aureus *between 1995 and 2000?*
- ***d*** *To test if a patient is infected with MRSA used to take 3 days. A test was developed in 2005 which can be completed in 2 hours. Explain why a test that can be carried out quickly is an advantage.*

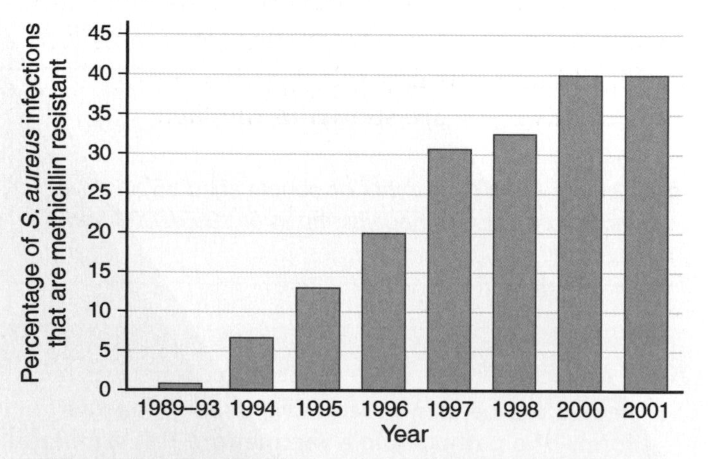

▲ This graph shows how the incidence of MRSA infections has increased over recent years.

NHS
National Patient Safety Agency
the bug stops here
Contaminated hands spread infections.
Improvement begins with you.
Clean your hands.
cleanyourhands campaign

Cleaner hospitals

Hand washing is an important way to prevent pathogens passing from one hospital patient to another. The first person to recognise the importance of hand washing was a Hungarian doctor called Semmelweiss. He worked in hospitals in the 1840s. At that time doctors had little knowledge of the cause of disease, so no attempt was made to prevent the spread of pathogens. Even though he did not know about bacteria and viruses, Semmelweiss argued that as doctors went from one patient to another they could be spreading life-threatening diseases. He made all the doctors working with him wash their hands after an operation and before visiting a new patient. Deaths on the hospital wards where Semmelweiss was in charge fell from 12% to just 1%.

Hand hygiene

A study of the importance of washing hands as a method of preventing the spread of MRSA was carried out in the 1990s. Trained staff observed doctors and nurses at timed periods to monitor if they washed their hands and how thoroughly this was done. They also measured the amount of hand-washing solution that was used. The results of this study are shown in the table.

Year	Amount of hand-washing solution used (litres per 1000 patient-days)	MRSA infections per 100 admissions
1993	3.5	0.50
1994	4.1	0.60
1995	6.9	0.48
1996	9.5	0.32
1997	10.9	0.25
1998	15.4	0.26

Questions

- **e** *Which method of measuring hand washing used in the study is the most reliable? Explain your answer.*
- **f** *Explain the relationship between the amount of hand-washing solution used and the number of new MRSA cases.*

Key point

- **The mutation and resistance of bacteria and viruses makes treatment of illnesses difficult and can lead to epidemics and pandemics, for example bird flu or MRSA.**

End of unit questions

1 The table is about the functions of parts of the body.

Match words from the list with each of the numbers **1–4** in the table.

A blood **B** gland

C kidney **D** skin

	Function
1	produces a fluid that helps to regulate body temperature
2	produces hormones
3	transports hormones
4	produces urine

2 This is part of the report of an investigation done by a student.

The purpose of this experiment was to determine the effect of light intensity on visual acuity. My hypothesis was that the best amount of light to see in would be a medium-bright light, about 245 lux, because I find it easiest to see when a light isn't too dim, but it's not so bright that I can't work very well.

To measure the responding variable each subject read the letters on a visual acuity wall chart.

I did the experiment in a windowless room.

I put a chair 6.1 metres away from the wall chart.

I set up a rheostat to dim the lights.

I used a light intensity meter to measure the amount of light reflected off the wall chart in each light level.

I got 20 classmates to volunteer.

I adjusted the light so it reflected from the chart at 780 lux.

A student sat in the chair and read the letters on the wall chart as far as possible.

The experiment was repeated at 260 lux, 90 lux, 30 lux and 10 lux.

The experiment was then repeated for all remaining students.

My results are shown in the table.

Student	Number of errors made by each student at each light intensity				
	759 lux	243 lux	81 lux	27 lux	9 lux
1	2	1	2	5	5
2	3	3	7	7	8
3	0	0	1	2	7
4	3	4	2	6	7
5	0	1	3	5	12
6	3	2	1	6	10
7	1	2	1	2	3
8	7	7	8	8	15
9	5	4	3	5	8
10	2	0	1	1	3
11	3	3	4	7	8
12	0	1	0	0	4
13	4	4	5	6	13
14	3	2	2	3	2
15	3	4	5	4	13
16	2	1	3	3	4
17	2	4	1	5	10
18	4	5	6	6	12
19	0	1	0	0	3
20	0	1	4	4	5

a What kind of variable is

i reflected light intensity? *(1 mark)*

ii number of errors? *(1 mark)*

b i Name **one** factor that the student controlled. *(1 mark)*

ii Name **one** factor that the student did not control. *(1 mark)*

c The student used 20 volunteers. Why was this better than using 5 volunteers? *(1 mark)*

d Work out the average number of errors at each light intensity. Write your results in a table. *(2 marks)*

e Describe **one** pattern you can see in the results. *(1 mark)*

f Describe **one** way of showing these results graphically. Say whether you would use a bar chart or a line graph, and say what you would plot on each axis. *(3 marks)*

g Is there any evidence to suggest that the student's hypothesis is correct, that the best amount of light to see in would be a medium-bright light, about 245 lux? Explain your answer. *(2 marks)*

3 The passage contains information about the 'morning after' pill.

> **What does the pill do?**
> The 'morning after' pill stops you from becoming pregnant. It's not 100% effective, but the failure rate is quite low – probably about 10%, and rather better than that if you take it as early as possible.
>
> The pill is believed to work principally by preventing your ovaries from releasing an egg, and by affecting the womb lining so that a fertilised egg can't 'embed' itself there.
>
> In Britain and many other western countries, it is not legally regarded as an abortion-causing drug, but as a contraceptive.
>
> **Who is the pill for?**
> It's now very widely used by women (especially young women) who have had unprotected sex. And in particular, it has proved of value to rape victims, couples who have had a condom break and women who have been lured into having sex while under the influence of drink or drugs.
>
> **Is it dangerous to use?**
> You might feel a little bit sick after taking it, but only about 1 woman in every 60 actually throws up. Uncommon side effects are headache, tummy ache and breast tenderness.
>
> **If the pill didn't work, and I went on and had a baby, could the tablet damage it?**
> We simply don't know the answer to this question. At present, no one has shown any increase in abnormalities among babies who have been exposed to the morning after pill. But past experience does show that other hormones taken in early pregnancy have harmed children.

a Some people regard this pill as an abortion-causing drug. Explain why. *(2 marks)*

b **i** Some people think that this pill should only be available on prescription. Suggest why they think this. *(1 mark)*

ii Others say it should be freely available 'over the counter'. What do you think? Give reasons for your answer. *(2 marks)*

c Scientists are uncertain whether the pill might cause abnormalities among unborn children. Suggest why. *(2 marks)*

4 A laboratory technician was cleaning out a cupboard. Dust from the cupboard made her sneeze.

a In this response dust is
A the coordinator **B** the effector
C the receptor **D** the stimulus.

b In this response the receptor is in
A the brain **B** the eye
C the nose **D** the spinal cord.

c In this response the coordinator is
A the brain **B** the nose
C the spinal cord **D** a synapse.

d Chemical transmitters are involved in
A sending impulses along sensory neurones
B sending impulses across the gap between a sensory neurone and a relay neurone
C sending impulses from one end of a relay neurone to the other
D sending impulses from a motor neurone to a relay neurone.

5 The graphs show how the concentrations of the hormones that control the menstrual cycle vary over 28 days.

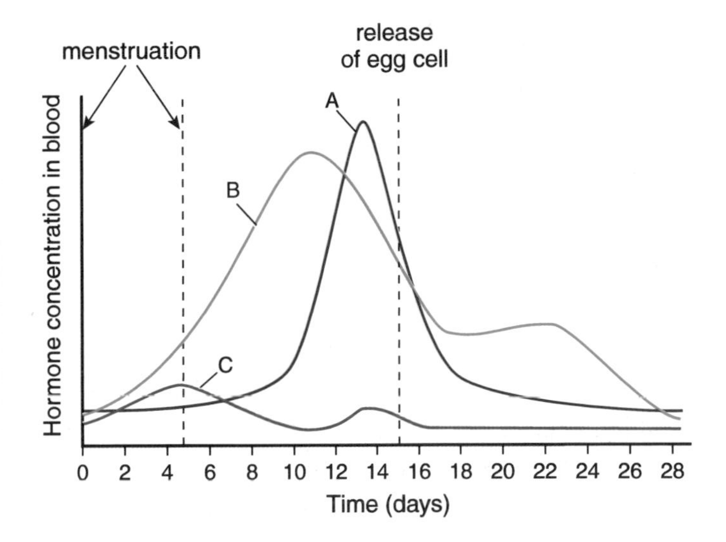

a Name
i hormone **A** *(1 mark)*
ii hormone **B** *(1 mark)*
iii hormone **C**. *(1 mark)*

b Explain why hormone **C** can be used as a fertility drug. *(2 marks)*

c Hormones similar in their effect to hormone **B** can be used as contraceptive drugs. Explain why. *(2 marks)*

6 The graph shows the average amount of cholesterol in the blood of people at different ages.

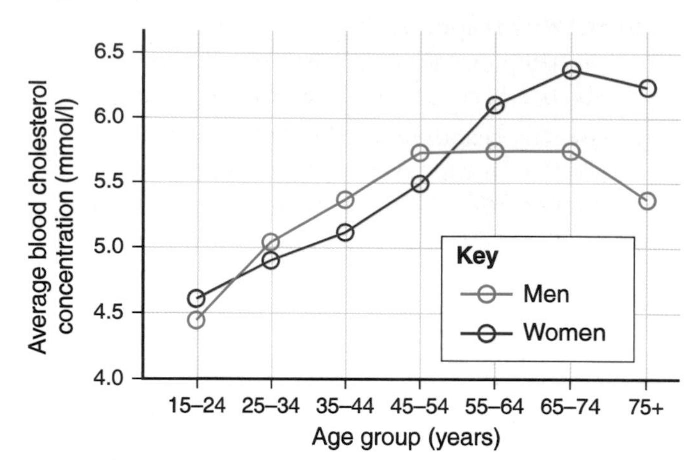

a The average blood cholesterol level for a 60-year-old woman is
A 5.5 **B** 5.7 **C** 6.2 **D** 6.6

b Using the data from the graph, which of the following has the highest risk of developing heart disease?
A 30-year-old men
B 50-year-old women
C 65-year-old men
D 60-year-old women

c The cholesterol in the blood is
A absorbed from the gut
B made in arteries
C made in the heart
D made in the liver.

d The amount of cholesterol in the blood increases when people eat
A salt **B** saturated fats
C starch **D** unsaturated fats.

7 Adults should eat no more than 6 g of salt a day. You can work out how much salt and fat there is in foods by reading the label. The amount of salt is usually given as the amount of sodium.

Amount of salt = amount of sodium × 2.5

The label on a take-away meal has the following information:

	100 g provides
Fat	8.0 g
of which saturates	6.7 g
polyunsaturates	0.3 g
Sodium	0.6 g

The mass of the whole meal is 300 g.

a Calculate the total amount of
i saturated fats *(1 mark)*
ii salt in this meal. *(1 mark)*

b Why is eating a lot of salt bad for your health? *(1 mark)*

c Explain why eating too much saturated fat increases the risk of heart attack. *(1 mark)*

8 The table is about the effects some chemicals have on the body.

Match words from the list with each of the numbers **1–4** in the table.

A alcohol **B** carbon monoxide
C nicotine **D** a solvent

Chemical	Effect on body
1	affects behaviour when inhaled
2	combines irreversibly with haemoglobin
3	makes it difficult to give up smoking
4	slows down the transmission of nerve impulses

9 New drugs must be tested before use. A form of ultrasound is being used by scientists to test the effectiveness of drugs designed to break down potentially life-threatening blood clots. Scientists from King's College of Medicine in London claim the technique provides a more reliable measure of the effectiveness of drugs than was previously available, and could remove the need to test new drugs on animals.

They have used the technique to test the effectiveness of a new drug – GSNQ – which dissolves blood clots. This reduces the risk of strokes. GSNQ was compared with the standard treatment of aspirin and heparin in a group of 24 patients who underwent surgery to clean a major blood vessel in the neck. Patients treated with GSNQ were found to have significantly lower numbers of clots during a 3-hour period after the operation.

A member of the research team said: 'Before this technique assessing a drug meant either doing animal tests, or taking blood from people and studying it under the microscope. Neither was a very good measure of what would actually happen when the drug was used in people.'

New drugs will still have to be thoroughly assessed in large-scale clinical trials, but the new technique will help scientists to decide which products should go to a full trial.

a Explain why new drugs have to be tested before they go on sale. *(1 mark)*

b How did the scientists measure the effectiveness of GSNQ? *(1 mark)*

c Give **three** advantages of the above method of testing GSNQ over traditional drug-testing methods. *(3 marks)*

d Explain why GSNQ will still need to be assessed in large-scale clinical trials before it is approved. *(2 marks)*

10 The graph shows the percentage of infections caused by MRSA in a hospital intensive care unit, between 1989 and 2003.

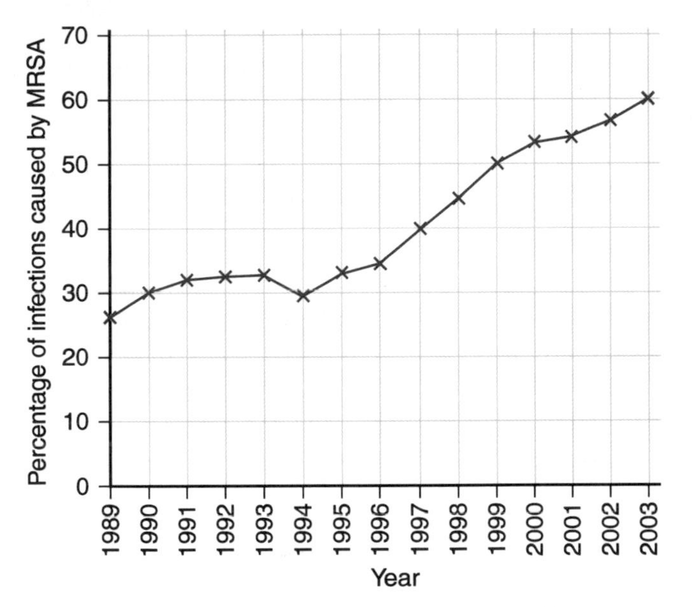

a The percentage of infections caused by MRSA became greater than 50% in

A 1998 **B** 1999 **C** 2000 **D** 2001

b MRSA stands for methicillin-resistant *Staphylococcus aureus*. Methicillin is an antibiotic. Antibiotics are used

A to kill viruses
B to kill bacteria
C to develop immunity
D as painkillers.

c Some pathogens are resistant to methicillin. This means

A a lower strength of methicillin is needed
B methicillin kills more pathogens
C no type of antibiotic will kill the pathogens
D some pathogens are not killed by methicillin.

d Which of the following would be the least effective in stopping the spread of MRSA in hospitals?

A keeping patients with MRSA infections away from other patients
B washing hands after visiting each patient
C wearing a fresh pair of clean gloves when treating patients
D testing patients for MRSA

11 The bar chart shows the number of cases of influenza in a large city in the UK.

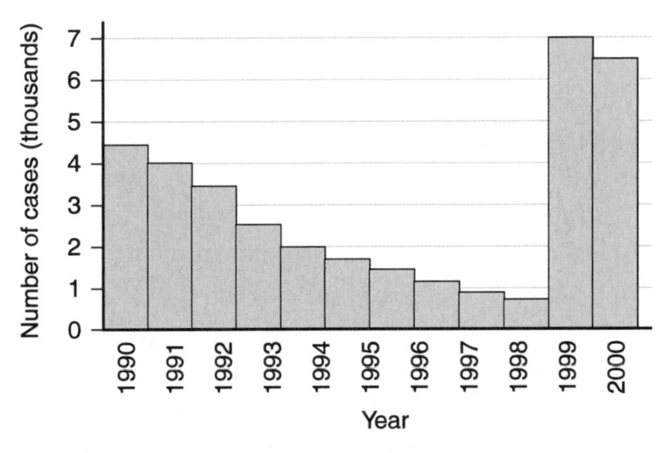

a The decrease in the number of cases of influenza between 1990 and 1999 occurred because more people

A became immune to the flu virus
B died from flu
C received antibiotics
D became infected with other diseases.

b The large number of cases in 2000 was likely to have occurred because

A a different type of flu virus was produced by mutation
B more people became immune to the flu virus
C more people were vaccinated against flu
D there were more old people.

c Most people who get flu recover in a few weeks. This is because their white blood cells destroy the virus by producing

A antibodies **B** antitoxins
C enzymes **D** vaccines.

d When people are vaccinated against flu they receive a vaccine which contains

A antibiotics **B** antibodies
C inactive viruses **D** resistant bacteria.

B1b 3 Variation and evolution

We often read about people trekking across freezing Arctic environments. But humans still find it difficult to survive in extreme cold without special clothing. Clothes like these keep the trekker warm and dry in places where the temperature can drop to –30 °C.

Survival kits

Many animals live in Arctic areas all year round – they are adapted to cold surroundings. Other animals are adapted for survival in very hot, dry environments. **Adaptations** are ways in which organisms have become specialised to survive in a particular habitat. These adaptations result from changes in genes. Competition is another factor that affects survival. Some adaptations enable an organism to compete more successfully. The genes for successful adaptations are passed on to future generations. This is how organisms evolve.

▲ Essential kit for Arctic explorers.

Changing genes

Genetic engineering, or genetic modification, techniques enable scientists to change the genetic make-up of organisms. This may involve removing faulty genes and adding healthy genes, or even developing new varieties of plants and animals with genes transferred from a completely different species. Genetically modified bacteria are used in the manufacture of drugs, such as insulin which is used to treat diabetes. Genetically modified crops (GM crops) are grown to increase food production.

Most scientists believe that food produced from GM crops is a way of increasing food production, especially in developing countries. However, many people believe that GM food is unsafe and that it is wrong to alter the genetic make-up of organisms.

◀ Growing rice plants containing an added gene to make them resistant to insect pests has led to a 10% increase in yield.

PUBLIC OPINION AGAINST GM FOODS

The public has hardened its stance against genetically modified foods. In a survey of almost 1000 people, carried out on behalf of the consumer magazine *Which?*, 61% were concerned about the use of GM material in food production.

Malcolm Coles, editor of *Which?*, said, 'Consumers clearly don't want GM foods and are hardening their stance against it.' Around a quarter support the growing of GM crops in the UK. Two years ago this figure was 32%.

Preventing inherited disease

In future it may be possible to use genetic engineering to prevent babies being born with inherited diseases. This would involve replacing faulty genes in sperm, eggs or in newly fertilised eggs. Many people believe that using genetics in this way is too dangerous and could lead to people choosing other characteristics of their children – creating 'designer babies'. Deciding whether something is right or wrong is an ethical question. Genetic engineering is controversial because it raises many ethical questions.

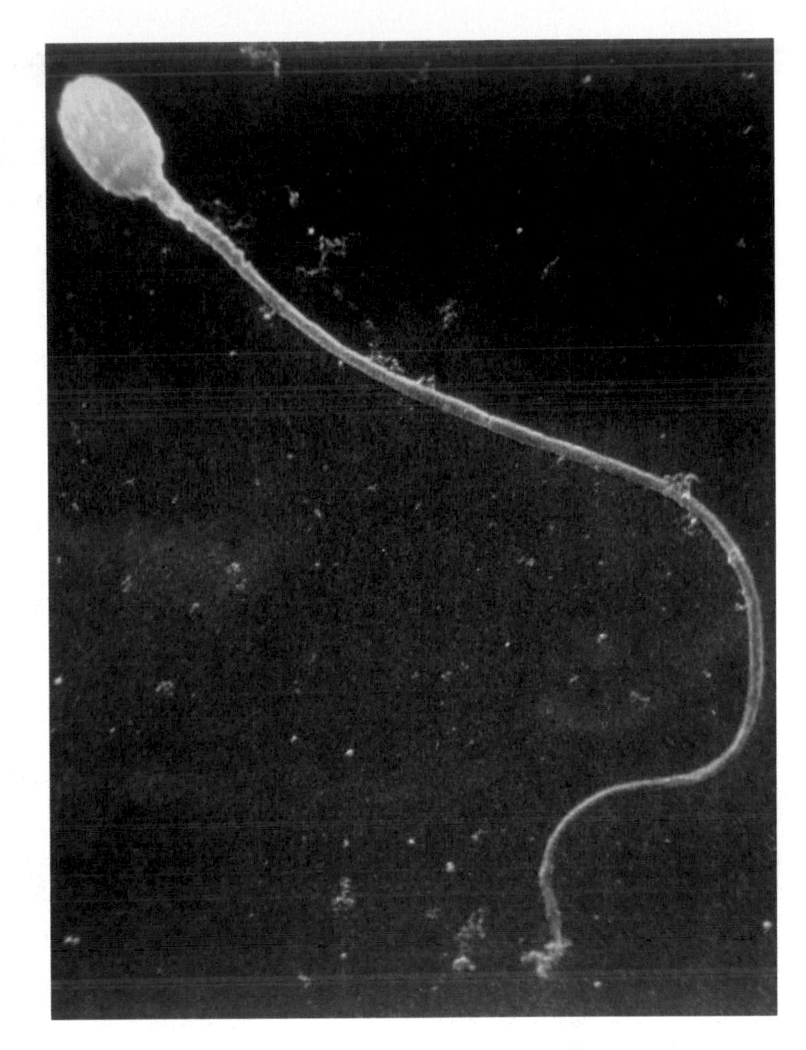

▶ Genes can be changed in sperm cells and fertilised eggs so that organisms develop with desired characteristics.

Creating the stuff of life

The tragic case of the Hashmi family has fuelled the debate over what is meant by 'designer babies'. Raj and Shahana Hashmi want to have a baby by in vitro fertilisation (IVF or test-tube baby). They want the baby to be free of inherited disease, and they want the baby's tissues to match the tissues of their 6-year-old son Zain, who suffers from a very rare blood disorder. By having a tissue match Zain would be able to have cells transplanted to save his life. The family's case is being considered by the House of Lords.

Think about what you will find out in this section

How are organisms adapted to their surroundings?	How are characteristics inherited?
How can we produce plants and animals with the characteristics we prefer?	How can the genetic make-up of plants and animals be changed?
Why have some species of plants and animals died out, and new species developed?	What are the advantages of cloning tissues and embryos?
What are the arguments for and against genetic modification?	Why are there different theories to explain evolution?

Competing for food

Sable Island is a remote island about 300 kilometres off the coast of Canada. The island is an attractive breeding ground for Grey and Harbour seals.

The Grey seals of Sable Island have been thriving, but Harbour seal populations have plummeted in recent years. Sharks like to eat both species of seal. Harbour seal pups and adults are smaller than Grey seal pups and adults. Grey seals and Harbour seals both eat sand lances, tiny fish that live just off the shores of Sable Island. To catch these fish, Grey seals swim close to the ocean floor and dig their snouts into the sand to find hiding fish. Harbour seals follow schools of sand lances and catch fish that wander from the school.

▲ Harbour seal.

▲ Grey seal.

Question

a *Suggest reasons why the Harbour seal is being out-competed by the Grey seal on Sable Island.*

Competing for territory

The jaguar is a solitary animal and will avoid face-to-face confrontation by marking out territory boundaries. It does this by leaving 'scats', which are small piles of **faeces**, and by patrolling the boundaries and 'calling', which can be heard up to 1.6 km away. The size of a jaguar's territory depends on food availability. In an area where food is plentiful, such as a forest, a jaguar can survive in a circular area of about 5 km in diameter. Where food is scarce, it may need to roam over an area of 500 km^2.

Question

b *What is the advantage to a jaguar of marking and patrolling a territory?*

Competing for mates

A population of red deer live on the Scottish Isle of Rhum. Males spend much of the year living amicably in large bachelor groups. In late summer the males move to traditional mating areas. These are areas of rich grassland which support a large number of females. The males have grown new antlers and they use these to fight for the prime spots in the mating areas. The females will only mate with the males who can defend the prime sites. After the mating season the males shed their antlers and return to their bachelor herds.

Fighting for mates.

Question

c *Suggest the advantage to the deer population of males fighting for mates.*

Planting crop seeds

Farmers who grow crops from seeds need to know how close together they can grow crop plants to get the best results. Automated seed planters can be programmed to sow a set number of seeds per square metre.

Cereal crops are grown to produce seeds. The graph shows the results of an investigation to find the optimum number of cereal seeds to sow.

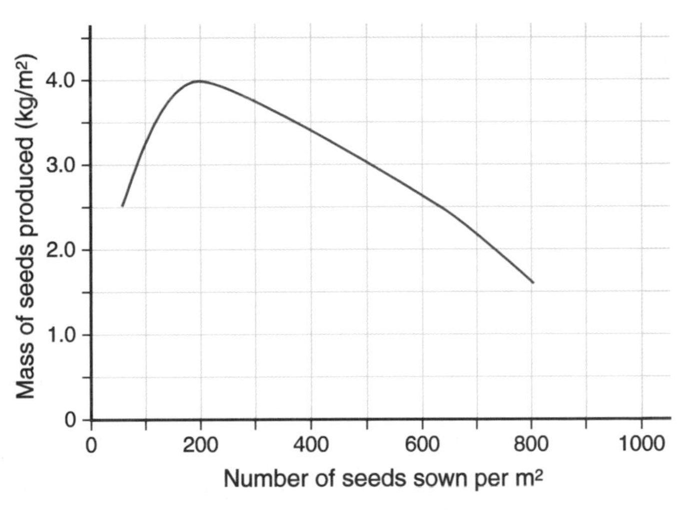

The seed yield of a crop depends on the number of plants grown per square metre.

Plants produce their own food by photosynthesis. They use energy from light to convert carbon dioxide and water into carbohydrates. They need other nutrients from the soil to make proteins.

Question

d *(i) Copy and extend the graph to estimate the seed yield if 1000 seeds are planted per square metre.*
(ii) How many plants would you advise the farmer to plant per square metre?
(iii) What do cereal plants compete with each other for?

Key points

- Animals compete with each other for food, territory and mates.
- Plants compete with each other for light, water and nutrients.

The bigger the better

Animals lose heat at the surface of their bodies. The larger an animal, the smaller is its surface area compared to its volume. This means that if two animals are shaped identically, the larger one will maintain its body temperature more easily.

This explains why there are no small birds in Antarctica. Even the smallest penguin, the Rockhopper, has a mass of 2.5 kg and Emperor penguins can grow to the size of a 10-year-old child.

The nineteenth century biologist Carl Bergmann observed that birds and mammals of the same species tended to be larger and heavier when they lived in colder climates. These observations led to 'Bergmann's rule' – that there is a correlation between body mass and average annual temperature.

Since then biologists have made many observations to test Bergmann's rule. The graph shows some of their results.

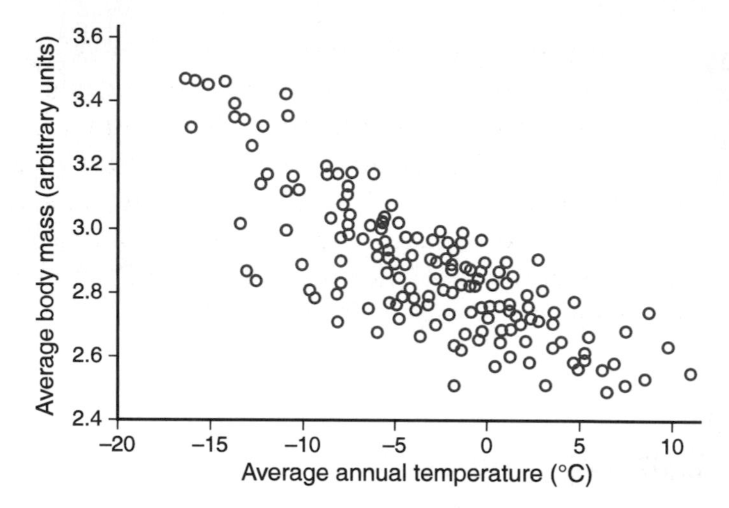

Question

a *Do these data support Bergmann's rule? Explain the reasons for your answer.*

Body shape

Joel Allen was a scientist who gathered data relating climate to variation in animals. Allen suggested that the ratio between body weight and surface area of warm-blooded organisms increases in colder climates in order to minimise heat loss. The extremities of organisms (limbs, tails and ears) are longer in warm climates than in cold climates because they act as heat-radiating organs.

The photographs show the Kit fox and the Arctic fox. The Arctic fox lives in cold conditions; the Kit fox lives in hot conditions.

▲ Kit fox.

▶ Arctic fox.

Question

b *Do these animals comply with Allen's rule? Explain the reasons for your answer.*

Insulation

Penguins have a thick layer of fat called blubber under the skin. Blubber is a very good insulator. Penguins don't use their feathers for flight; they use them to keep warm. Their feathers are short and consist of fine woolly hairs. The hairs trap air next to the body. When it gets very cold the penguins puff their feathers out to trap even more air. Air is a poor **conductor** of heat. That is why double glazing works – the two panes of glass trap an insulating layer of air. Feathers from birds like geese and ducks can be used to fill duvets to keep us warm at night.

Huddling

In winter the Antarctic temperature drops to –30 °C and the wind speed can reach 200 km per hour. In these conditions, penguins huddle together in large groups. Most of the penguins are then sheltered by a layer of penguins. They all take a turn on the outside! The penguins keep warm because the surface area of the whole group is reduced – it is like having one very large penguin.

Penguins huddle together for protection against the wind.

Caribou

Caribou have two layers of fur covering their bodies. They have fine, crinkly underfur and a thick coat of guard hairs on top. The guard hairs are hollow and filled with air.

Well insulated against the cold.

Question

c *Explain how guard hairs help the caribou to survive in the Arctic.*

The polar bear

Polar bears are well adapted to life in the Arctic.

Question

d *Look at the photograph of the polar bear. Suggest four ways in which polar bears are adapted to life in the Arctic.*

Key point

- Animals may be adapted for living in cold places by reducing heat loss by:
  - reducing their surface area, e.g. smaller ears and shorter limbs
  - insulation, e.g. long hair or a thick layer of fat
  - behaviour, e.g. huddling to reduce the total surface area of all the animals.

Surviving heat and attacks

Death Valley

Death Valley in California, USA, is one of the hottest and driest places on the Earth. The temperature can rise to 55 °C. There is rarely more than 2 cm of rain per year. Death Valley got its name from the number of gold prospectors who died there – mainly from thirst. But many species of animals and plants live and breed there. These organisms are adapted to survive in the dry environment.

One of the most inhospitable places on Earth.

The Joshua tree is one of the few tree species able to live in Death Valley.

The Joshua tree

The main problem facing desert plants is dehydration. Heat from the Sun evaporates water from their surfaces. The key to surviving in the desert is to collect as much water as possible when it rains and store it. To do this the Joshua tree has two sets of root systems. One set is a shallow root system; the shallow roots only reach down to 50 cm, but they spread out over a wide area to catch rainwater. The other set of roots stores any surplus water in undergound bulbs. The bulbs are buried up to 10 m under the soil. A bulb can reach 1 m in diameter and have a mass of 20 kg.

Question

a *Suggest two advantages of the bulbs growing up to 10 m underground.*

The creosote bush

The creosote bush has several adaptations for surviving in the desert:

- its small leaves are covered by a waxy substance
- the leaflets fold together to decrease surface area
- during extremely dry periods, the leaves are shed.

Question

b *Explain how each of these adaptations helps the plant to survive in the desert.*

Spring leaves and summer leaves

Some plants produce different types of leaves in the wet season and the dry season. The table shows the dimensions of the leaves in one of these plant species.

Leaf dimension	Spring leaf (wet season)	Summer leaf (dry season)
length (mm)	30	50
maximum width (mm)	10	1
surface area (mm^2)	300	150
volume (mm^3)	60	60

Question

c *(i) Calculate the ratio of the surface area to the volume of the two types of leaf.*
(ii) Explain the advantage to the plant of producing summer leaves.
(iii) Give one disadvantage to the plant of producing summer leaves.

The jack rabbit

We cool ourselves down by sweating. But sweating uses up precious water. The jack rabbit uses its big ears to cool down. Its ears act like the radiator of a car. A car radiator receives hot water from the car engine, gives out heat to the environment and returns cooled water to the engine. In the same way, the jack rabbit's ears receive warm blood from the body and radiate heat to the environment. The cooled blood then returns to the body. This method of cooling conserves water.

▲ Ears are not just for hearing.

Defence mechanisms

Plants have many ways of deterring animals from eating them. These include structural adaptations and chemical adaptations.

Question

d *How is the honey locust adapted to protect itself from hungry herbivores?*

▲ Honey locust tree.

Hemlock is a poison produced by hemlock plants. The plant produces the poison to deter animals from grazing on it. To warn animals off, the plant smells like rotting vegetables.

▲ Hemlock.

▶ Danger signal.

The bright stripes on the monarch butterfly larva advertise its presence. The larva feeds on milkweed. Milkweed contains chemicals that are poisonous to birds, but not to the larva. Any bird that eats a larva will find it very unpleasant and will not eat another larva.

Key points

- Plants may be adapted for surviving in hot places by having a thick, waterproof covering to reduce water loss, reducing the area of their leaves to reduce water loss, storing water and having long roots to collect water.
- Animals may be adapted for surviving in hot places by increasing their surface to increase heat loss, e.g. having large ears.
- Animals and plants may be adapted to cope with specific features of their environment, e.g. thorns, poisons and warning colours to deter predators.

Collecting and analysing ecological data

How many woodlice live in a wood?

A quick look at the photograph will tell you that the woodlice are not distributed evenly across their habitat. Because of this it is not easy to find how many woodlice there are in this part of the wood. Instead of trying to count all the woodlice, we can use sampling. We count the numbers in a small area, then use this number to estimate the total number.

Question

a *What assumption do we make when calculating the total population size in this way?*

Woodlice live on the floor of a wood.

Quadrats

The most common method of sampling animals and plants is using the quadrat. This is a square frame, usually of side either 50 cm or 1 m. Quadrats are often subdivided into 10 cm squares.

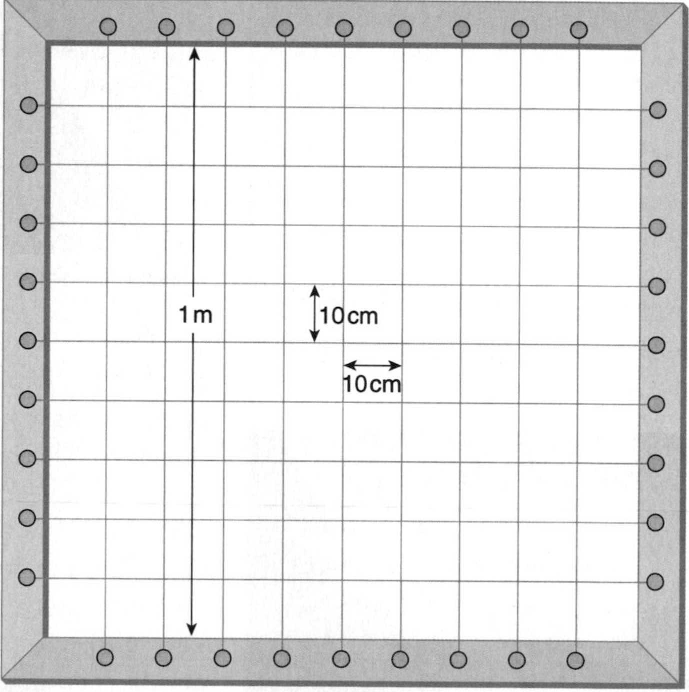

A 1 m quadrat divided into 10 cm squares.

A group of four students each placed a 10 cm quadrat on the floor of the wood. The diagram shows their quadrats.

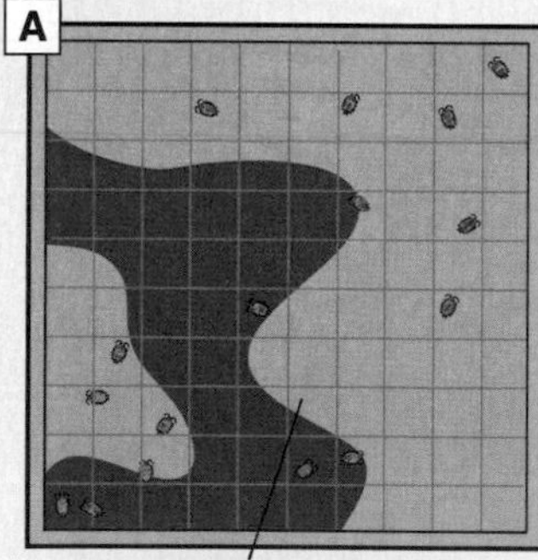

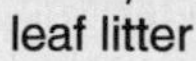

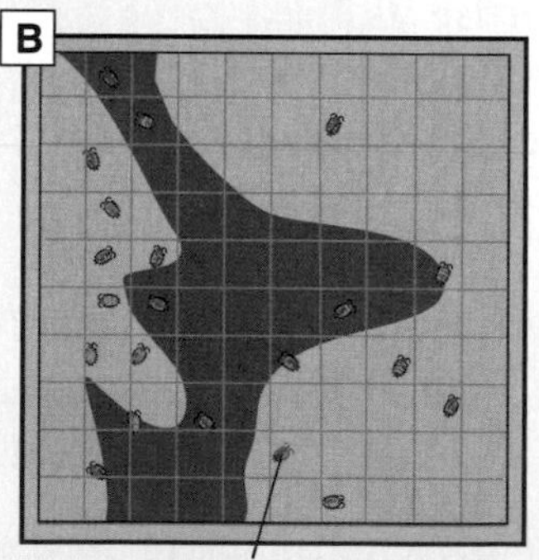

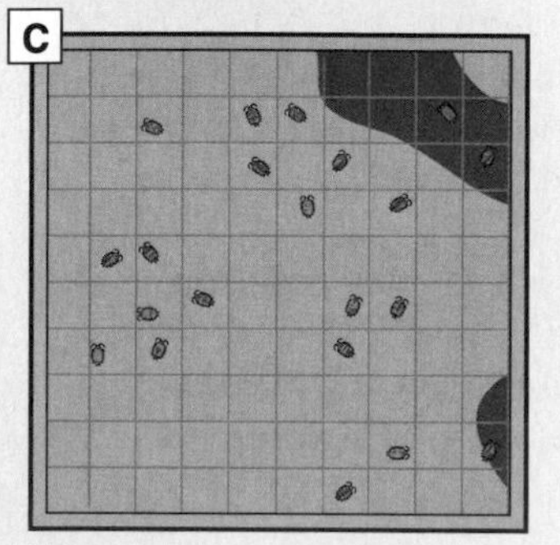

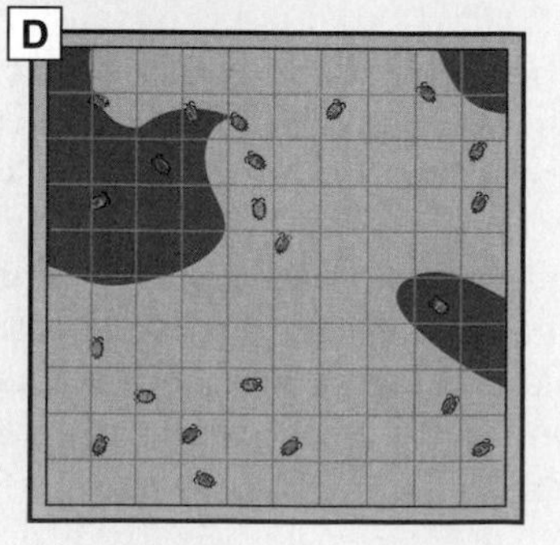

Question

b *(i) Count the number of woodlice in quadrat A. Use this number to estimate the number of woodlice in 1 m² of woodland.*
(ii) Now calculate the average number of woodlice in quadrats A, B, C and D. Use this number to estimate the number of woodlice in 1 m² of woodland. Compare your answer with that in part (i). What does this tell you about using quadrats to estimate population size?

Uneven distribution

The woodlice are not evenly distributed on the floor of the wood.

Question

c (i) Where are most of the woodlice found?
(ii) Suggest a hypothesis to explain this distribution.
(iii) Design an investigation to test your hypothesis. You should include: the independent variable, the dependent variable, how you intend to measure the dependent variable, the control variables.

Clover growing amongst grass.

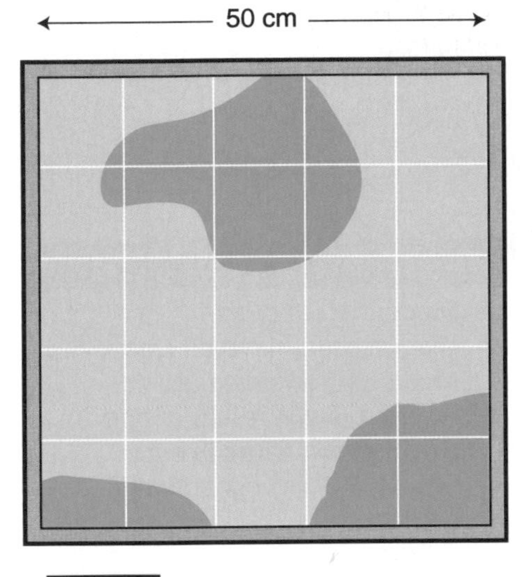

Estimating cover

Clover grows in clumps amongst grass. A student wanted to find out the area of a lawn that was covered by clover. She placed a 50 cm quadrat on the lawn as shown in the diagram.

Question

d (i) To estimate the area of lawn covered by clover, count any square more than half covered by clover as a clover square. Use your answer to calculate the percentage of the quadrat covered by clover.
(ii) How could you modify this method to get a more accurate estimate of the area of the quadrat covered by clover?

Moorland plants

Bracken and bilberries both grow on moorland. A group of students wanted to find out if the distribution of bracken and bilberries was affected by the pH of the soil. Using a large number of quadrats, they calculated the percentage area covered by each species in soils with different pH. The table shows their results.

pH	Average percentage cover	
	Bilberry	Bracken
4.8	82	0
5.0	78	12
5.2	66	26
5.4	55	39
5.6	38	56
5.8	16	65
6.0	8	71

Question

e (i) Plot the results on **one** *bar chart to show how pH affects the distribution of the two species.*
(ii) Use the bar chart to describe patterns in the data.
(iii) Suggest an explanation for the distribution of the two species.

Key points

- We can use quadrats to sample the distribution of organisms in a habitat.
- The larger the number of quadrats, the more reliable the results.
- The distribution of organisms in a habitat is affected by many factors including light, shade, temperature, water, ions and pH.

Passing on information

There's no-one like you!

There may be many people in your school but it is easy to tell them apart. Even though there are millions of people in the world, no two are exactly the same. Your height, eye colour and the shape of your earlobes are just three **characteristics** that vary from person to person – and distinguish you from everyone else.

The similarity between parents and their offspring is because some characteristics are **inherited** – they are passed from parents to their offspring. Other characteristics are affected by environmental factors. For example, your strength and speed are not just inherited characteristics – they are also affected by many environmental factors including diet and training. Similarly, the size of leaves on a plant is affected by light, temperature and nutrients.

A group of students recorded some of the different characteristics in their class. The results are shown in these graphs.

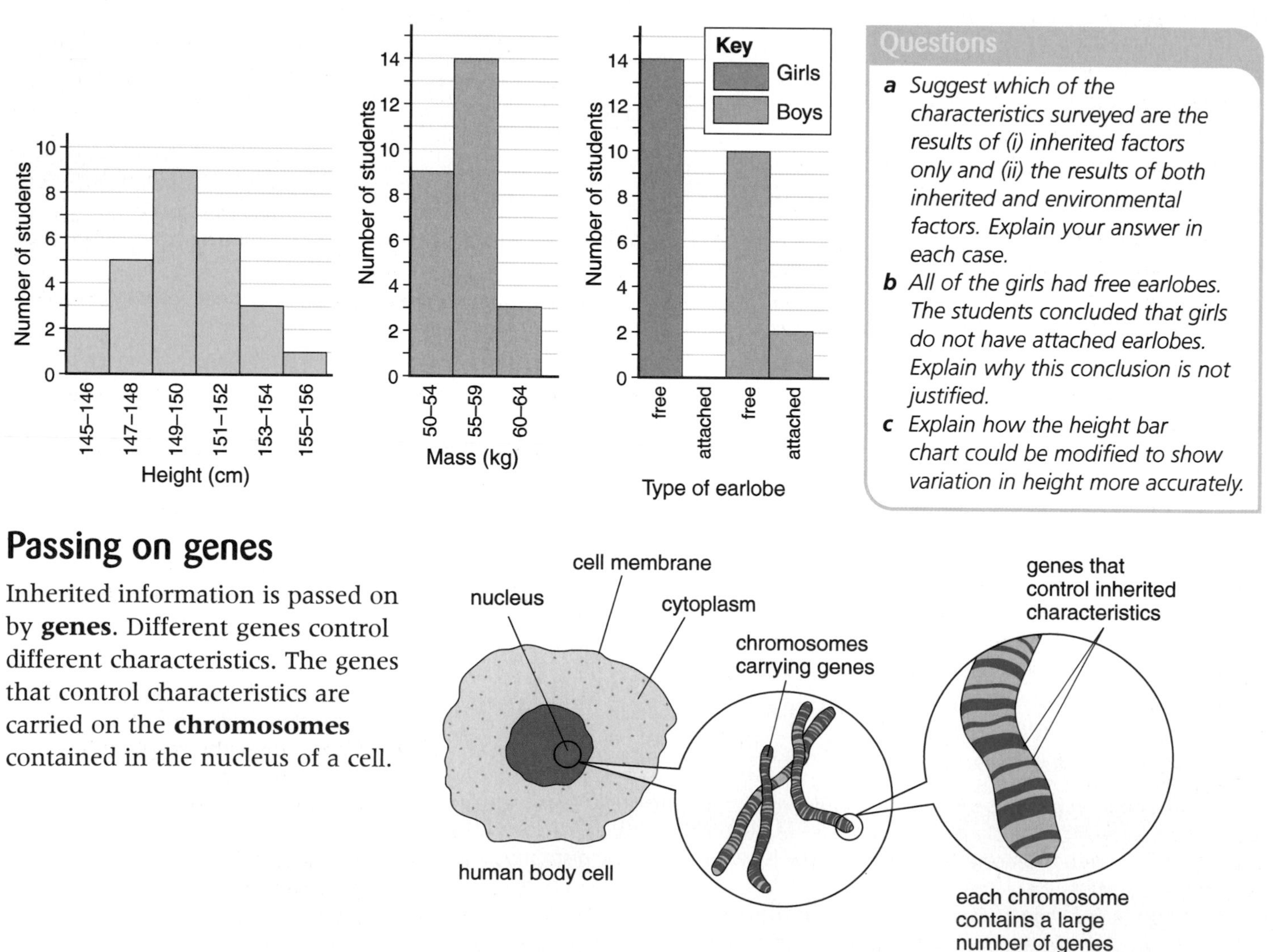

Questions

a *Suggest which of the characteristics surveyed are the results of (i) inherited factors only and (ii) the results of both inherited and environmental factors. Explain your answer in each case.*

b *All of the girls had free earlobes. The students concluded that girls do not have attached earlobes. Explain why this conclusion is not justified.*

c *Explain how the height bar chart could be modified to show variation in height more accurately.*

Passing on genes

Inherited information is passed on by **genes**. Different genes control different characteristics. The genes that control characteristics are carried on the **chromosomes** contained in the nucleus of a cell.

The start of life

Your life started when a **sperm** cell from your father fertilised an **egg** cell from your mother. This is why you have inherited half of your genes from your father and half from your mother.

Sperm and egg cells are sex cells called **gametes**. The joining or fusion of male and female gametes is called **sexual reproduction**. This type of reproduction produces individuals who have a mixture of genetic information from two parents.

Question

d *Plant seeds are formed by sexual reproduction. Explain why a plant grown from a seed will have characteristics of both parent plants.*

Reproducing from one parent

Some plants and animals can also reproduce by **asexual reproduction**. In this type of reproduction there is no fusion of cells and only a single parent is needed. The diagram shows couch grass – a common weed found in gardens. This grass spreads by producing new plants from an underground stem. The plants growing from the underground stem are produced by asexual reproduction. Individuals produced by asexual reproduction have exactly the same genes as the parent. Organisms with identical genetic information are known as **clones**.

The photograph shows bean aphids feeding on broad bean plants. In the autumn male and female aphids mate and produce eggs. The aphids that hatch from these eggs in spring are all wingless females, known as 'stem mothers'. These stem mothers produce offspring of both sexes without needing any males.

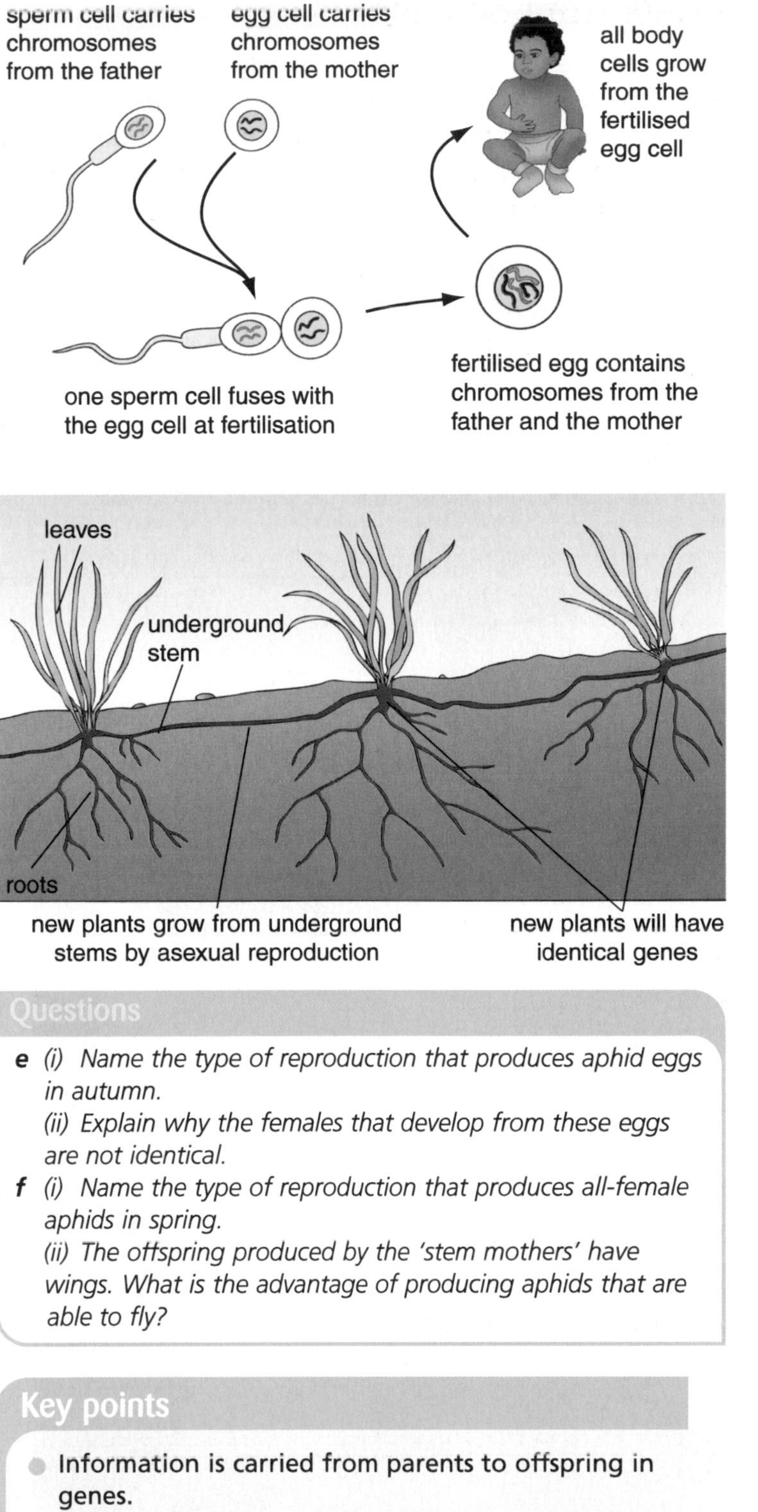
sperm cell carries chromosomes from the father
egg cell carries chromosomes from the mother
all body cells grow from the fertilised egg cell
one sperm cell fuses with the egg cell at fertilisation
fertilised egg contains chromosomes from the father and the mother
leaves
underground stem
roots
new plants grow from underground stems by asexual reproduction
new plants will have identical genes

Questions

e *(i) Name the type of reproduction that produces aphid eggs in autumn.*
(ii) Explain why the females that develop from these eggs are not identical.

f *(i) Name the type of reproduction that produces all-female aphids in spring.*
(ii) The offspring produced by the 'stem mothers' have wings. What is the advantage of producing aphids that are able to fly?

Key points

- Information is carried from parents to offspring in genes.
- Genes are carried on chromosomes found in the nucleus of cells.
- Sexual reproduction produces a mixture of genetic information from two parents.
- There is no mixing of genetic information in asexual reproduction.

B1b 3.6 High-tech breeding

Plants from cuttings

Young plants can be grown from older plants by taking cuttings. A new plant can be produced quickly and cheaply from each cutting. Plants grown from cuttings have identical genes to the parent plant and to each other.

Questions

a *Why is taking cuttings an example of a cloning technique?*

b *A new variety of plant was developed by a gardener. Would the first plant of this new variety have been grown from a seed or from a cutting from another plant? Explain your answer as fully as you can.*

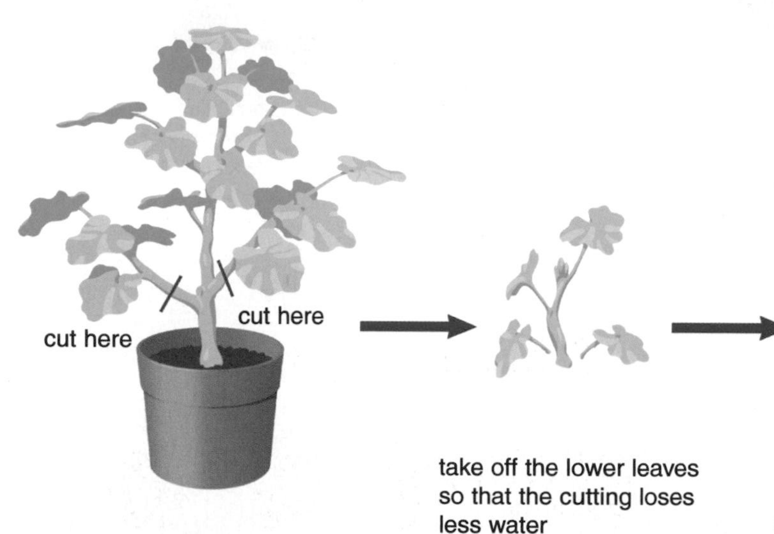
cut here
cut here
take off the lower leaves so that the cutting loses less water

plant the cutting in compost and water it
cover it with a polythene bag

roots form after a few weeks

Test-tube plants

Plant **tissue culture** is a modern cloning technique. Tiny pieces of plant tissue are used to grow whole plants. The pieces of tissue are grown on a special growth medium containing nutrients and hormones. Roots, stems and leaves grow from the cells in the piece of tissue. Using tissue culture, plant breeders can grow large numbers of identical plants from just a small piece of tissue.

Tissue culture is useful when large numbers of identical plants are needed and plants cannot be produced using cuttings.

Speeding up breeding

Farmers use only animals with the most useful characteristics for breeding. For example, dairy farmers want to breed cows which produce large amounts of milk.

By using **embryo transplants** breeders can produce a large number of genetically identical calves from a single fertilised egg. This involves removing the developing embryo from a pregnant cow that has a high yield of milk. The cells of the embryo are split apart before they become specialised.

Each separated cell can then be grown in the laboratory to form multiple embryos. As the separated cells have the same genes, the embryos that are formed will also be genetically identical. Each new embryo is implanted into a host mother and continues to develop.

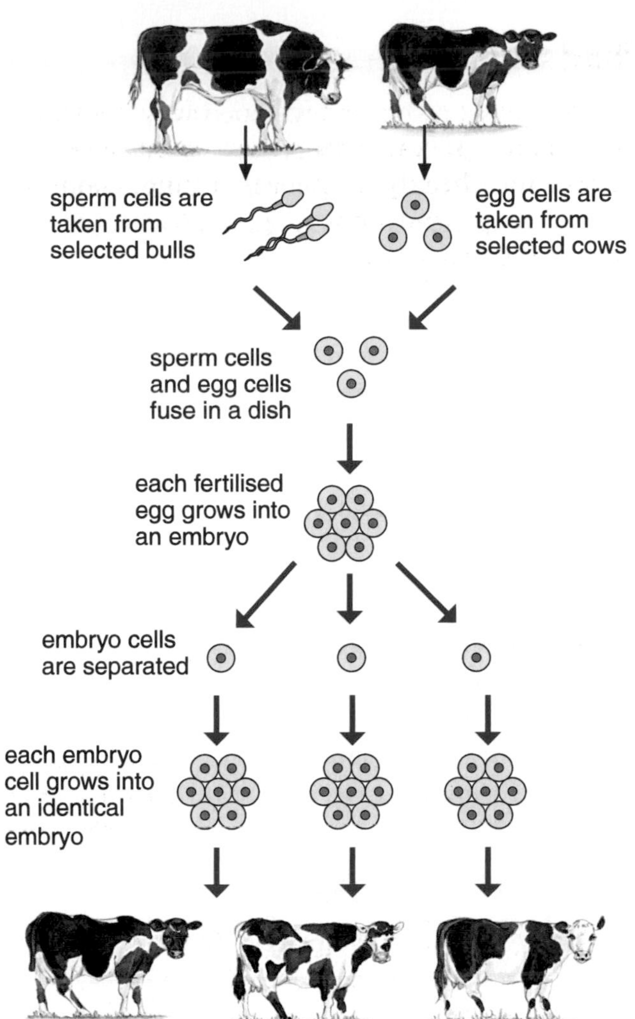
sperm cells are taken from selected bulls
egg cells are taken from selected cows
sperm cells and egg cells fuse in a dish
each fertilised egg grows into an embryo
embryo cells are separated
each embryo cell grows into an identical embryo
each embryo is implanted into a different host mother

Question

c Breeding normally involves selecting the most suitable bull and cows and allowing them to mate. Each cow then gives birth to a single calf after a few months of pregnancy. Explain the advantages of using embryo transplants instead of normal breeding.

Pet cloning

The diagram shows the techniques that were used to clone Snuppy.

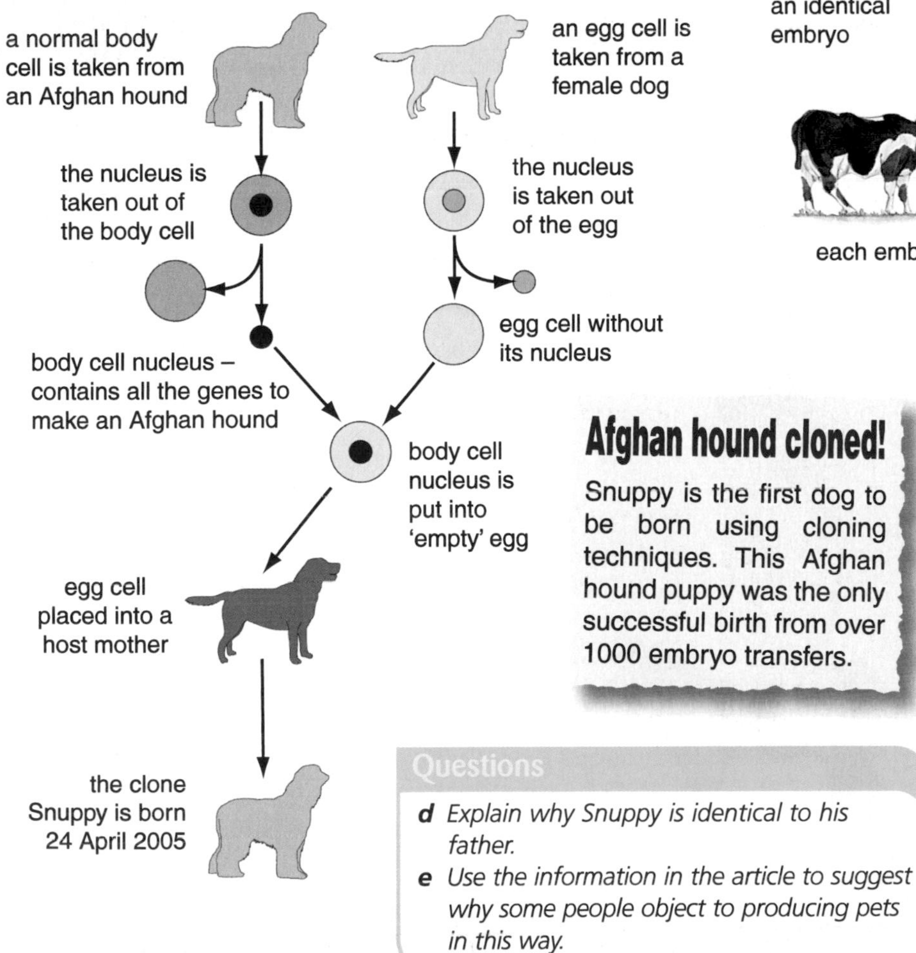
a normal body cell is taken from an Afghan hound
an egg cell is taken from a female dog
the nucleus is taken out of the body cell
the nucleus is taken out of the egg
body cell nucleus – contains all the genes to make an Afghan hound
egg cell without its nucleus
body cell nucleus is put into 'empty' egg
egg cell placed into a host mother
the clone Snuppy is born 24 April 2005

Afghan hound cloned!

Snuppy is the first dog to be born using cloning techniques. This Afghan hound puppy was the only successful birth from over 1000 embryo transfers.

Questions

d Explain why Snuppy is identical to his father.

e Use the information in the article to suggest why some people object to producing pets in this way.

Key points

- New plants can be produced quickly and cheaply from cuttings.
- Identical organisms can be produced using modern cloning techniques, including tissue culture, embryo transplants and adult cell cloning.
- Producing plants and animals using cloning techniques raises ethical issues.
- People need to be able to make informed judgements about the ethical issues concerning cloning.

Swapping genes

Gene transfer

Scientists have recently developed ways to transfer genes from one organism to another. This process is called genetic engineering or **genetic modification**. Genetic modification allows plants, animals and microorganisms to be produced with specific characteristics.

People have been breeding plants and animals for many hundreds of years to develop varieties with specific characteristics. Crop plants such as wheat and potatoes, and farm animals such as cattle and sheep, have been developed over many years of breeding. Traditional breeding methods involve mixing many genes over many years. Using genetic modification, the genetic make-up of an organism can be changed very quickly.

Genetic modification allows just one individual gene to be inserted into a plant or animal so that it develops a specific characteristic. When genes are inserted at an early stage in development, plants and animals will grow with the chosen characteristic. Compared to traditional breeding, with GM technology new varieties can be produced much more quickly.

Question

a *Traditional breeding involves sexual reproduction in selected plants and animals. Explain why traditional breeding involves mixing many genes.*

Plant breeders have produced varieties of potato that are fast-growing and disease-resistant. This has involved selecting which plants to use for breeding over many years.

Using GM technology

Genetic modification is now used to manufacture foods and medicines and to produce new varieties of crop plants. One of the first applications of GM in food production was the manufacture of an enzyme called chymosin. This enzyme is used to make cheese from milk. Chymosin used to be obtained from the stomach lining of newborn calves. Calves produce chymosin to digest milk. The enzyme is now produced from genetically modified bacteria or yeast. Today about 90% of hard cheese is made using chymosin produced with GM technology.

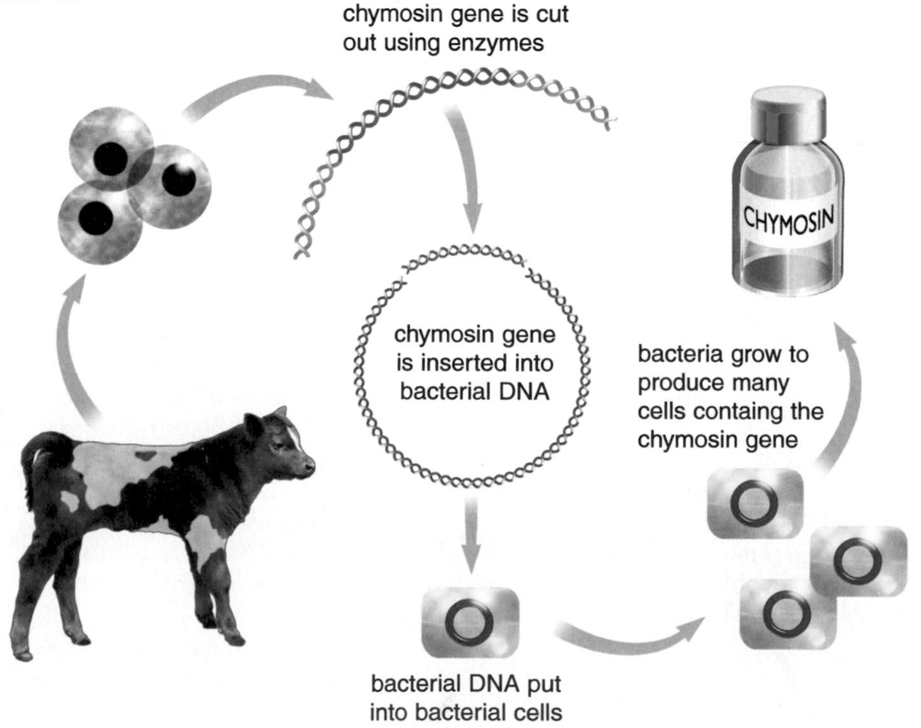
chymosin gene is cut out using enzymes
chymosin gene is inserted into bacterial DNA
bacterial DNA put into bacterial cells
bacteria grow to produce many cells containg the chymosin gene
CHYMOSIN

Questions

b *Suggest why adult cows do not produce chymosin.*

c *Explain why genetically engineered chymosin is popular among vegetarians.*

Manufacturing medicines

GM technology is also used in medicine to manufacture substances that used to be obtained from human or animal tissues. For example, people who have haemophilia, an inherited disease, do not produce the proteins needed for their blood to clot. A bruise or a cut can result in severe bleeding. Haemophiliacs are now treated using clotting proteins which are produced by genetically modified bacteria.

Question

d *Proteins to promote blood clotting used to be extracted from human blood. Suggest the benefits of producing these proteins using GM organisms.*

GM crops

New genes can also be transferred to crop plants which are grown to produce food. Crops which have had their genes changed in this way are called genetically modified crops (**GM crops**). For example, some bacteria produce a protein that is poisonous to insects. The gene that makes this poison can be cut out of bacteria and added to crop plants. Because the GM crop plant will now produce the poisonous protein, the farmer will not need to spray the crop with **insecticide**. But other insects, such as bees, may also be killed when they feed on the GM crops.

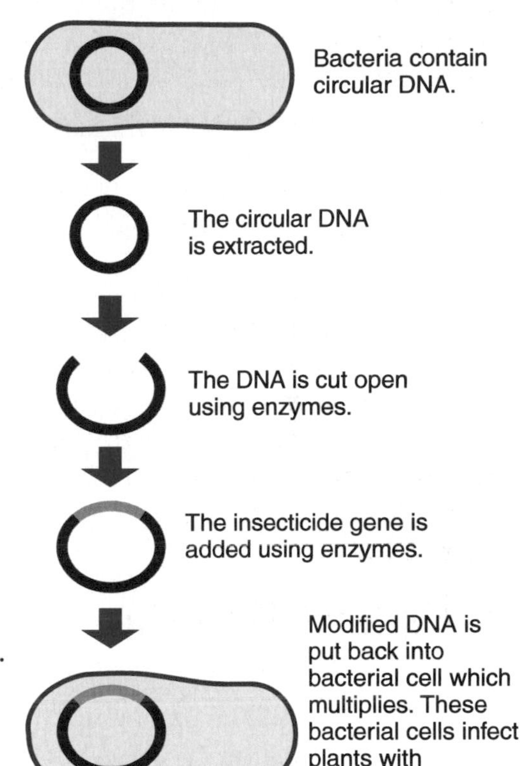
Bacteria contain circular DNA.
The circular DNA is extracted.
The DNA is cut open using enzymes.
The insecticide gene is added using enzymes.
Modified DNA is put back into bacterial cell which multiplies. These bacterial cells infect plants with insecticide genes.

GM crops have also been produced so that crop plants are resistant to herbicides (weedkillers). Farmers use herbicides to kill weeds which compete with the crop plants for light, water and nutrients. By inserting a gene which makes the plant resistant to herbicides, farmers can spray over an entire field, killing all plants apart from the GM crop.

Question

e *Suggest two advantages and two disadvantages of growing insect-repelling GM crops.*

Key points

- In genetic engineering, genes can be 'cut out' and inserted into other organisms.
- Plants and animals will develop with specific characteristics when genes are inserted at an early stage in their development.
- Modifying the genetic make-up of plants and animals raises ethical issues.
- People need to be able to make informed judgements about the ethical issues concerning genetic engineering.

Social and ethical issues

Public concerns

Many people are concerned about the use of the use of genetically modified products. The photographs show the action some people have taken to stop GM products getting into the food we eat.

Here are some of the views of the people involved.

The protestors' views

We don't know the long-term effects. GM products could be damaging the environment and our health – but we don't know yet.

GM crops containing an insecticide could create 'superbugs' – pests which are resistant to insecticides.

GM crops could reproduce with wild plants and introduce GM genes into other plants. We don't know what effect this could have.

There is no need to increase food production in this country – we have more than enough food already.

The food scientist's view

GM products are carefully tested before being sold.

Using GM crops means that fewer harmful chemicals such as pesticides are used in the environment.

There is no evidence that GM foods have a harmful effect on the environment. It is just bad publicity from biased groups that gets people worried.

We need to increase food production on a worldwide scale to feed an ever-increasing population.

GM crops are very useful in developing countries where pests and disease are much more common than in the UK.

Question

a *Imagine that a company wants to grow GM crops near your school. The local press want people to write in with their views. Do you support GM technology or are you against it? Give as many reasons to support your view as you can.*

Designer babies

About 6000 babies are born in the UK each year as a result of *in vitro* fertilisation (IVF). A recent report states that about one in three couples have difficulty conceiving. IVF is one way of aiding conception. The technique involves taking eggs from a woman and fertilising the eggs in the laboratory – producing a 'test-tube baby'. The fertilised egg develops into a small ball of embryo cells which is then implanted into the woman.

Prior to implantation, a single cell from an embryo conceived by IVF can be removed and genetically tested before parents decide whether to implant or discard the embryo. Testing embryos using genetic screening is a major step forward in preventing inherited disease but it raises some difficult ethical questions.

The ethical issues include:

Should parents be allowed to test for other characteristics? Embryos developed from IVF treatment are tested for genetic disorders. At present tests are not carried out for characteristics such as intelligence and sporting ability. Deciding the characteristics of a child is often reported as producing a 'designer baby'.

Should embryo testing be available for all pregnant women? At present women having IVF treatment are allowed different tests from those who are pregnant, because the embryo can be tested at a very early stage in development.

Should all testing include all genetic disorders? Some people want genetic testing for genetic disorders to be offered to all parents so that they have the information. Other people say that some parents would use a result showing even a very slight disorder as a reason for abortion. This is like rejecting people with even slight disabilities.

▲ This small ball of cells is an embryo at a very early stage of development. The cells can be tested to see if they have genes that could cause an inherited disease.

Questions

b *Should parents and doctors be allowed to select any of the genetic characteristics of children? Give reasons to support your answer.*

c *Should genetic testing of embryos include all genetic disorders? Give reasons to support your answer.*

Key points

- Applications of science in medicine and food production can raise ethical issues.
- People need to be able to make informed judgements about economic, social and ethical issues concerning cloning and genetic engineering.

Evidence about the past

Evolving organisms

There are many different types of animals and plants. Different types of organism are called **species**. There is evidence that all species have evolved from simple organisms that lived on Earth more than three billion years ago. Similar species are descended from a common ancestor by a process of gradual change.

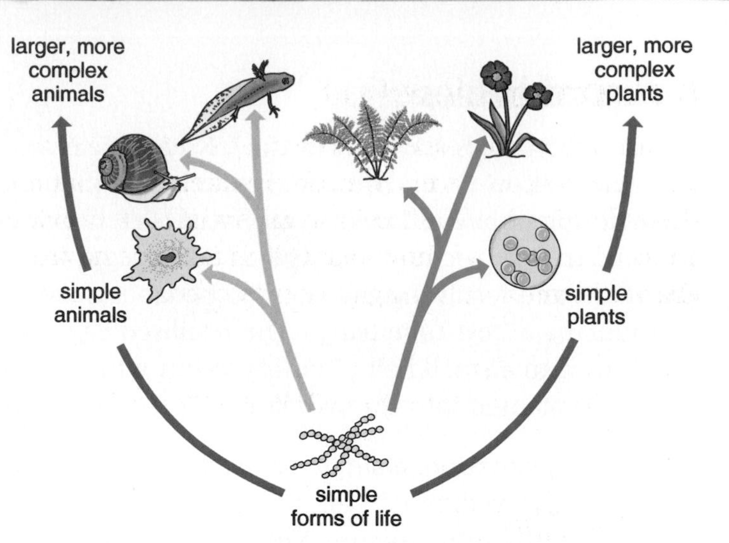

Simpler forms of life evolved to form larger and more complex organisms.

Evidence from the past

Evidence of the way organisms have evolved can be seen in **fossils**. Fossils are the remains of plants and animals that are found in rocks. Fossils can be formed from:

- hard parts of animals and plants that do not decay easily such as bones, teeth, shells and the woody remains of plants
- animals and plants that have not decayed because the conditions for decay were not present (moisture, oxygen and warmth are needed for decay to happen)
- traces of animals such as footprints.

Fossils are only rarely formed. This is why there are gaps in the fossil record, and scientists can only suggest how one kind of organism evolved from another. Not enough fossils have been found to show exactly what happened to each kind of organism that lived in the past. Fossils of simple organisms, such as those made from just one cell, are extremely rare. This is one reason why scientists cannot be certain about how life began on Earth. The evidence for evolution can be interpreted in different ways. This is why evolution is a theory and not a fact, and why there are conflicting theories.

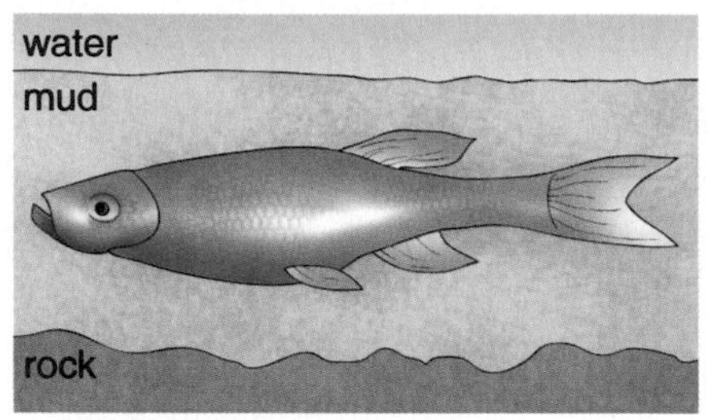

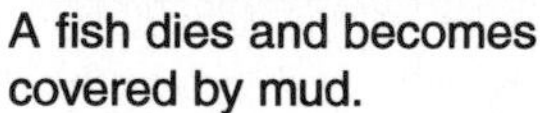
A fish dies and becomes covered by mud.

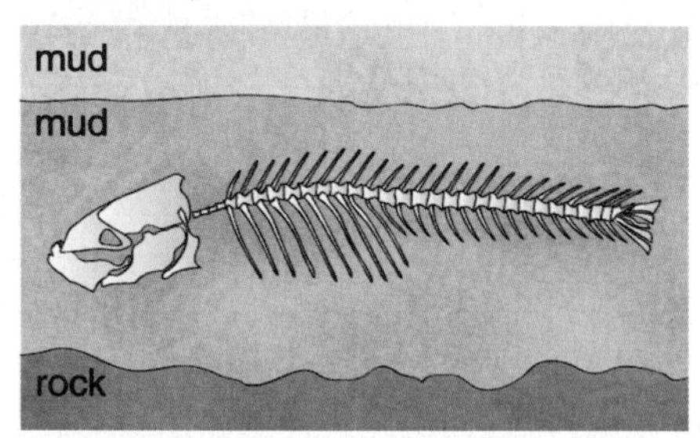

The soft tissues of the fish decay. The only part that remains is the skeleton.

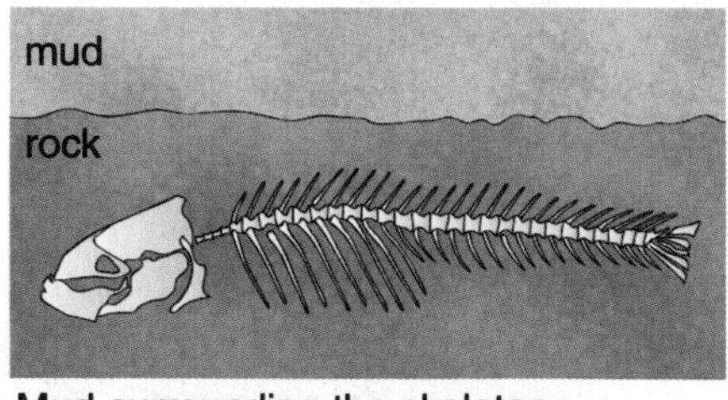

Mud surrounding the skeleton turns into rock.

Questions

a *Whole bodies of woolly mammoths have been found in frozen soils. These mammoths were almost perfectly preserved – their stomachs were full of 30 000-year-old grass and their flesh was still attached. Explain why fossil remains usually show only the remains of hard tissues such as bone.*

b *Some of the mammoths were found with stone arrowheads in their bodies. Explain what this evidence tells us about how humans lived at the time that mammoth fossils were formed.*

A record of evolution

The location of a fossil also provides evidence of its age. The deepest layers of rock are likely to contain the oldest fossils. The fossils in layers of rock provide a record that shows how plants and animals have changed over a very long period of time.

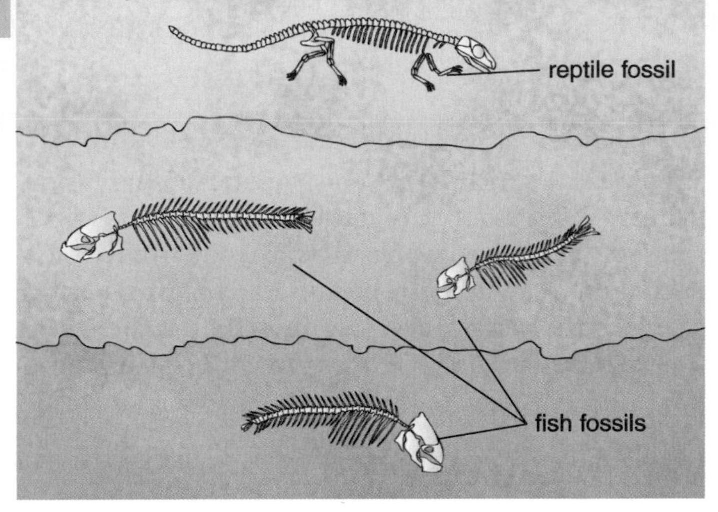

More layers of rock are formed on top of the older rock
reptile fossil
fish fossils

Question

c (i) Which fossil is the oldest? (ii) What is the evidence that this is the oldest fossil?

Extinct species

The **fossil record** shows that some species lived in the past but are now extinct. About 100 million years ago there were many more species of reptiles than there are today, including many dinosaur species.

Species become extinct because:
- the environment in which they live changes
- new predators or disease kill them
- they cannot compete with other species.

The 'Terror Birds' are an example of a species that is now extinct. These large, flightless birds roamed the grasslands of South America between 2 million and 62 million years ago. They were ferocious predators, able to chase their prey at speeds of 40 miles per hour. Scientists have put forward two theories to explain why these birds became extinct. One theory is that as the climate became warmer the grassland habitat was replaced with dense tropical forest. An alternative theory is that the Terror Birds could not compete with carnivorous mammals.

Terror Birds are now an extinct species.

Question

d South America used to be separated from North America. At that time carnivorous mammals were only found in North America. Explain why the Terror Birds became extinct after the two continents became joined.

Similarities and differences

Comparing the features of different species provides evidence of how they may have evolved. Species that share many similar characteristics are likely to be closely related. Species that are not closely related in evolutionary history are more likely to have very different characteristics.

The diagram shows the bones present in the fin of a fish, the fin of a whale and the wing of a bat.

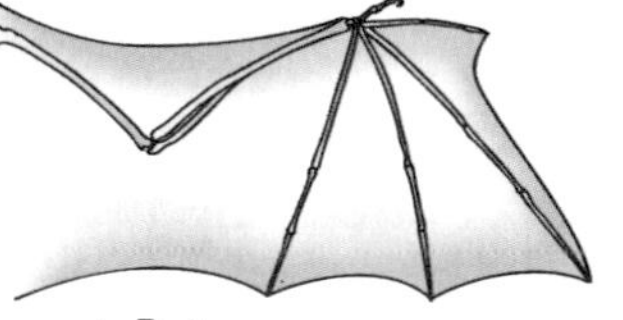
Bat.

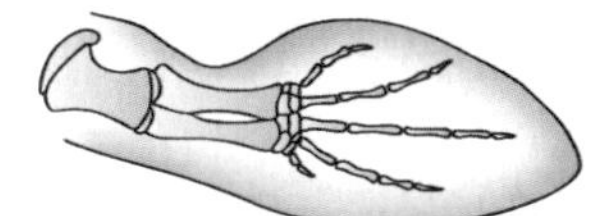
Whale.

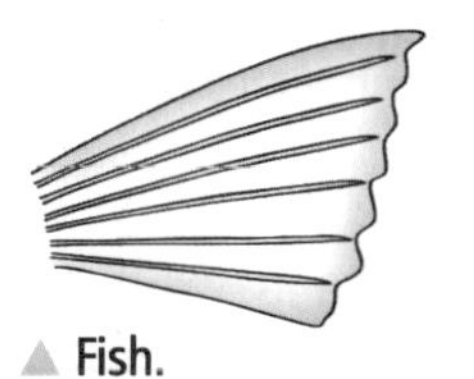
Fish.

Questions

e Describe how the fins of fish and whales differ.
f Describe the evidence that suggests that whales are more closely related to bats than they are to fish.

Key points

- Fossils provide evidence of how different organisms have changed.
- Similarities and differences between species indicate their evolutionary and ecological relationships.
- Some species have become extinct.
- Scientists cannot be certain about how life began on Earth.

B1b 3.10 Forming new species

Changing ideas

People used to think that all species had always remained the same. They also thought that species had been created at the same time that the Earth was made. As scientists learned more and more about plants and animals they concluded that species are changing and that new species have been formed. The first explanations of how evolution occurs were put forward by Lamarck and later by Darwin.

Lamarck's theory

One of the earliest theories of how evolution takes place was proposed by a scientist called Lamarck. According to Lamarck, a species changes over a period if time because it passes on to its offspring changes it acquires during its lifetime. The diagram shows how Lamarck's theory explains how species of wading birds developed long legs.

Questions

a *What are the advantages of wading birds having long legs?*

b *Use your knowledge of how characteristics are inherited to suggest a major weakness of Lamarck's theory.*

To reach fish in deeper water, wading birds stretch their legs. This makes their legs slightly longer.

Having slightly longer legs is passed on to the next generation. Birds in this generation also stretch their legs.

Over many generations, the wading birds' legs become much longer.

▲ Lamarck's theory of evolution.

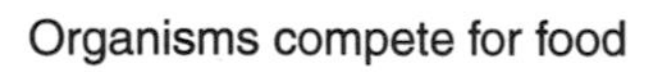

Organisms compete for food

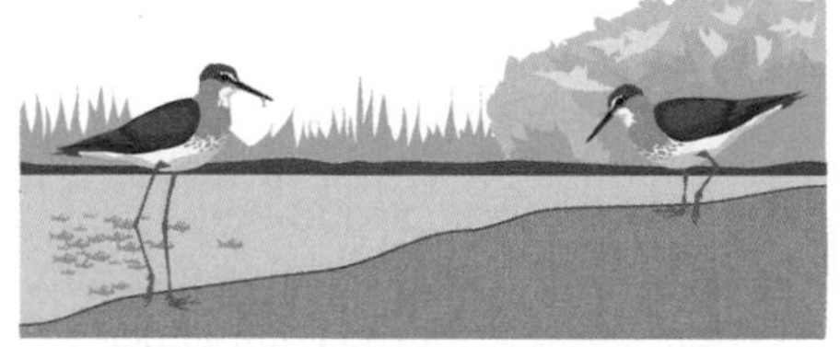

Individuals of the same species may have different characteristics, such as slightly longer legs.

Individuals struggle to survive. Some die because of lack of food or may be eaten by predators.

Individuals with useful characteristics are more likely to survive, and pass on their characteristics to the next generation.

◀ Darwin's theory of natural selection.

Darwin and natural selection

Scientists' present ideas about the way species evolve are based on the theory of natural selection, which was first put forward by Charles Darwin. Natural selection brings about changes in species because:

- changes in genes (**mutations**) produce new forms of genes
- differences in the genes of individual organisms produce different characteristics
- individuals with characteristics most suited to the environment are more likely to survive and breed
- the genes which have enabled individuals to survive are then passed on to the next generation.

Natural selection can only occur when individuals of the same species show different characteristics. These differences occur because mutation produces new forms of genes. This is why the change in a species may become more rapid as new forms of genes are produced.

Question

c *Darwin's theory of how species change is very different from Lamarck's theory. Explain how wading birds would have developed long legs by the process of natural selection.*

Different theories

Evolution has taken place over billions of years. There is not enough evidence from fossils and from the comparison of species to prove what caused species to change over this very long time period. This is why scientists cannot be certain about how life began on Earth and why there are conflicting theories to explain evolution.

Question

d *What are the differences between Darwin's and Lamarck's theories?*

Surviving and breeding

The effects of natural selection can be seen in a species called the peppered moth. There are two varieties of this species – a light variety and a dark variety. Both varieties feed at night and rest on trees during the day. The photograph shows the two varieties on a tree in a city.

Question

e *Which variety of moth is more likely to be eaten by insect-eating birds in a city? Explain your answer.*

In an investigation:

- Large numbers of light and dark varieties of moth were caught in a trap.
- The moths were marked with a spot of paint on the underside of their bodies and then released.
- Equal numbers of moths were released into a polluted woodland and into a unpolluted woodland.
- After a few days the moths were trapped again and the number of marked moths was counted.

The results are shown in the bar graph.

Marked moths recaptured (% of total recaptured)
0
10
20
30
40
50
light moths
dark moths
Polluted woodland
light moths
dark moths
Non-polluted woodland

Questions

f *Suggest why the moths were marked with paint on the underside of their bodies.*

g *What percentage of light moths was recaptured in: (i) the unpolluted woodland; (ii) the polluted woodland?*

h *Explain why more dark moths than light moths were recaptured in the polluted woodland.*

i *Suggest why only a small percentage of both varieties of moth were recaptured.*

Key points

- Some theories of evolution conflict with Darwin's theory.
- You should be able to interpret evidence relating to the theories of evolution.
- Darwin's theory is based on the natural selection of individuals most suited to their environment.
- There may be a more rapid change in a species when genes mutate.
- An investigation determines if a relationship exists between two variables.
- Scientists identify patterns and relationships to make suitable predictions.

Challenging ideas

The theory of evolution

When Charles Darwin presented his theory of evolution over 100 years ago people were astonished and even hostile. Darwin's ideas were only gradually accepted because:

- the theory of evolution undermined the idea that God made all the animals and plants that live on the Earth
- people believed that organisms had always been as they appeared.

People found it hard to believe that humans could have evolved. They laughed at the idea that humans as a species are closely related to apes such as chimpanzees. Although most people now accept that humans have evolved, scientists do not agree about how humans evolved, and how closely related we are to other species.

CHIMPS BELONG TO HUMAN BRANCH OF FAMILY TREE

A recent report says that chimps are so closely related to humans that they should be placed in the same branch of the evolutionary tree as us.

Evolutionary relationships

Studying the similarities and differences between species provides evidence of their evolutionary relationships. The diagrams show the skeletons of a human, an ape and a monkey. The features of these skeletons provide evidence to show how closely related these species are.

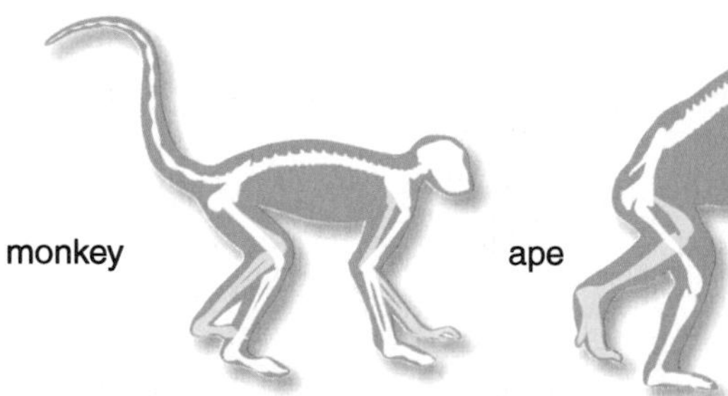

Questions

a *Describe two features of the ape and monkey skeletons that are not found in the human skeleton.*

b *Describe two features of the monkey skeleton that are not found in the human and ape skeletons.*

c *Which species, apes or monkeys, is more closely related to humans? Give reasons to support your answer.*

Tracing human evolution

The diagram shows the evolutionary relationship between humans and other animals. Your family tree shows that you are more closely related to your parents than to your aunts and uncles. Similarly, species that have evolved from the same ancestors are related. The theory of evolution does not suggest that chimpanzees 'turned into' humans. Chimpanzees are themselves the result of evolution. Evidence suggests that an ancestral ape species gave rise to both chimpanzees and humans.

Question

d *(i) Which species are most closely related to humans? (ii) Use the information in the diagram to explain your answer.*

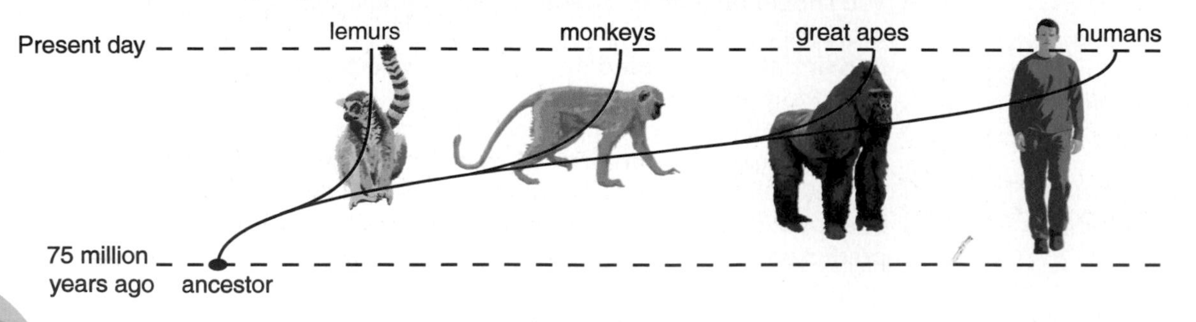

Fossils of extinct species

One very important fossil discovery was the bones of a human-like creature that lived 3.2 million years ago. The scientists who found the fossil skeleton called it 'Lucy'. The skeleton has many features similar to chimpanzees, but the structure of the leg and hip bones show that Lucy walked on two legs, like us. Walking on two legs is a very important difference between humans and apes. This is why Lucy's species, called 'Southern Ape', is thought to be an extinct human ancestor.

Working out relationships

Scientists believe that humans share a common ancestor with extinct species such as Southern Ape and with chimpanzees. However, scientists are not sure how this common ancestor evolved into these three species. The diagram on the right shows how these three species could be related.

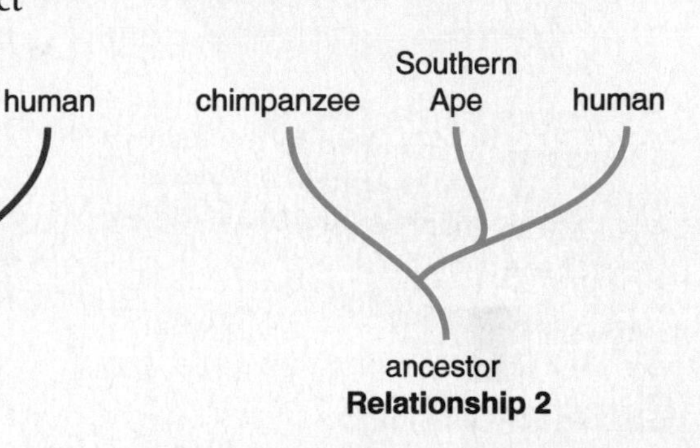

The table below shows some of the features of the skeletons of these three species.

Characteristic	Human	Southern Ape	Chimpanzee
thumb length	short	short	long
legs and arms	arms shorter than legs	arms shorter than legs	legs shorter than arms
brain size	large (1500 cm^3)	small (620 cm^3)	small (440 cm^3)

Questions

e *Which of the relationships is shown by (i) brain size, and (ii) the length of legs and arms?*

f *What evidence in the table suggests that the Southern Ape walked upright on two legs?*

Charting human evolution

Scientists have pieced together the evidence from different fossils of extinct species to chart how human evolution may have occurred. The bars on the chart show when a species appeared and when it became extinct.

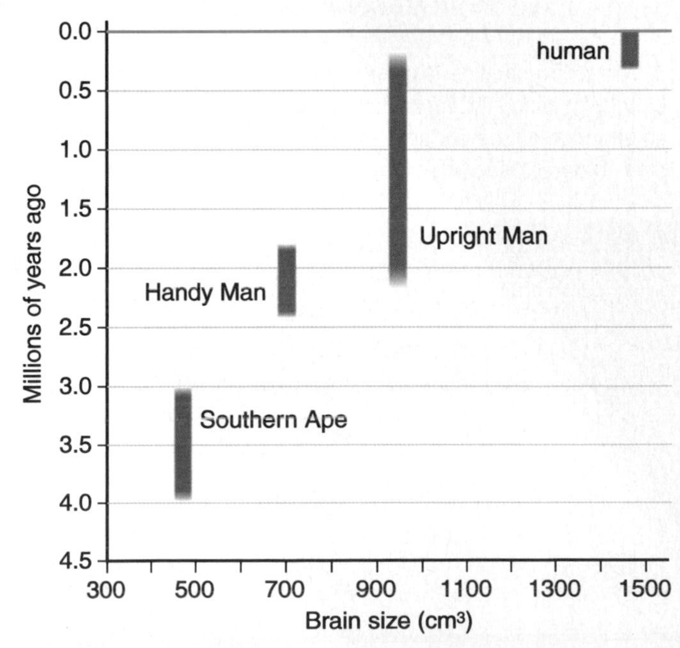

Questions

g *How many years ago did 'Southern Ape' first appear?*

h *For how long did the species 'Handy Man' survive?*

i *Which species in the chart was the most intelligent? What evidence supports your answer?*

Key points

- Darwin's theory of evolution was only gradually accepted.
- Scientists cannot be certain about how life began on Earth.
- The similarities and differences between species may provide evidence of their evolutionary and ecological relationships.

B1b 4 Taking care of the planet

In 2003 Europe had the hottest summer ever recorded. The newspaper cuttings describe some of the effects of the heatwave. Was the summer of 2003 a sign that the Earth's climate is changing?

Parisians thronged the bank of the River Seine which has been turned into an urban beach with sand, cafes, deckchairs and palm trees as the temperature in the capital neared 40 °C (104 °F) again yesterday.

Amsterdam zoo fed its chimpanzees iced fruit and sprayed ostriches with cold water to keep them cool as temperatures in the Dutch capital edged towards 30 °C (86 °F), the Dutch news agency ANP reported.

Polish fire crews battled 35 forest fires on Monday and about a quarter of the country's woodlands were at serious risk of fire after temperatures topped 30 °C (86 °F) for much of July, authorities said.

13 Spaniards have died in the heatwave, and 30 taken to hospital because of the heat in Cordoba, Seville and Huelva in Andalusia.

The death toll from Portugal's biggest wildfires in decades rose to 11 after two bodies were found in charred woodland, but cooler overnight temperatures enabled firefighters to contain all but three major blazes.

Climate change

The impact that humans are having on our planet is always in the news. We are already beginning to see climate changes that may bring disasters to many parts of the world. For example, severe flooding in Bangladesh gets more frequent with each decade. The Earth is not ours to use as we wish; we hold it in trust for future generations, so we must look after the Earth and its resources.

The photographs show some of the concerns that people have about the environment.

▼ This satellite photo shows air pollution over China covering an area larger than the UK. Many developing nations are increasing the size of their industries. This can bring about air pollution on a massive scale.

▼ Forests in many parts of the world are being cleared to provide grazing land for cattle. In India the tiger has become an endangered species as its native forests are relentlessly cut back. Deforestation affects the Earth's climate.

Climate change is bringing about more flooding. In recent years scenes like this have become much more common in many parts of the world.

A huge area of land was flooded when this hydroelectric dam was built, destroying more animal and plant habitats.

For many years, scientists have made careful measurements of changes to the environment. From these measurements they have made predictions about the world in the 2050s. The map shows what many scientists think will happen if there is 'business as usual', that is if we do not change the way we treat our planet.

The map shows that several parts of the world, such as SE Asia, will have more frequent and more severe storms. Deforestation will continue in South America and crop yields in many parts of the world will fall, including India and southern Africa.

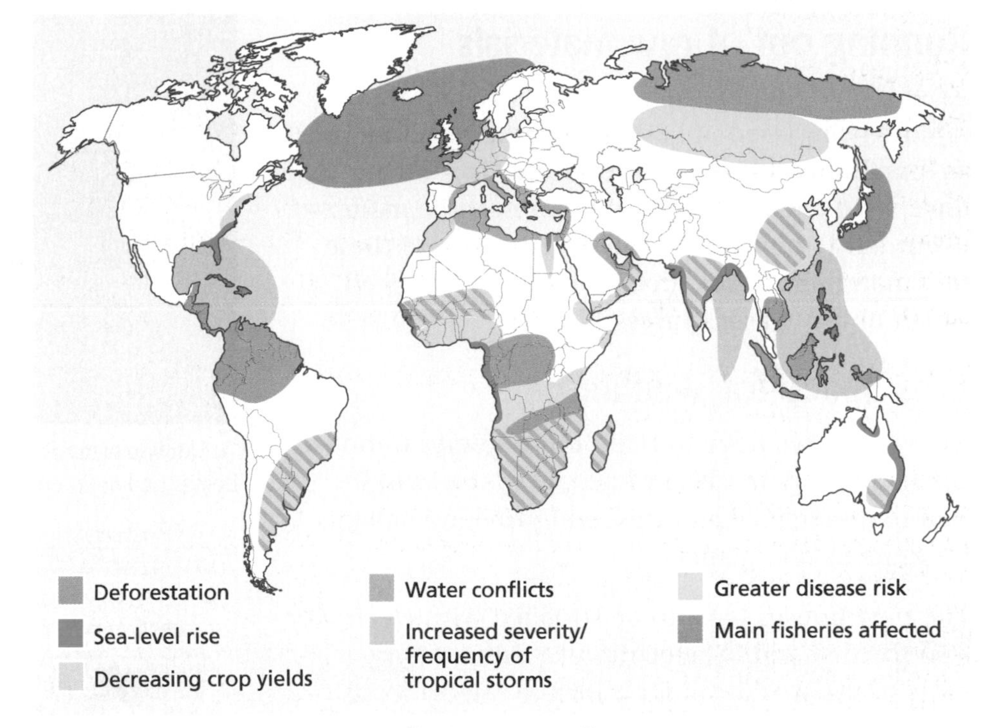

The world in the 2050s, assuming 'business as usual'.

Think about what you will find out in this section

How is the increase in human population affecting the surface of our planet?	How do scientists investigate the effect of human activities on the environment?
How are human activities affecting the Earth's atmosphere and climate?	How can science help us to plan sustainable development?
How are humans making it difficult for some species to survive?	

The world's population

In 2005 the world's population was estimated to be 6.5 billion.

Question

a *Look at the graph.*
(i) Describe how the world's population changed between 1750 and 2000.
(ii) What proportion of the estimated world population in 2050 will live in less developed countries?
(iii) Suggest reasons why the populations of less developed countries are rising faster than those of more developed countries.

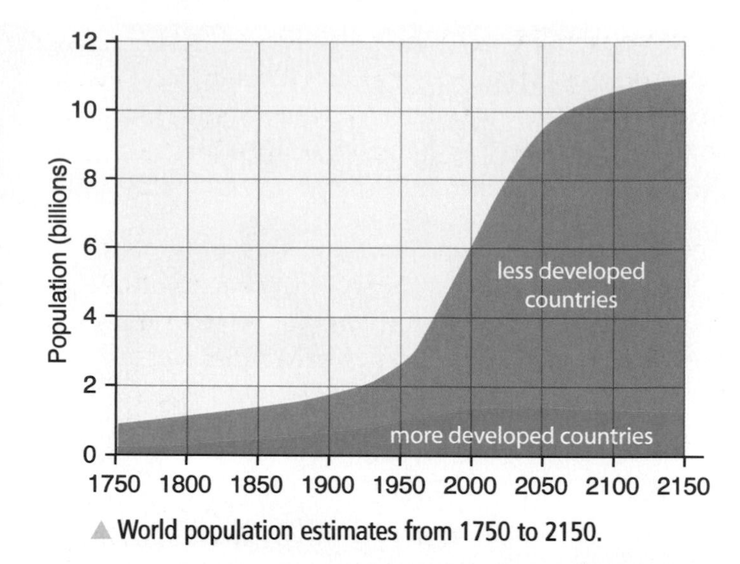

▲ World population estimates from 1750 to 2150.

Six thousand years ago, the population of the world was about 0.2 billion. People lived in small groups, and most of the world was unaffected by human activities.

Running out of raw materials

As world population increases, raw materials are rapidly being used up. Some of these resources, such as fossil fuels, are **non-renewable**. This means that once we have used them, no more can be made, so eventually supplies will run out. As we use these raw materials, a great deal of waste is produced. Much of this waste causes pollution.

▲ Satellite pictures at night show how much electricity is being used in Europe.

Less space for wildlife

Every year, we have to build more homes to house the increasing numbers of people in the world. Before building began, this land provided habitats for animals and plants.

The new houses have to be supplied with electricity – so the demand for electricity is rising. Many parts of the world cannot afford to import fossil fuels to produce electricity, so they build dams for hydroelectric schemes. More habitats for animals and plants are lost.

Most houses are built using brick and cement. Bricks are made from clay which is obtained from quarries. Cement is made from limestone. Quarrying can threaten the habitats of rare species. The inhabitants of each new house will produce rubbish. This is dumped, tipped at landfill sites or even burned. Still more habitats are lost to animals and plants.

Question

b *Give one disadvantage of each of the three methods of getting rid of rubbish.*

▼ The more people there are, the more rubbish we make.

Each household also produces sewage. In this country most sewage is treated before being passed into rivers. But a lot of raw sewage is still dumped out at sea. This can be washed ashore by the tide to pollute beaches.

The increase in population means that more land is used for agriculture. Using land for agriculture reduces the number of habitats for wild animals and plants. The **pesticides**, **herbicides** and fertilisers used by farmers can cause further damage to local animals and plants.

▶ When treated sewage is dumped into rivers, microbes feed on it and multiply. Their respiration uses up the oxygen in the water, causing fish to suffocate.

Question

c *Give three ways in which human activities reduce the amount of land available for other animals and plants.*

DDT – the wonder pesticide?

In 1939, a pesticide called DDT was used for the first time to kill mosquitoes that carry malaria. DDT was seen as an ideal pesticide because it was toxic to a wide range of insects but it seemed to be harmless to mammals, fish and plants. DDT was widely used during World War II and was credited with saving the lives of millions of people who would have died of malaria.

But as DDT became more widely used, some insects became resistant. Even worse, in areas where DDT had been sprayed widely, scientists observed that bird populations had declined as well as mosquito populations. The scientists found that DDT interfered with the birds' ability to produce eggshells. The eggshells were thinner so most of them cracked before the young birds were hatched.

DDT is now banned in developed countries, but the World Health Organization estimates that 22 of the world's poorest countries rely on DDT to fight malaria because of its effectiveness and affordability. Malaria affects more than 300 million people and causes at least one million deaths annually. More than 3000 people, mostly children, die from malaria each day in Africa alone.

Question

d *Scientists have given politicians lots of data about DDT. Should DDT be banned in every country in the world? Explain your reasons.*

▲ A mosquito taking a meal.

Key points

- Human population growth means that raw materials are being used up quickly and more waste is being produced, causing more pollution.
- Humans are destroying animal and plant habitats by building, quarrying, farming and dumping waste.
- Some species are finding it difficult to survive.

Asian brown cloud

We need energy to power factories, homes and motor vehicles. But producing this energy can have serious effects on both humans and the environment.

The 'Asian Brown Cloud', a 3 km-thick blanket of pollution over South Asia, may be causing the premature deaths of half a million people in India each year, deadly flooding in some areas and drought in others, according to one of the biggest scientific studies of the phenomenon.

The grimy cocktail of ash, soot, acids and other damaging airborne particles is as much the result of low-tech polluters such as wood- and dung-burning stoves, cooking fires and forest clearing as it is of industry, the UN-sponsored study found.

More than 200 scientists contributed to the study. They used data collected by ships, planes and satellites to study Asia's haze.

The scientists say more research is needed but some trends are clear. Respiratory illness appears to be increasing along with the pollution in densely populated South Asia, researchers said, suggesting that the pollution plays a role in the 500 000 premature deaths that occur annually in India.

Scientists say it's too early to draw definite conclusions about the impact of the cloud and of similar hazes over East Asia, South America and Africa.

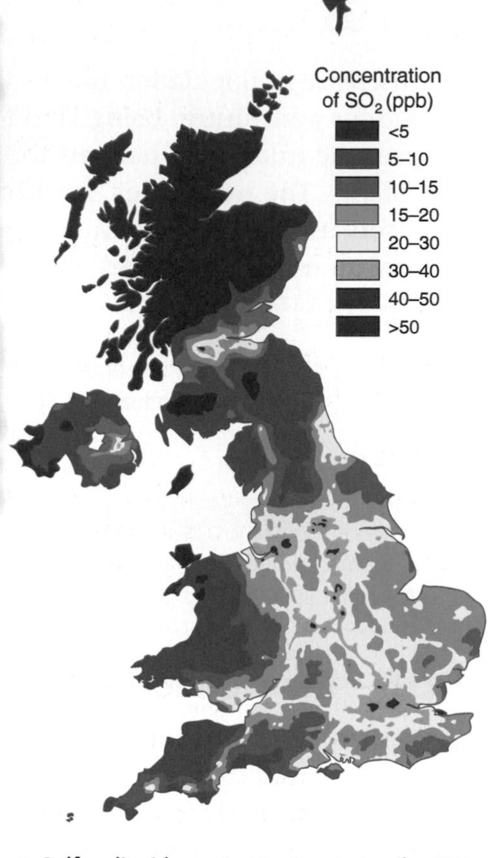

▲ Sulfur dioxide concentrations over the UK.

Question

a *(i) Suggest what kinds of data the scientists collected in this study.*
(ii) Suggest why it is 'too early to draw conclusions' from this study.
(iii) What could scientists do to help 'low-tech' polluters reduce pollution?

One effect of air pollution is the production of acid rain. Acid rain is formed mainly from gases released from car exhausts.

Petrol contains fuel molecules called hydrocarbons. These are oxidised when the fuel burns to release energy. The main gas released is carbon dioxide. Because petrol also contains nitrogen compounds and sulfur compounds, nitrogen oxides and sulfur dioxide are also formed. These gases pass into the atmosphere.

Question

b *What evidence is there from the map that most sulfur dioxide in the atmosphere comes from motor vehicles?*

How can scientists help?

You can now buy low-sulfur petrol and diesel from petrol stations. This reduces the amount of sulfur dioxide emitted by cars.

Nitrogen oxide gases are thought to contribute to asthma. Scientists have developed catalytic converters to convert the nitrogen oxides into nitrogen and carbon dioxide. The catalyst in the converter is the precious metal platinum. The diagram shows where the catalytic converter is fitted into the car.

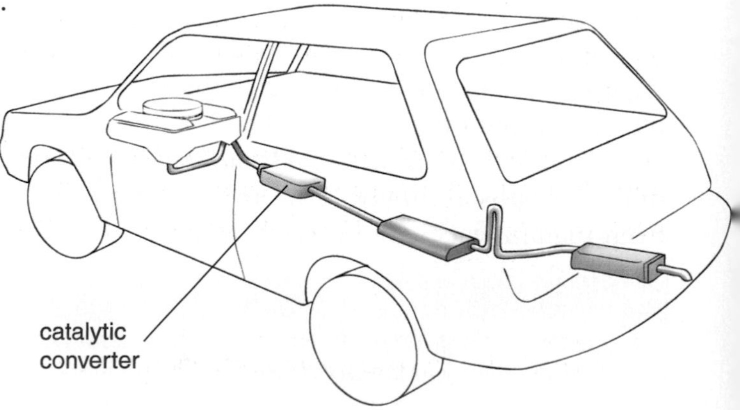

Question

c *Not all cars in the UK are fitted with catalytic converters. Suggest a reason for this.*

Acid rain

Carbon dioxide and sulfur dioxide in exhaust fumes from cars and lorries dissolve in rainwater and make it slightly acid. Acid rain has a pH of 4 or less. It has very damaging effects on the environment.

▶ Coal-burning power stations release sulphur dioxide into the atmosphere.

How does air pollution affect the environment?

Living organisms can be used to monitor the effects of air pollution and acid rain.

Lichens are organisms that live on the bark of trees and on stones. The table shows the number of species of lichen on the bark of trees at different sulfur dioxide concentrations.

	Concentration of sulfur dioxide in air (micrograms per m³ of air)					
	5	30	35	50	70	150
Number of species of lichen	14	10	8	7	5	2

Question

d *Could you use the number of lichen species to give an accurate prediction of the sulfur dioxide concentration in the air? Explain your answer.*

The larvae (young forms) of many types of insect live in fresh water. They are sensitive to the pH of the water. The table shows how the pH of the water affected the number of species of insect larvae and plants in five different ponds.

Pond	pH of water	Number of species of insect larvae	Number of species of plants
1	4.4	4	8
2	4.8	5	11
3	5.7	9	16
4	6.6	19	23
5	8.1	14	21

Question

e *Give two alternative explanations for the effect of pH on the number of species of animals found.*

Trees are also damaged by acid rain. One of the first signs of acid rain damage is that the young leaves near the top of the tree die. This is called 'crown-loss'.

Crown-loss is not the only damage caused by acid rain. The acid affects the covering of the older leaves. This makes it easier for disease organisms to attack the tree.

Question

f *How will the loss of its young leaves affect the tree?*

Key points

- Air is polluted by exhaust fumes from motor vehicles.
- Sulfur dioxide from exhaust fumes dissolves in rain to produce acid rain.
- Acid rain damages living tissues and some types of stones.
- Lichens can be used as indicators of the concentration of sulfur dioxide.
- Aquatic animals can be used as indicators of the acidity of fresh water.

The Brazil issue

Most forest clearance across the world is the result of slash and burn farming. In Brazil the government grants people land to farm for a short period of time. Once the land has become infertile, the farmers then move a little deeper into the forest and clear the next section of land.

Cattle ranchers also clear forests to provide grazing land. Over 74% of the beef consumed in Europe comes from Brazil. This is 80% of the beef Brazil produces, and most of it is produced on land that was once forest.

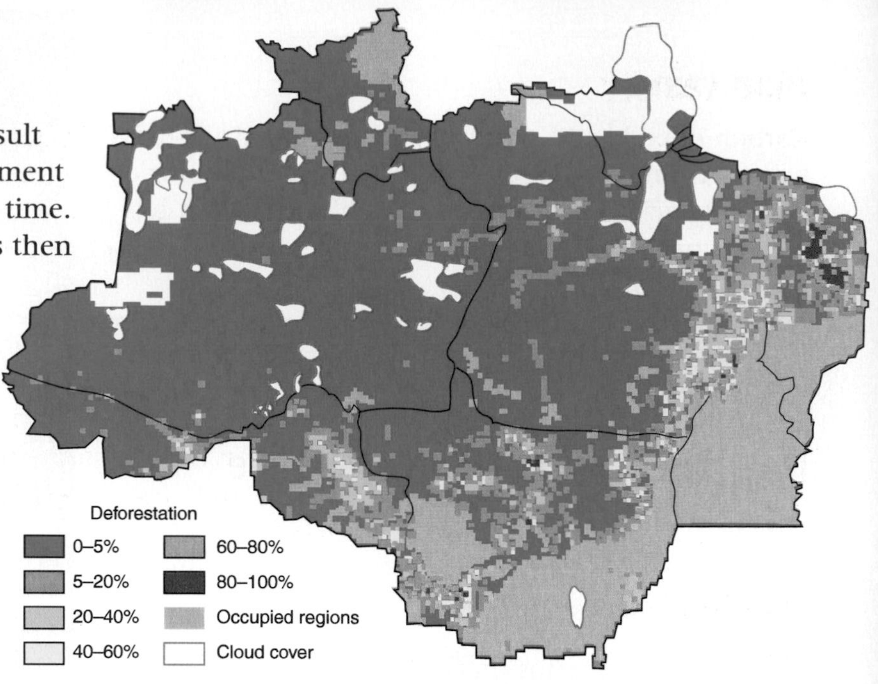

▲ Forests are disappearing at an alarming rate in Brazil.

Greenhouse gases

Forest clearance, or deforestation, affects the atmosphere. Global warming is now regarded as a major problem facing the world. Some studies suggest that there has been an overall rise in temperatures by 0.3 °C every decade across the planet. This rise is from an increase in the levels of greenhouse gases being released into the atmosphere. One of the most abundant of these gases is carbon dioxide. Levels of this gas in the atmosphere have risen 25% in the last 50 years. This is mostly due to industry in both developed and developing countries and the increased levels of car ownership and air travel in the developed world.

However, burning of vegetation in the areas where deforestation is taking place accounts for around 25% of the carbon dioxide released into the atmosphere. This equates to 2000 million tonnes of carbon dioxide every year.

Question

a *Brazil is a developing country. Suggest why its government seems to be encouraging deforestation.*

CO_2 in balance

Plants remove carbon dioxide from the atmosphere during photosynthesis, and most living organisms pass carbon dioxide back to the atmosphere when they respire. These two processes balanced each other for thousands of years. But in the last few centuries humans have interfered with this balance in two major ways: combustion of fossil fuels and deforestation.

When trees are cut down, the branches are burned. This releases carbon dioxide into the atmosphere. The roots of the trees die and are decomposed by microorganisms. When microorganisms respire they release yet more carbon dioxide into the atmosphere.

Cutting down trees also means that the amount of photosynthesis going on in the world is reduced. Trees take in millions of tonnes of carbon dioxide every year.

Questions

b *Give two reasons why cutting down forests increases the amount of carbon dioxide in the atmosphere.*

c *In the USA large areas of forest are mature – the trees have stopped growing. How will these forests affect the composition of the atmosphere?*

Biodiversity

Nobody knows how many species there are in the world. Scientists' estimations vary between 5 million and 80 million. Most scientists agree, though, that about half these species live in tropical rainforests. Another estimate is that deforestation is losing the world 150 species every day. Scientists call this a 'loss of biodiversity'.

Why does this matter? Approximately 40% of all prescription medicines are based on natural compounds found in microorganisms, plants and animals. Nature provides us with penicillin, aspirin, morphine and steroids. New medicines to fight breast cancer come from the bark of the pacific yew tree. The rosy periwinkle plant is used to fight Hodgkin's disease and childhood leukaemia.

> **Question**
>
> ***d*** *Explain the possible consequences of loss of biodiversity.*

Methane

The number of cattle in the world is rising rapidly, particularly in places like Brazil. Cows have a four-chambered stomach. Microbes live in these chambers and digest parts of the cow's food, producing methane. Methane is another greenhouse gas. Cattle produce about 100 000 000 tonnes of methane per year – 20% of the total methane in the atmosphere.

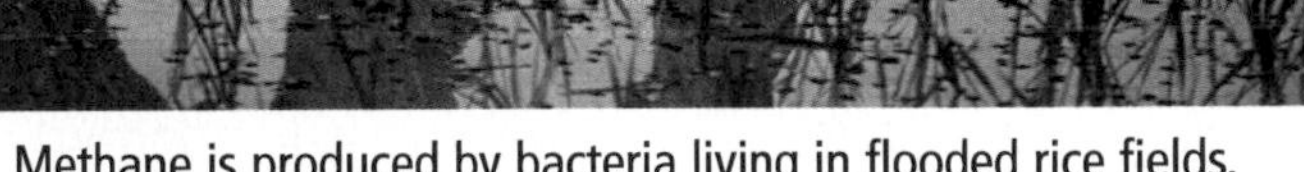

Methane is produced by bacteria living in flooded rice fields.

Most of the remaining atmospheric methane comes from rice fields. Rice fields are under water for long periods, so there is very little oxygen in the soil. Bacteria in these soils produce a lot of methane. As the world's population has increased, so has the total area of rice fields to provide food for all the people.

The increased concentrations of greenhouse gases are causing the temperature of the atmosphere to rise, slowly but surely. The diagram shows why this is happening.

This rise will cause changes to climates around the world – some places will become wetter, others drier. The warming of the atmosphere is causing melting of the polar icecaps. This will cause sea levels to rise. Some low-lying pieces of arid land will become flooded.

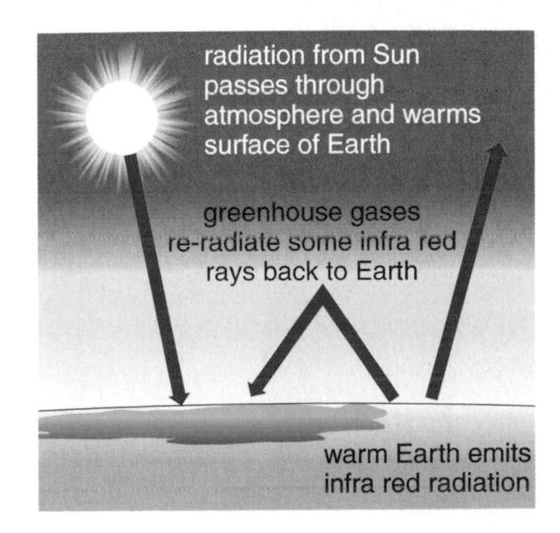
radiation from Sun passes through atmosphere and warms surface of Earth
greenhouse gases re-radiate some infra red rays back to Earth
warm Earth emits infra red radiation

> **Key points**
>
> - Carbon dioxide in the atmosphere is rising because of increased use of fossil fuels and increased deforestation.
> - Loss of biodiversity is also caused by deforestation.
> - Methane in the atmosphere is rising because of increased numbers of cattle and rice fields.

India moves forward

India is changing rapidly. It has a population of more than one billion. Its population grows by about 17 million each year (the population of the UK is about 60 million). Two-thirds of India's population lives in the countryside, but over the past 20 years cities have grown rapidly as industry has expanded and people have moved to find employment.

The increase in industrialisation and population has created problems:

- forests have been cut down for fuel and for farmland
- **water tables** have fallen and soils have become poorer
- use of pesticides and fertilisers has increased
- urbanisation has increased pollution of water, land and air
- rapid growth of the economy has led to increased numbers of motor vehicles and plastics, putting a strain on energy supplies.

There is no quick fix to India's problems. Different people have different views about the best way to tackle it.

▲ Millions of people in the world live in poor conditions.

Developing countries should protect their forests.

We only use a fraction of the energy you use per head of population.

Your pollution is ruining the planet.

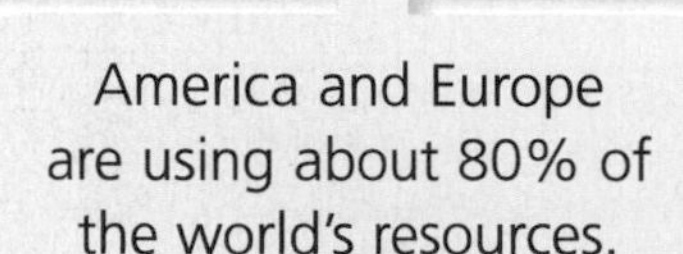

Why shouldn't we try to improve our standard of living?

Your increasing population is taking too much of the world's resources.

America and Europe are using about 80% of the world's resources.

Conserving natural resources

Most energy used in industrialised countries comes from non-renewable energy resources. Industrial countries are using up non-renewable energy resources far more quickly than developing countries. The ways in which we can conserve non-renewable energy resources include:

- making our homes more energy efficient
- making less use of private motor vehicles by walking or by using public transport.

The Ecohouse has been designed to use energy in a sustainable way.

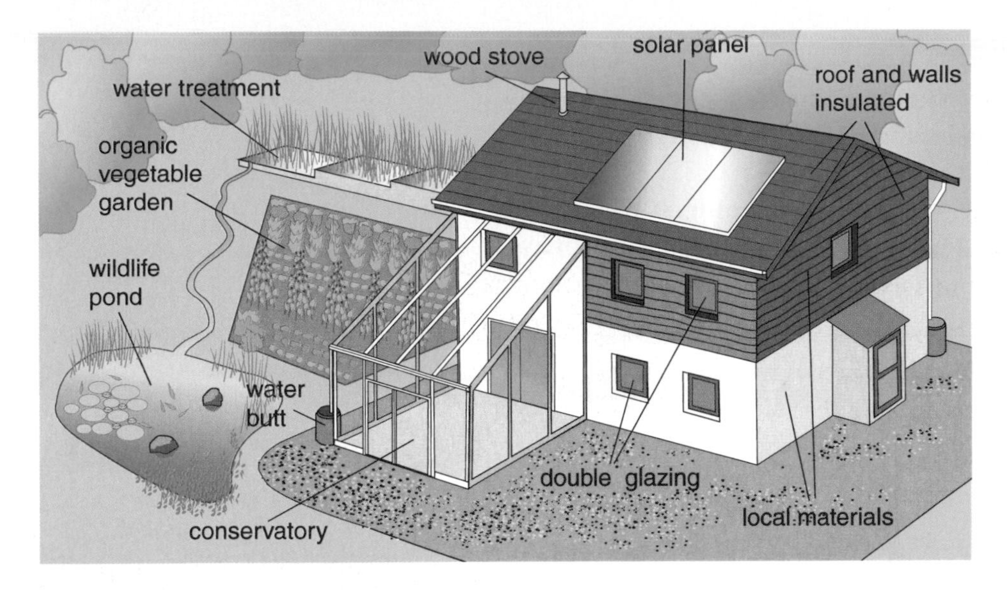

Ecohouse – the sustainable house.

Question

a Explain how each of the labelled features helps the house to be sustainable.

Waste and recycling

The chart shows the amount of waste produced per head of population in different countries.

Municipal waste is the waste that is collected by refuse lorries from homes and shops.

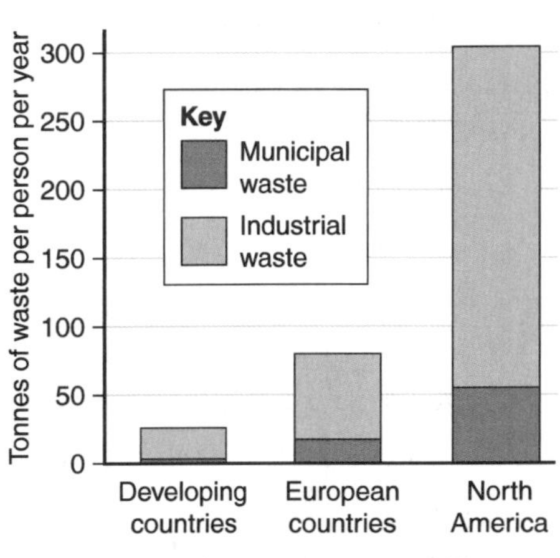

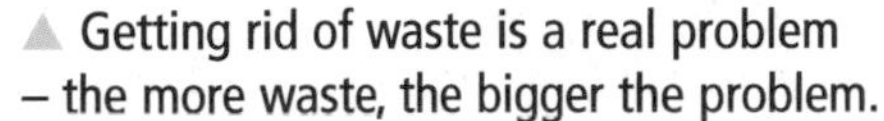

Getting rid of waste is a real problem – the more waste, the bigger the problem.

Question

b (i) Estimate the ratios of municipal waste to industrial waste for developing countries and for North America.
(ii) Suggest an explanation for the difference in the ratios.

One way of conserving precious natural resources is to recycle them. We can all contribute to this by using recycling bins. Many local councils now provide households with two bins – one for rubbish that can be recycled and one for waste that cannot be recycled. Many products are now designed so that the materials in them can be recycled.

The Government has given local councils recycling targets to meet. This means that a much higher proportion of waste is now being recycled rather than dumped. Recycling paper means that fewer forests are cut down. Recycling glass and metals means that fewer quarries are dug.

Recycling materials helps to conserve natural resources.

Question

c How does recycling each of the following save natural resources? (i) newspaper; (ii) bottles; (iii) aluminium cans.

Key points

- Sustainable development means improving the quality of life without compromising the lives of future generations.
- This means using natural resources and energy carefully and in planned ways.
- Natural resources can be conserved by recycling paper, glass and metals.

Global warming – fact or fiction?

Investigating climate change

People have differing views of climate change and whether global warming is really happening. Read the report below from a 'Greenforce' conference.

> Climate change is with us. A decade ago, it was conjecture. Now the future is unfolding before our eyes. Canada's Inuit see it in disappearing Arctic ice and permafrost. The shantytown dwellers of Latin America and Southern Asia see it in lethal storms and floods. Europeans see it in disappearing glaciers, forest fires and fatal heat waves.

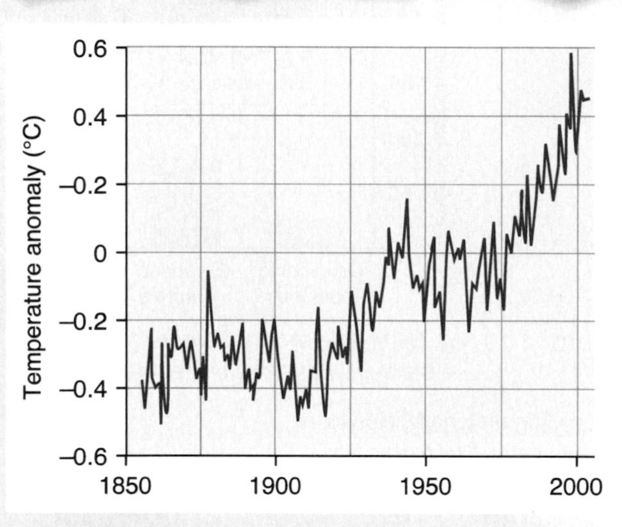

▲ How global temperatures have varied since 1860.

Question

a *Are the above statements supporting the idea of global warming based on valid and reliable evidence, or on non-scientific observations?*

The graph shows the how the global temperature each year differed from the mean temperature for the period. This difference is called the temperature anomaly.

Question

b *(i) Describe the trend in global temperatures between 1860 and 2000. (ii) What is the advantage in recording 5-year averages rather than annual averages? (iii) Reliable measurements of the global temperature only go back to the 1860s. Suggest a reason for this.*

▼ Every chunk of ice core tells the story of the atmosphere it was formed in.

To find out about the climate before 1860, scientists have to rely on 'proxy' records rather than direct measurements with instruments. For example, the width of tree rings is related to temperature. Other techniques that have been used include examining the time of crop harvests and other historical records.

Scientists needed much more reliable data if they were to prove that global temperature was rising, and to establish a link between global warming and carbon dioxide. To do this they obtained evidence from ice cores.

The photograph shows a scientist examining part of a 3 km-long ice core drilled from the Antarctic. As the ice was formed, air bubbles were trapped in it. Scientists can analyse these bubbles to find the composition of air trapped hundreds of thousands of years ago. They can also estimate the temperature of the atmosphere when the bubbles were trapped. The graph on the next page shows some of their results.

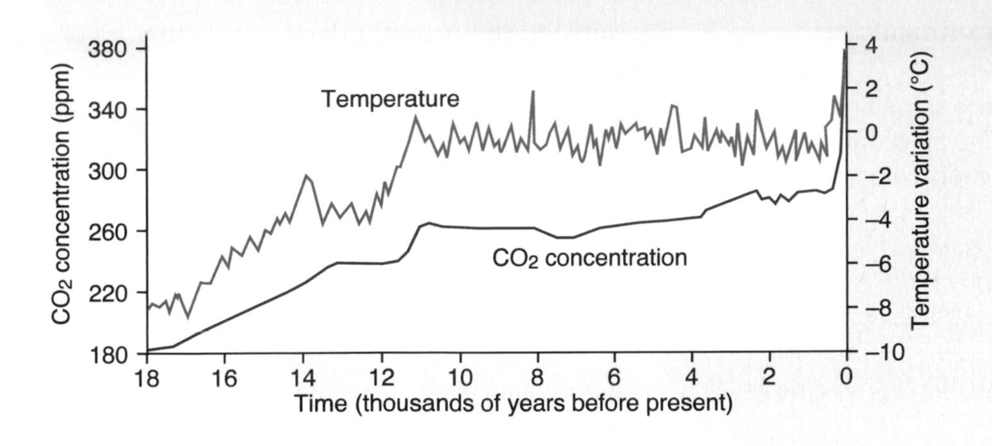

Question

c *(i) What was the mean air temperature in Antarctica 14000 years ago?*
(ii) Describe how the carbon dioxide concentration in the atmosphere changed from 18000 years ago to the present.
(iii) Explain how the data from the graph gives evidence that carbon dioxide is a greenhouse gas.
(iv) Explain why the graph does not prove that carbon dioxide is responsible for global warming.

Different views

Read the newspaper cuttings about the heat wave of 2003 again (p 66). Some scientists are not convinced that the **greenhouse effect** was to blame. Below are some statements by these scientists:

> 'We must be wary of trying to link naturally varying weather events like fires and floods to simplistic events like 'global warming'. The highest recorded temperature for Australia was in 1889, for North America in 1922 and for Asia in 1974. If we junked every car, closed every factory, and shut down every power station, climate would still change, and we would still suffer heat waves and ice ages.'

The views of such scientists influence politicians. In 2004, Russia finally agreed to sign the **Kyoto Treaty** that calls for industrialised nations to reduce their greenhouse gas emissions by 5% by 2012. This meant that nations responsible for 55% of greenhouse emissions had agreed to take action. But the USA, which is responsible for 25% of greenhouse gas emissions, still refuses to sign.

At the G8 summit in 2005, the US President George Bush said 'To a certain extent, [climate change] is [man-made], obviously.' But he defended US rejection of the Kyoto treaty, saying 'the Kyoto treaty would have wrecked our economy … my hope is to move beyond the Kyoto debate and to collaborate on new technologies that will enable the United States and other countries to diversify away from fossil fuels.'

Question

d *Summarise evidence from this chapter that: (i) supports the views of the scientists above; (ii) supports the global warming theory for climate change.*

Question

e *50% of electricity in the USA is generated by coal-fired power stations and this is expected to rise to 80% over the next 5 years.*
(i) Does President Bush now accept the scientific evidence that links global warming to carbon dioxide emission?
(ii) Why do you think that the USA will not sign the Kyoto treaty?

Key points

- Scientists collect data to provide evidence for environmental change.
- A correlation between two variables does not always mean that one is causing the other.
- We need to plan future development carefully to make sure it is sustainable.
- Different people have different ideas about how this should be done.

End of unit questions

1 A class investigated how temperature affected the number of earthworms in soil.

Each soil sample was taken from the same field and was 1 m square and 15 cm deep. The air temperature was taken and earthworms were counted on the same day each month.

The results are shown in the table below.

	Jan	Feb	Mar	Apr	May	Jun	Jul	Aug	Sep	Oct	Nov	Dec
Air temp. (°C)	3	1	1	5	8	15	20	16	12	9	8	6
No. of worms	20	5	8	33	75	12	9	15	35	43	75	53

a **i** What were the control variables in the investigation? *(2 marks)*

ii Give **two** ways in which the investigation could have been improved to give more reliable results. *(2 marks)*

b **i** Plot the data on one graph. *(2 marks)*

ii In which month was the least number of earthworms found? Suggest an explanation for this. *(2 marks)*

c The class decided that another factor might also be affecting the number of earthworms. They decided to find out the rainfall for the area. Suggest where they might find this information. *(1 mark)*

The table below shows the monthly rainfall for the area.

	Jan	Feb	Mar	Apr	May	Jun	Jul	Aug	Sep	Oct	Nov	Dec
Total rainfall (mm)	45	33	28	55	75	25	8	12	35	45	60	55

i Add this data to the graph. *(1 mark)*

ii In which two months were most earthworms found? *(2 marks)*

iii Which conditions do earthworms prefer? *(1 mark)*

iv Outline an investigation you could do in a laboratory to find if earthworms prefer these conditions. *(2 marks)*

2 The table is about methods of producing offspring.

Match words from the list with each of the numbers **1–4** in the table.

A asexual reproduction
B genetic engineering
C sexual reproduction
D tissue culture

	Function
1	transferring genes from one species to another
2	produces offspring with no fusion of gametes
3	producing offspring from a small group of cells
4	produces offspring with a mixture of characteristics from two parents

3 The diagram shows the technique involved in adult cell cloning.

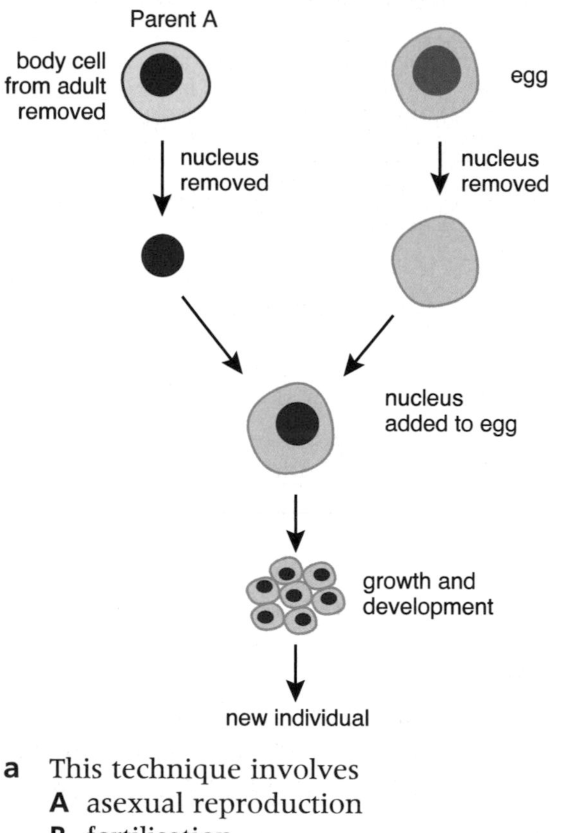

a This technique involves

A asexual reproduction
B fertilisation
C mutation
D sexual reproduction.

b The new individual is identical to Parent A because they

A have the same genes
B are formed by the fusion of gametes
C have developed in the same conditions
D have the same enzymes.

c Some people object to producing animals in this way. This is most likely to be because cloning raises issues which are

A economic **B** ethical

C scientific **D** social.

d Which one of the following is **not** contained in a nucleus?

A cells **B** chromosomes

C DNA **D** genes

4 Plant tissue culture is a method used to produce new plants. The flow diagram shows one method of plant tissue culture.

Small piece of tissue is removed from a plant, e.g. a piece of root or stem tissue
↓
The tissue is transferred to a culture medium
↓
New tissue develops containing unspecialised cells
↓
New tissue is transferred to a new culture medium
↓
New specialised shoots and root cells develop
↓
Developing plants are separated and grown under optimum conditions

a Name the type of reproduction involved in plant tissue culture. *(1 mark)*

b Describe **two** advantages of producing plants using this method rather than from seed. *(2 marks)*

c Explain why a disease is more likely to destroy a whole batch of plants grown by plant tissue culture than a batch of plants grown from seeds. *(1 mark)*

d Suggest **two** factors which need to be controlled to create the optimum growing conditions for the developing plants. *(2 marks)*

5 Fossils have been used to provide evidence of human evolution.

a i Explain why bones and teeth are normally the only fossils found. *(2 marks)*

ii Usually, only a few bones from a single skeleton are found scattered around a wide area. Suggest reasons for this. *(2 marks)*

b Scientists use fossil evidence to work out how humans evolved. The diagram shows the skulls of a gorilla and a human and Neanderthal Man – an extinct species.

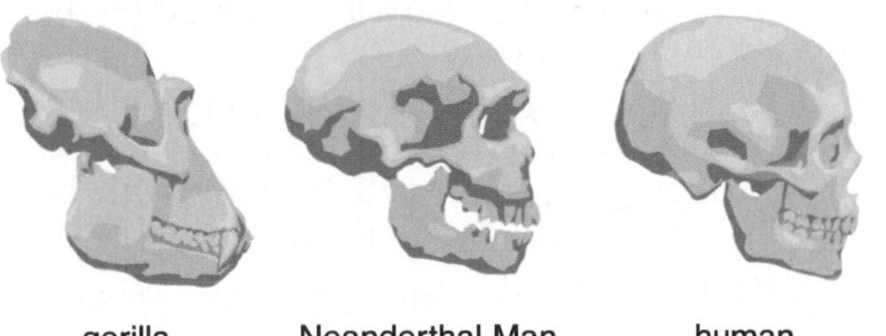

Describe **two** features of the skull which suggest that Neanderthal Man is more closely related to humans than gorillas. *(2 marks)*

6 The land snail is found in grasslands and in woodlands. The shells of snails differ in colour and in the number of dark bands. The colour of the shells may be yellow or brown. Some shells have no bands and some have lots of bands.

Birds such as thrushes feed on snails.

The scattergraph shows the percentage of yellow unbanded snails in two different habitats.

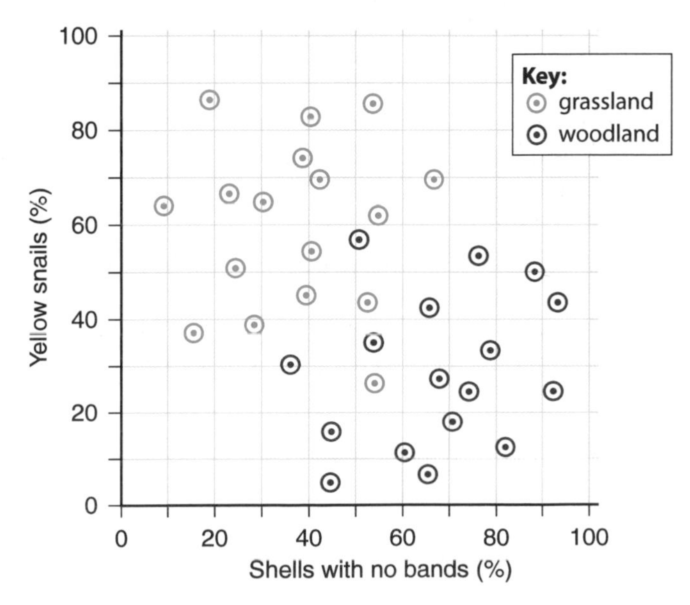

a Which of the following statements best describes the relationship between habitat and the colour and banding of the land snail?

A There is no relationship.

B The percentage of yellow snails with bands is higher in woodlands.

C The percentage of yellow snails with no bands is higher in woodlands.

D The percentage of yellow snails with no bands is higher in grasslands.

b Which of the following statements provides the best explanation of this relationship?

A Snails which are more easily seen are more likely to be eaten by thrushes.

B There are more thrushes in woodlands.

C Thrushes only eat snails with banded shells.

D Snails with yellow shells are better camouflaged.

c The differences in the appearance of the shells of the land snail is an example of

A competition

B cloning

C natural selection

D variation.

d Snails which are better suited to their habitat are more likely to survive and pass their genes on to their offspring. This is an example of

A competition

B mutation

C natural selection

D predation.

7 Human activities affect the environment.

Match words **A**, **B**, **C** and **D** with the spaces **1–4** in the paragraph.

A carbon dioxide

B fertiliser

C methane

D sulfur dioxide

Water may be polluted by ___1___. Burning wood produces ___2___. Acid rain is produced mainly by ___3___ dissolving in water. Rice fields add ___4___ to the atmosphere during the day.

8 Organisms affect or are affected by the composition of the atmosphere.

Match words **A**, **B**, **C** and **D** with the spaces **1–4** in the sentences.

A cattle

B lichens

C microorganisms

D trees

___1___ can be used as indicators of air pollution.

___2___ break down dead materials and return carbon dioxide to the atmosphere.

___3___ give off large amounts of methane into the atmosphere.

___4___ lock up large amounts of carbon dioxide for many years.

9 **a** Give **two** reasons why tropical rainforests are being cut down at a large rate. *(2 marks)*

b Cutting down rainforests reduces biodiversity.

i Explain what is meant by 'biodiversity'. *(1 mark)*

ii Give **one** consequence for humans of a reduction in the biodiversity of tropical forests. *(1 mark)*

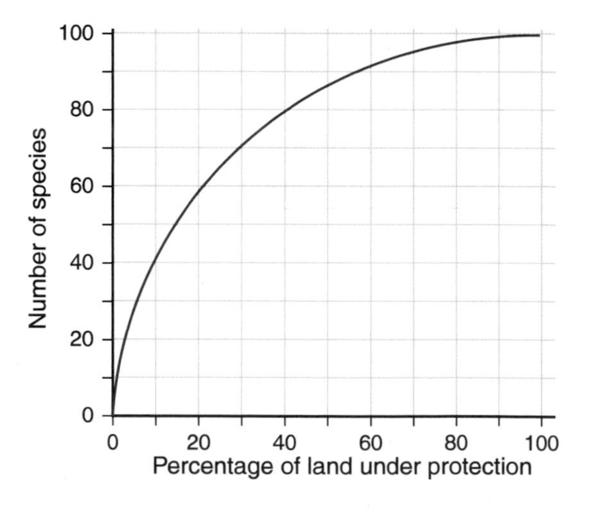

c Biodiversity can be preserved by protecting parts of rainforests.

i Describe, in as much detail as you can, the relationship between the percentage of land protected and number of species preserved. *(3 marks)*

ii Explain why small-scale protection projects can be very effective. *(1 mark)*

10 In 1952 in London there was a thick fog for several days in December. This fog trapped air pollutants. The graph shows the concentration of sulfur dioxide and smoke particles in the atmosphere. It also shows the number of deaths per day.

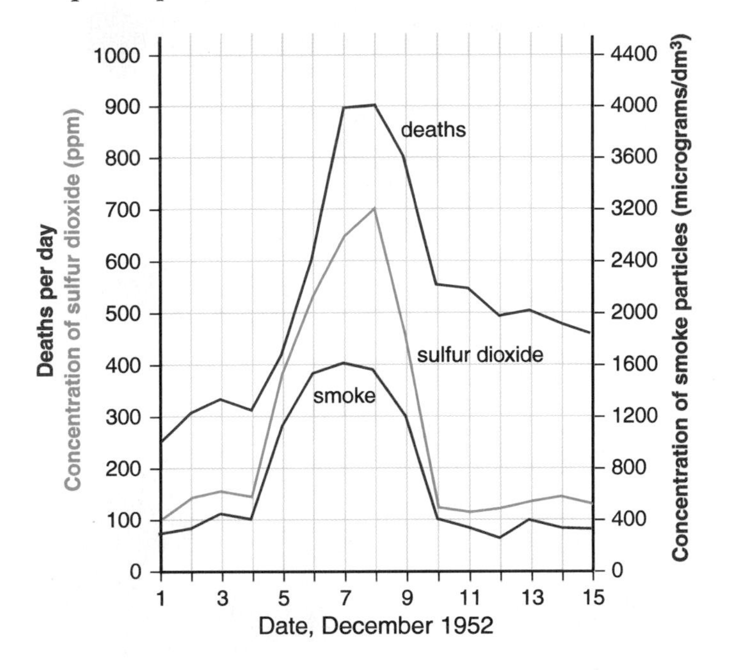

a What was the maximum number of deaths per day?
A 400 **B** 700 **C** 900 **D** 3600

b What was the concentration of smoke particles on 7 December?
A 400 micrograms per cubic decimetre
B 650 micrograms per cubic decimetre
C 900 micrograms per cubic decimetre
D 1600 micrograms per cubic decimetre

c The data
A proves that sulfur dioxide caused the large number of deaths
B shows an exact correlation between sulfur dioxide concentration and the number of deaths
C shows partial correlation between sulfur dioxide concentration and the number of deaths
D shows that that the number of deaths depends on smoke particle concentration.

d Sulfur dioxide dissolves in water to produce
A an acid solution
B an alkaline solution
C a neutral solution
D sewage.

11 a Explain what is meant by 'sustainable development'. *(1 mark)*

b China has the largest population in the world. Demand for electricity there is increasing rapidly. Suggest **two** reasons for this increase in demand. *(2 marks)*

c The table shows how China plans to change its methods of generating electricity by 2020.

Year	Percentage of electricity obtained from energy source					
	Hydroelectricity	Coal	Oil	Gas	Nuclear	Other
2000	24.8	69.3	4.8	0.3	0.7	0.1
2002	27.1	58.6	1.6	7.5	4.2	1.0

Will the changes to China's electricity production help the environment? Use data from the table to explain your answer. *(3 marks)*

12 An 'ecological footprint' is the measure of how much land and water a human population needs to produce the resources required to sustain itself and to absorb its wastes.

The average American uses 25 hectares to support his or her current lifestyle.
The average Canadian uses 18 hectares.
The average Italian uses 10 hectares.

a Explain why Americans have larger ecological footprints than Italians. *(2 marks)*

b Suggest **three** ways in which ecological footprints can be reduced. *(3 marks)*

Cells and photosynthesis

Lifting weights takes a lot of energy. The energy comes from the weight-lifter's food. But how did the energy get into the food; and how does the weight-lifter release energy from food?

All energy transfers in living organisms take place in tiny parts of cells called organelles. You can see some organelles in cells through a light microscope. But most of our understanding of how organelles work comes from using electron microscopes which can make out much more detail than light microscopes. The organelles that transfer energy are chloroplasts, found in some types of plant cell, and mitochondria, found in all living plant and animal cells.

It takes a lot of energy to do this.

The energy used to lift the bar over the weight-lifter's head comes from the millions of mitochondria in the cells in her muscles. These transfer the chemical energy in her food into energy that makes her muscles contract.

But where does the chemical energy in food come from? Chloroplasts in plant cells absorb energy from sunlight and use this to produce sugars containing chemical energy. So all the weight-lifter's energy comes from sunlight.

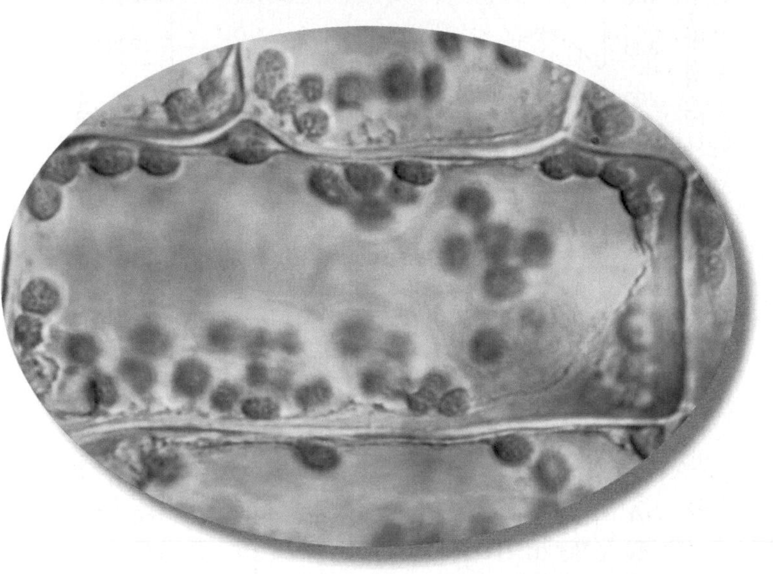

Tiny chloroplasts produce all the Earth's food.

Food for the future

To go on long space voyages to planets such as Mars, astronauts will have to grow their own food on board. They will also need to renew the oxygen on board the spacecraft. Plants could do this for them. But will plants grow and reproduce in space?

Tiny mitochondria provide you with all of your energy.

In 2002, the space shuttle Atlantis carried seeds of the *Brassica rapa* plant to the Russian space station Mir. Scientists in the space lab used the seeds to carry out experiments on plant growth. They are using *Brassica rapa* because this plant produces edible oils, which are difficult to provide in a space-grown diet. The oils come from the seeds, and the young leaves can also be eaten in salads. The unopened flower heads, which are very similar to broccoli, can be eaten as well.

▲ A modern food factory.

Plants make their food by photosynthesis. Understanding photosynthesis is critical for future long-duration space missions. In addition to providing food for the astronauts, plants can generate oxygen, remove carbon dioxide and purify water. So, living plants could help maintain proper spacecraft atmosphere, and recycle water. This research will also have direct application to future production of crops that the astronauts could eat, such as lettuces, radishes or onions.

A growing world population needs more food. But the natural growing season in countries like the UK is only a few summer months. Now we can grow crops all the year round in greenhouses. But growing crops like this is expensive, so scientists research the conditions which will make plants grow fastest at the least cost. This means understanding what goes on in plant cells during photosynthesis and respiration.

Photosynthesis takes place in leaves. You can see that there are many different types of cell in the leaf. These different leaf cells are adapted to their functions in the photosynthesis process.

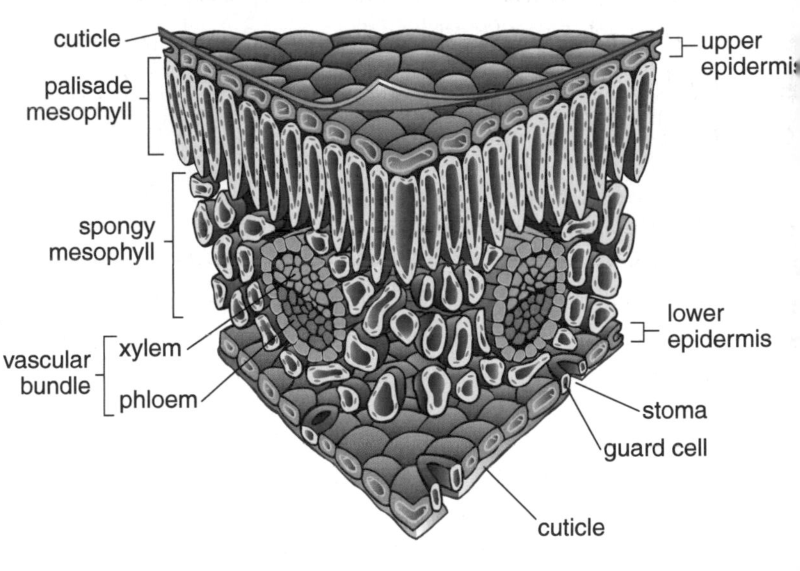

▲ Inside a leaf.

Think about what you will find out in this section

How is the structure of different types of cells related to their function in a tissue or an organ?	What happens to the carbohydrates made by photosynthesis?
How do substances move into and out of cells?	How can growers increase the rate of photosynthesis in their plants?
How do different factors affect the rate of photosynthesis?	What are the benefits and disadvantages of growing plants in artificial conditions?

Palisade cells

Most photosynthesis takes place in **palisade cells** in leaves. We can't see much of the structure of these cells through a light microscope but the more powerful electron microscope shows us a lot more detail.

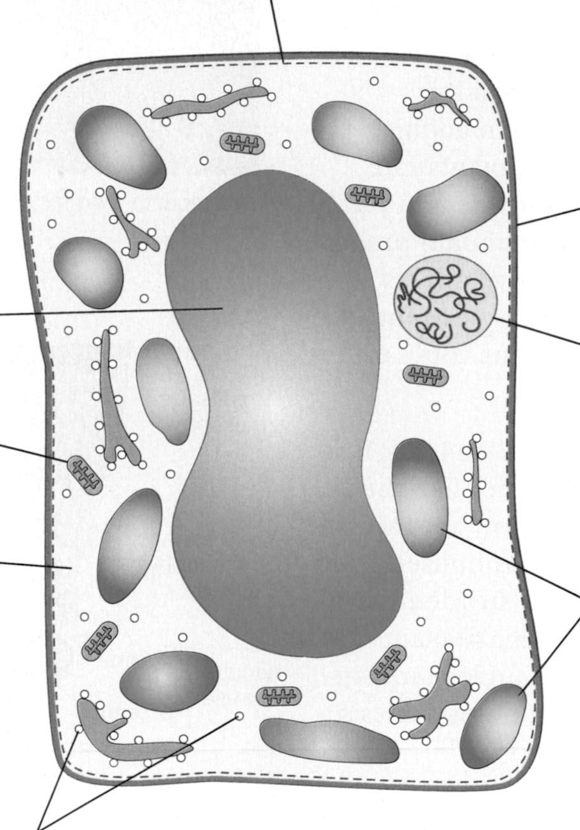
The **cell membrane** has tiny pores. These allow small molecules like water and gases to diffuse through, but keep large molecules inside the cell.
The **cell wall** is made up of tiny fibres. Together these are very strong, so the cell wall supports the cell.
The **vacuole** is filled with a solution of ions called **cell sap**. When it is full the vacuole supports the cell.
The **nucleus** controls the activities of the cell. It contains genetic information for making proteins. Some of these proteins are enzymes. Enzymes catalyse chemical reactions in the cell.
mitochondrion (see opposite)
The **cytoplasm** is where many of the chemical reactions in the cell occur. These reactions are catalysed by enzymes.
Chloroplasts contain a green pigment called **chlorophyll**. This absorbs light energy. Chloroplasts convert this light energy into chemical energy in carbohydrates.
The enzymes concerned with the synthesis of proteins are found close to organelles called **ribosomes** in the cytoplasm. A copy of the genetic information needed to produce a protein is made from the DNA in the nucleus. This copy becomes attached to a ribosome. Enzymes then use this information to join amino acids in the correct sequence to produce the protein.

▲ Palisade cells are the starting point for the manufacture of almost all the world's food. A palisade cell is a typical plant cell.

Question

a *Draw a table with two columns. In the left-hand column, write the names of the parts of a plant cell. In the right-hand column, give the function of each part of the cell.*

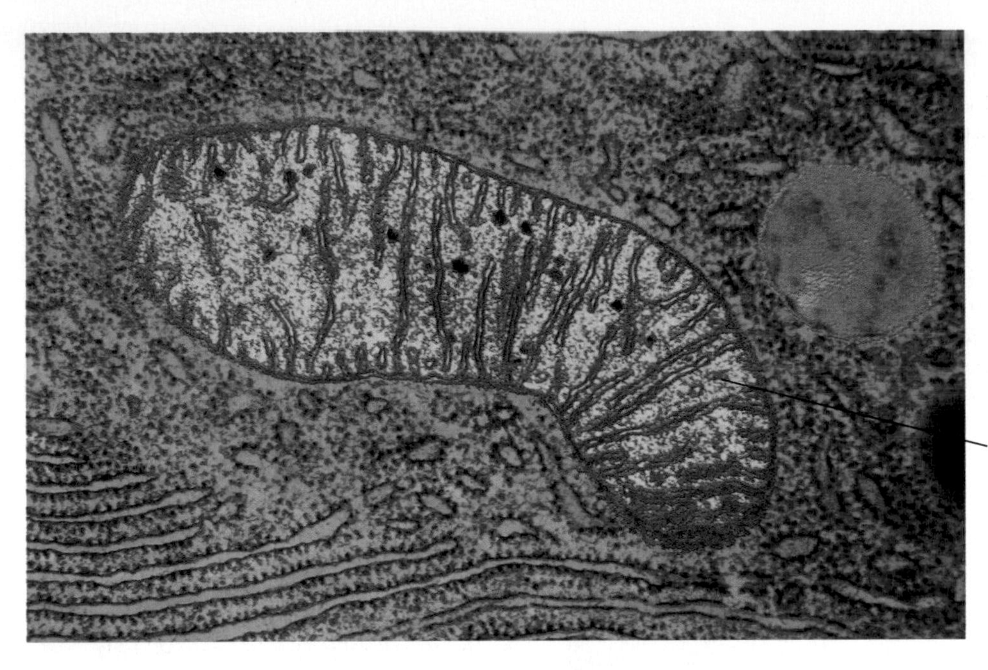

Mitochondria are where energy stored in sugars is released by chemical processes in the cell. The energy is released in a series of chemical reactions called respiration.

How do animal cells differ from plant cells?

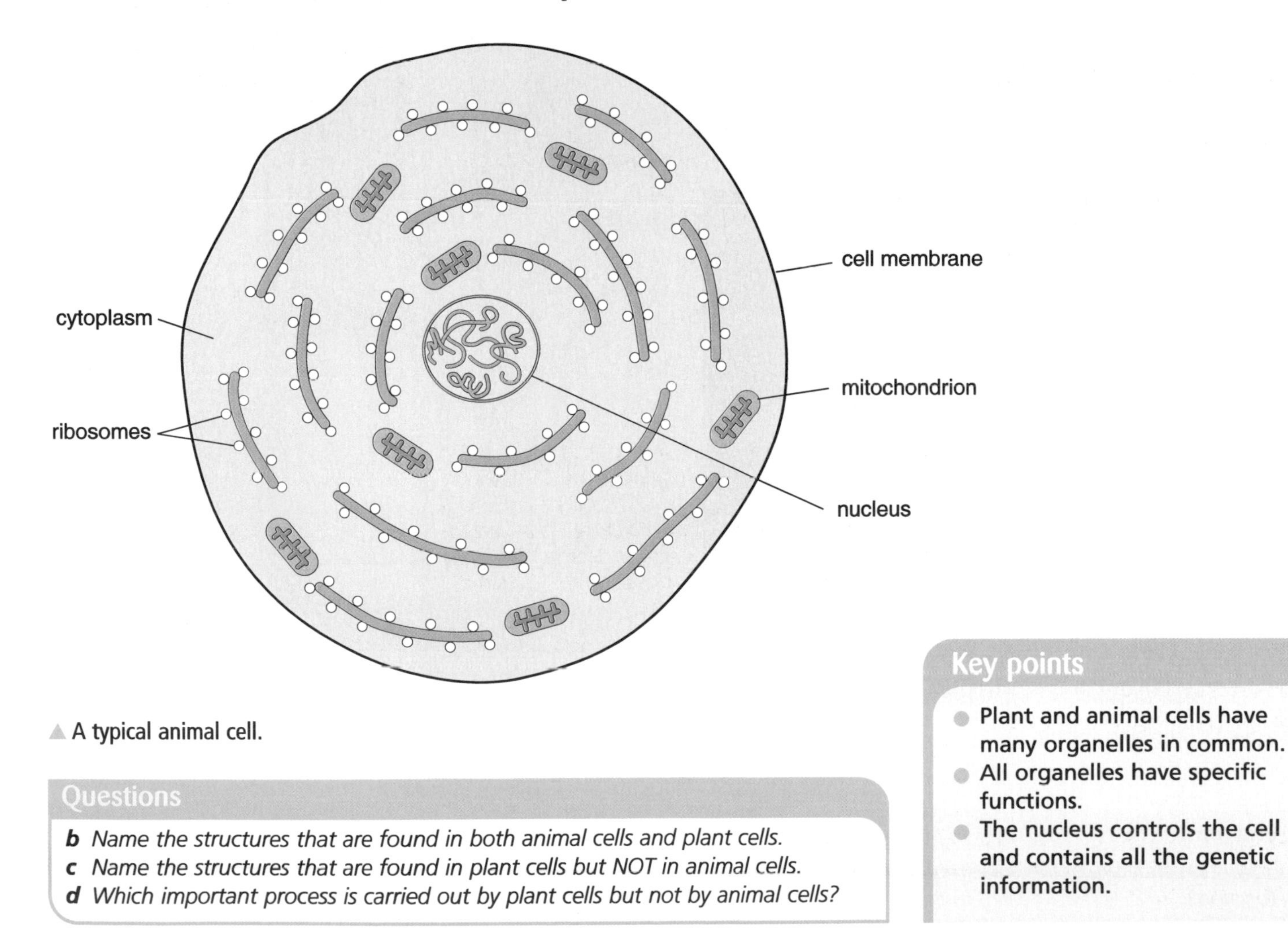

▲ A typical animal cell.

Questions

b Name the structures that are found in both animal cells and plant cells.
c Name the structures that are found in plant cells but NOT in animal cells.
d Which important process is carried out by plant cells but not by animal cells?

Key points

- Plant and animal cells have many organelles in common.
- All organelles have specific functions.
- The nucleus controls the cell and contains all the genetic information.

Specialised plant cells

There are many different kinds of cell in a leaf. Each type of cell is **specialised** or adapted to its function.

The function of the xylem is to bring water and ions from the roots to the leaves. Mature xylem cells are dead. Their cell walls are impregnated with a strong, waterproof material called lignin. The end walls of the cells break down to form continuous vertical tubes called xylem vessels.

Gases enter and exit the leaves via pores called stomata. Each stoma is surrounded by two guard cells. The changing shape of the cell wall of the guard cells creates a pore through which gas exchange can take place.

The photograph shows phloem cells from a leaf. The phloem cells carry the sugars made in photosynthesis to other parts of the plant.

Question

a *How is the structure of a xylem vessel adapted to its function?*

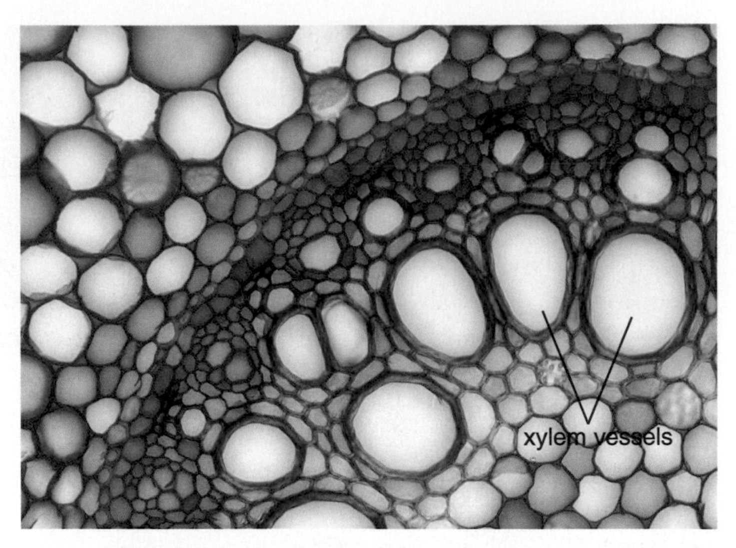

▲ The plant's plumbing system.

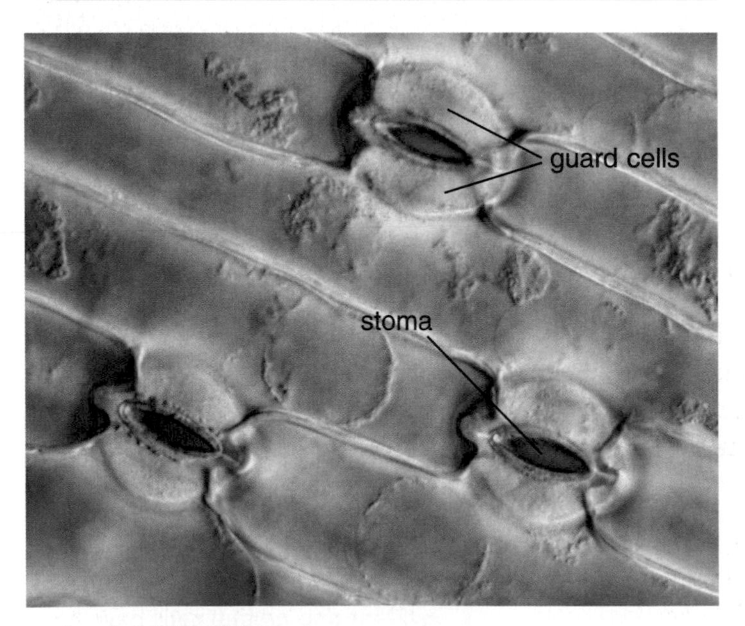

▲ The plant's breathing system.

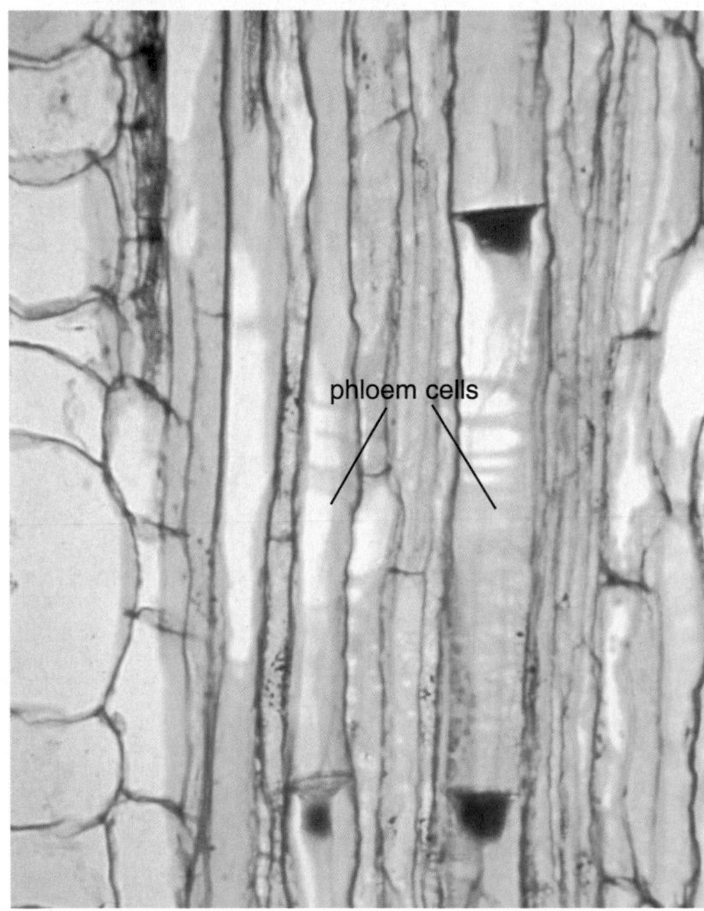

▲ The plant's food transport system.

Question

b *How is the structure of the phloem cells adapted to their function?*

Specialised animal cells

The oesophagus (gullet) is an organ, essentially a long, muscular tube, that carries food from the mouth to the stomach. The food is still mainly solid at this time. The different types of cell in the gullet help it to carry out its function.

The muscle cells contain thousands of protein fibres. These fibres can be made to slide over each other causing the muscle cells to contract. Food is moved down the gullet by contraction of the muscle cells squeezing the food along. This process is called peristalsis.

The gland goblet cells produce a slimy material called mucus. The mucus moves out of the cells onto the inner lining of the gullet. Here, the mucus acts as a lubricant, cutting down friction between the solid food and the lining of the gullet.

The cells lining the gullet are dead and hard, like the cells on the surface of the skin. These lining cells protect the living cells underneath from being damaged as hard food moves over them.

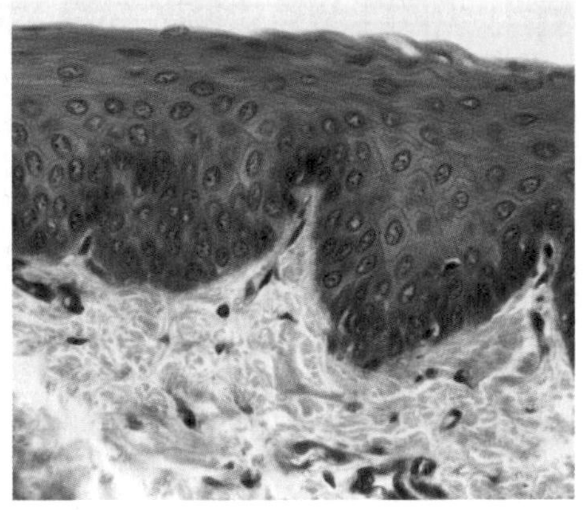

▲ Your gullet contains many different kinds of cell.

Sperm cells

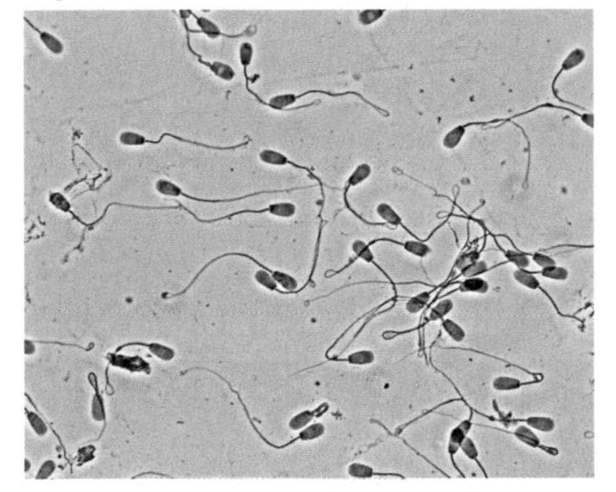

▲ Sperm cells.

A sperm cell has to be able to move to reach an egg.

Question

c *How does each of the following help a sperm cell to move:*
(i) tail
(ii) mitochondria
(iii) shape?

Nerve cells

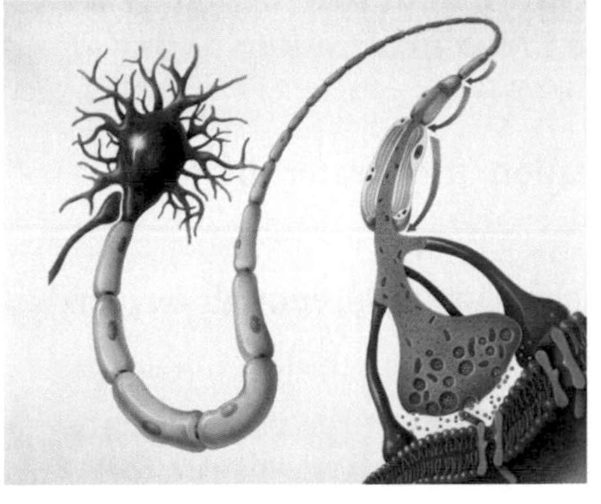

▲ A motor neurone cell.

A motor neurone has to be able to receive information from a relay neurone, then transmit an impulse over a long distance.

Question

d *Explain how a motor neurone is adapted for: (i) receiving information (ii) transmitting information over long distances.*

White blood cells

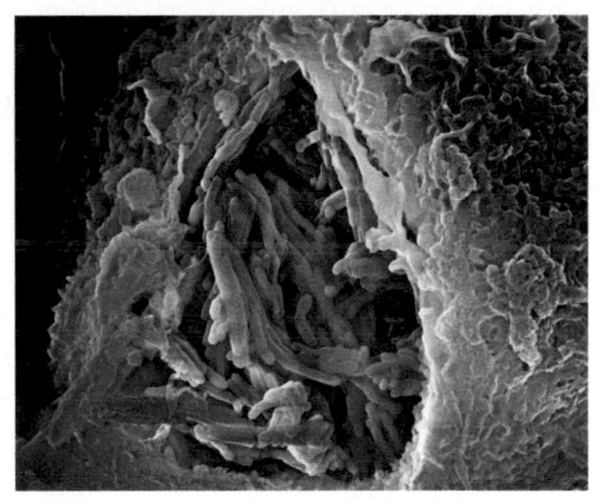

▲ A white cell engulfing bacteria.

Some white cells protect us by engulfing bacteria, then digesting them.

Question

e *Explain two ways in which this type of white blood cell is adapted for its function.*

Key points

- The structure of each type of cell is adapted to its function.
- It is important to link each structural feature to the way in which it helps the cell to carry out its function.

Why can't we drink sea water?

The main problem facing people in lifeboats on the open ocean is **dehydration** – even though they are surrounded by sea water. This is because sea water contains ions – about four times the concentration of the ions in our body fluids. Drinking one litre of sea water causes the ion concentration of the body **fluids** to rise by about 10%. The effect of this rise is to cause water to move out of the body cells, making the cells shrink as water moves out into the body fluids. To prevent too much damage to the cells, the kidneys produce more urine to get rid of the excess ions. The person loses more water in urine than was taken in as sea water. To explain why cells shrink in some solutions you need to understand two processes – diffusion and osmosis.

Diffusion

A few crystals of potassium permanganate are placed in a beaker of water. Six hours later, coloured particles have moved to all parts of the liquid. This is because particles in gases and in solutions move about all the time. Because they are moving about, the particles spread themselves out evenly. This process is called **diffusion**. The effect of diffusion is that particles move from where there is a high concentration to where there is a low concentration, so that equilibrium is reached on both sides.

The greater the difference in concentration, the greater the rate of diffusion.

Gas molecules diffuse rapidly. This enables us to get enough oxygen into the blood to transport to the body cells.

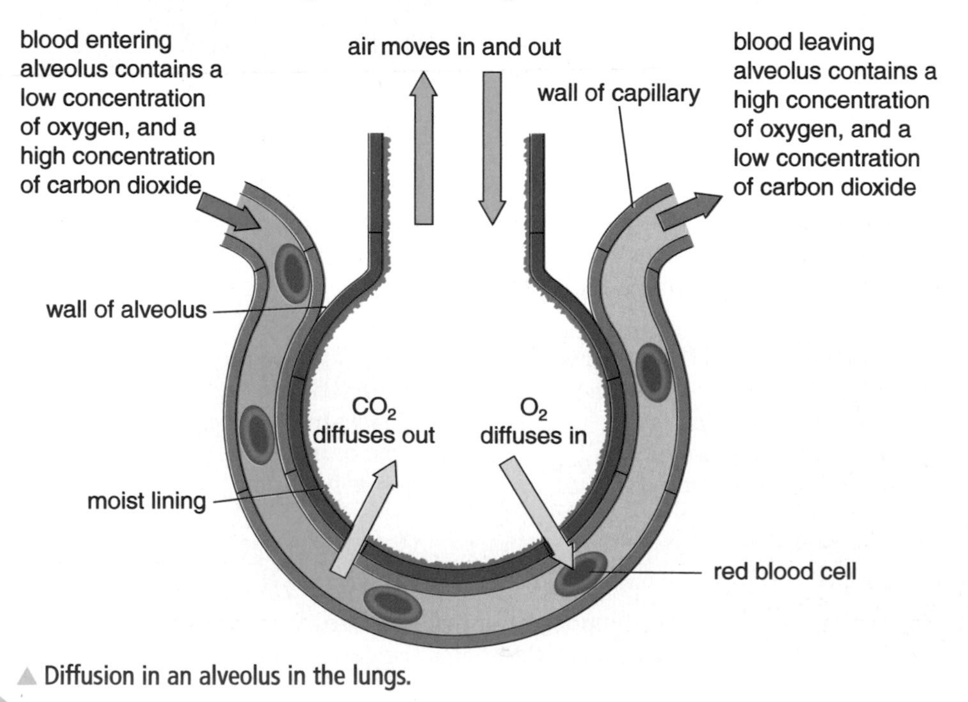
blood entering alveolus contains a low concentration of oxygen, and a high concentration of carbon dioxide
air moves in and out
wall of capillary
blood leaving alveolus contains a high concentration of oxygen, and a low concentration of carbon dioxide
wall of alveolus
CO_2 diffuses out
O_2 diffuses in
moist lining
red blood cell

Diffusion in an alveolus in the lungs.

Diffusion of coloured particles.

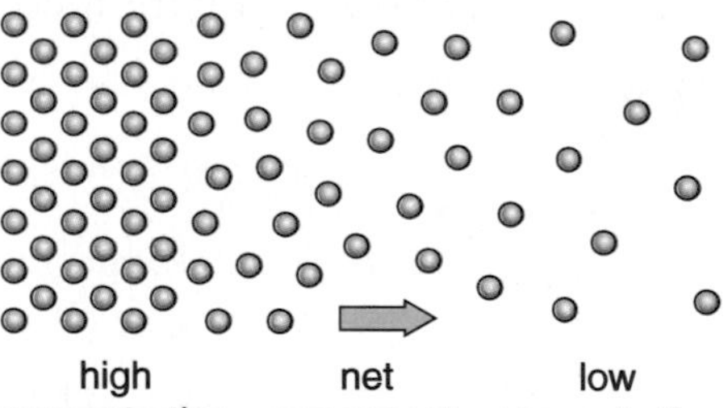
high concentration
net movement
low concentration

Particles move from areas of high concentration to areas of low concentration.

Carbon dioxide diffusing out of a jar.

Question

a *Use information from the diagram to explain why oxygen and carbon dioxide diffuse in opposite directions.*

Osmosis

Red blood cells shrink when placed in concentrated salt solution. Under normal conditions, the concentration of water molecules and salt particles is the same in both red blood cells and blood plasma. After drinking sea water there is a higher concentration of salt particles in the plasma. This means that there is a lower concentration of water molecules in the plasma than inside the red cells. The cell membrane of the red cells is partially permeable: water particles can diffuse through, but salt particles are too big. Water molecules therefore diffuse out through the cell membranes into the plasma and the red cells shrink. This movement of water molecules through a partially permeable membrane from a high concentration of water to a lower one is called **osmosis**.

▲ Red blood cells in salt solution.

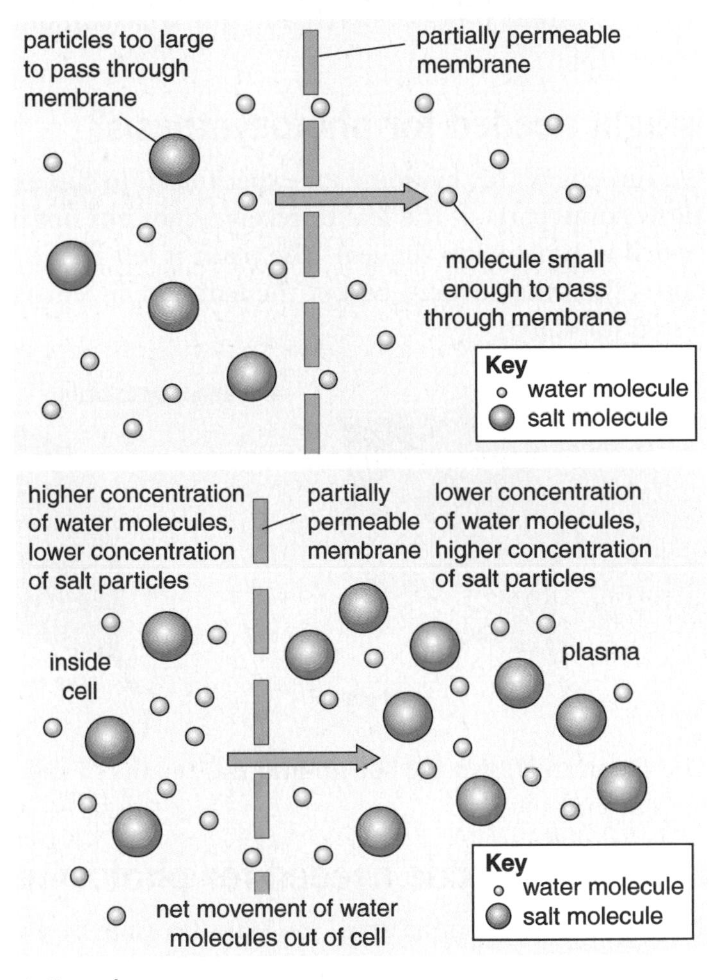

▲ Osmosis.

Questions

b *A person drinks a large volume of water very quickly. What effect will this have on the volume of the red blood cells? Explain your answer in terms of osmosis.*

c *If we place animal cells in pure water, they burst. But plant cells do not burst. Explain why, in terms of the difference in structure between animal cells and plant cells.*

Membranes save lives

Many lifeboats are now equipped with a reverse osmosis machine to provide drinking water from sea water. The diagram shows how it works.

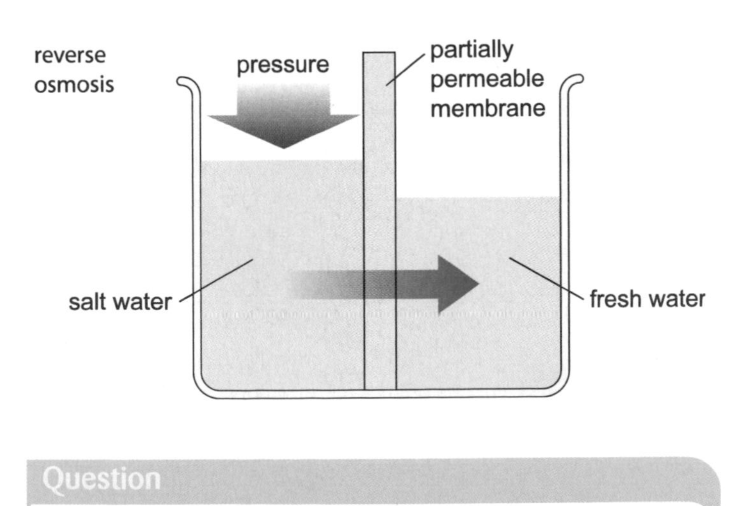

Question

d *Use information from the diagram to explain how the machine produces drinking water.*

Key points

- There are different ways for substances to pass into and out of cells.
- Diffusion is the movement of molecules or ions from a region of high concentration to a region of low concentration.
- Osmosis is the movement of water molecules through a partially permeable membrane from a high concentration of water to a lower one.

Investigating photosynthesis

Starch – the energy food

The first cornflakes were made in 1906 by Keith Kellogg when he was searching for a digestible bread substitute. Kellogg accidentally left a pot of boiled wheat to stand and the wheat grains softened. When Kellogg rolled the softened wheat and let it dry, each grain of wheat emerged as a large thin flake. The flakes turned out to be a tasty cereal. Kellogg had invented cornflakes.

One of the easiest ways to see if a plant is photosynthesising is to test it for **starch**. Any excess sugar produced during photosynthesis is stored as the insoluble product starch and this will stain blue-black with the iodine test.

▲ The first starch of the day.

Is light needed for photosynthesis?

We can check this by doing an experiment. In the experiment we allow some parts of the leaf to receive light but not others. A foil stencil is attached to the leaf. The plant is left in the light for a few hours then the covered part of the leaf and an uncovered part are tested for starch.

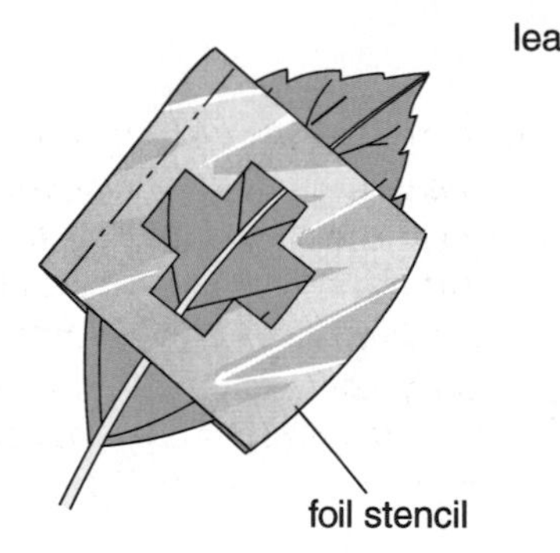

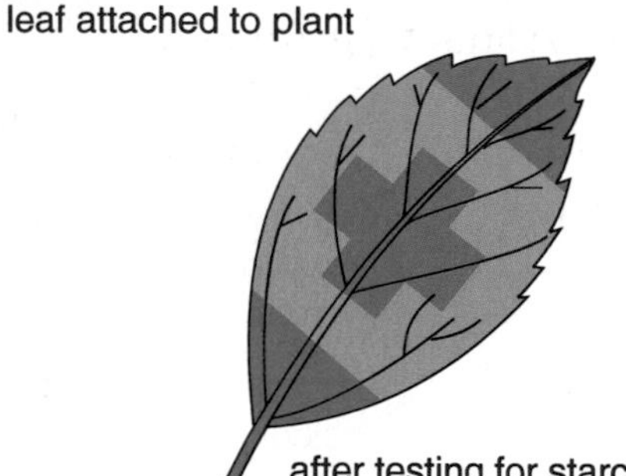

The diagram shows the set-up and the results of the iodine test for starch.

▲ The iodine test. If starch is present, the leaf turns blue-black.

Question

a *Explain the appearance of the leaf after the iodine test.*

Is carbon dioxide needed for photosynthesis?

To test for this we must stop one leaf of a plant getting carbon dioxide. The diagram on the right shows how this is done. After a few hours in the light, both the leaves are tested for starch using the iodine test.

The leaf in the flask did not turn blue-black in the iodine test, but a leaf which had not been in the flask turned blue-black.

Question

b *(i) What is the control in this experiment?*
(ii) Explain the results of the iodine test for starch.

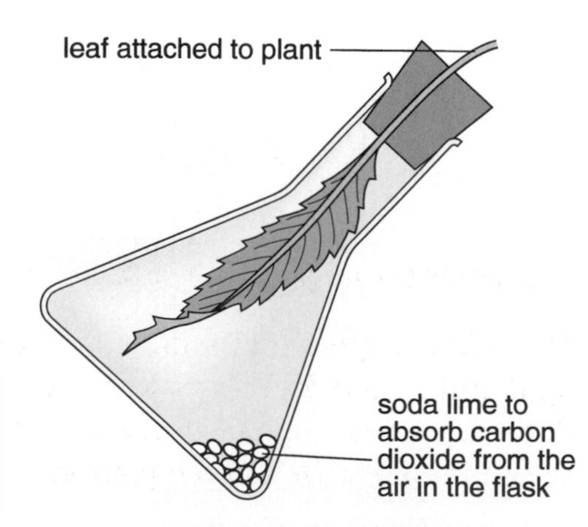

▲ Preventing a leaf getting carbon dioxide.

Is chlorophyll needed for photosynthesis?

We can't take chlorophyll out of a leaf without killing it. But some leaves have patches without chlorophyll. These leaves are called **variegated** leaves.

A plant with variegated leaves is left in the light for a few hours. One of its leaves is then tested for starch. The results of the iodine test for starch are shown in the diagram.

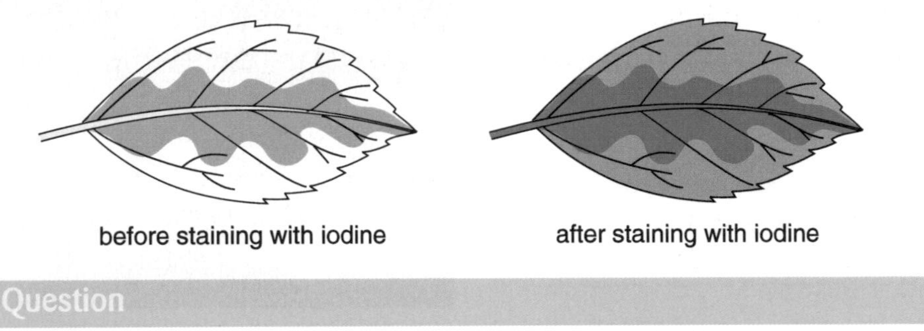

Question

c *Explain the results of the iodine test on the variegated leaf.*

Showing that plants produce oxygen

We can collect the gas given out by plants using an aquatic plant and the equipment shown. Aquatic plants are ones that grow naturally in water. They can obtain dissolved oxygen and dissolved carbon dioxide from the water. When the gas in the tube is tested we find that it is mainly oxygen.

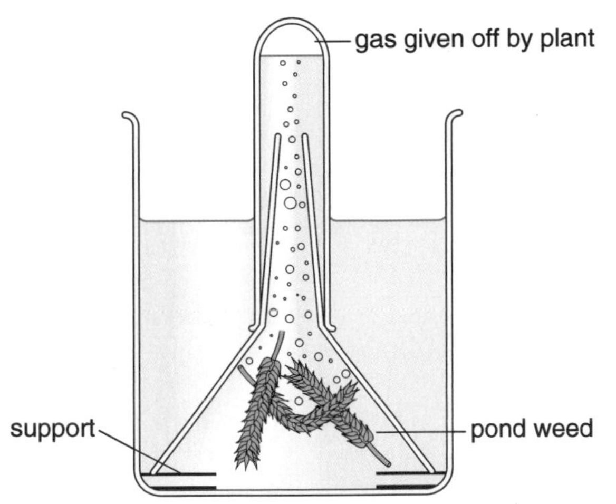

▲ Collecting the gas given out by an aquatic plant.

These four experiments confirm that green plants need carbon dioxide, light and chlorophyll to make starch; and that they produce oxygen. Photosynthesis consists of many chemical reactions, but these can be summarised by the word equation:

carbon dioxide + water + **light energy** → glucose + oxygen

Question

d *(i) How would you show that the gas in the tube was oxygen?*
(ii) The experiment was repeated but this time the apparatus was left in the dark. How would the composition of the gas in the tube be different from the gas collected when the tube was in the light?

Key points

- Photosynthesis is the means by which plants produce their own food using the Sun's energy.
- The process also requires carbon dioxide, water and chlorophyll.
- Oxygen is produced as a waste product of the reaction and glucose as an energy source.

What limits the rate at which crops grow?

Growing corn is big business. Corn plants produce food via photosynthesis, so growers need to know the conditions in which photosynthesis works fastest.

These conditions can be investigated in a school laboratory using simple apparatus. The photosynthesis equation tells us what conditions plants need to grow:

carbon dioxide + water + **light energy** → glucose + oxygen

But plants need heat as well as light energy in order to grow.

Gold on stems?

Question

a *If this was a reaction you were investigating in chemistry, give three ways in which you could increase the speed of the reaction.*

The speed of a reaction can usually be increased by increasing the concentration of the chemicals. Many reactions also go faster if the temperature is increased.

A **limiting factor** is something that slows down or stops a reaction even when other factors are in plentiful supply.

Measuring the rate of photosynthesis

The easiest way of measuring the rate of photosynthesis is to measure the rate at which oxygen is produced. The two sets of apparatus shown in the diagrams do this in different ways.

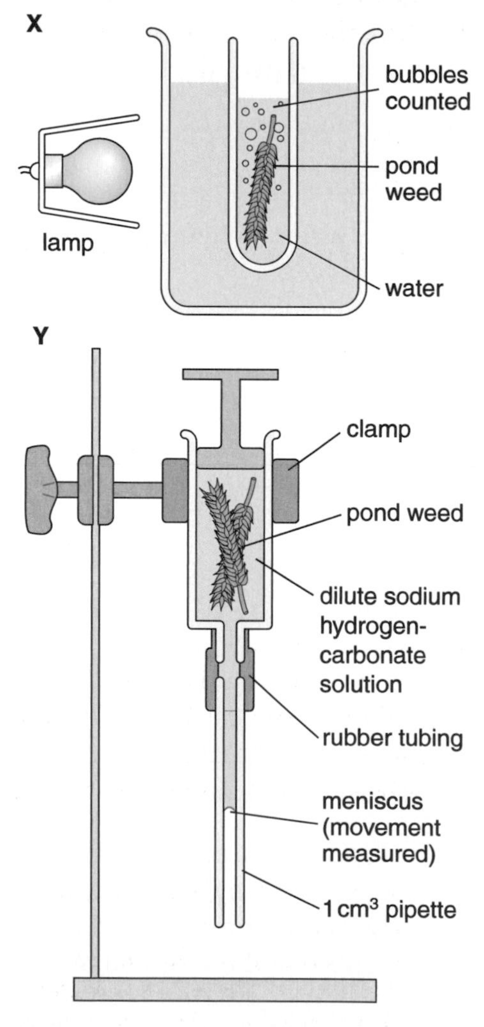

Measuring the rate of photosynthesis.

Question

b *(i) What two measurements do you need to make in order to calculate the rate at which oxygen is produced?*
(ii) Which set of apparatus in the diagram would give the more reliable data? Explain the reason for your answer.
(iii) What assumption is made when using either set of apparatus to measure the rate of oxygen production?

How light intensity affects the rate of photosynthesis

We know that plants need light to photosynthesise but, to advise growers, we need to investigate how variations in light intensity affect the rate of photosynthesis.

Questions

c *Outline how would you use apparatus* ***Y*** *to measure the effect of light intensity on the rate of photosynthesis.*

d *The table on the next page shows results obtained from this experiment. On graph paper, draw a graph of the results.*

When we do this we find that the rate of photosynthesis increases only to a set level. After that, increasing the amount of light does not increase the rate of photosynthesis any further. Some other factor such as shortage of carbon dioxide or low temperature is now limiting the rate.

Light intensity (arbitrary units)	Rate of movement of meniscus (mm/5 minutes)
0	0
2	3
4	6
6	9
8	12
10	12

Carbon dioxide as a limiting factor

The atmosphere has a very low concentration of carbon dioxide. It is about 0.03%. This limits the rate of photosynthesis.

Increasing carbon dioxide concentration increases the rate of photosynthesis up to point X on the graph. After this, increasing the concentration of carbon dioxide does not increase the rate of photosynthesis any further. Some other factor, such as shortage of light or low temperature, is now limiting the rate of photosynthesis.

Question

e How could the rate of photosynthesis at point X on each graph be increased?

Temperature as a limiting factor

Temperature speeds up the rate of photosynthesis in two ways:

- it speeds up the movement of the molecules which react inside the chloroplasts
- it speeds up the rate of diffusion of carbon dioxide from the atmosphere into the leaves.

Increasing temperature increases the rate of photosynthesis up to point X on the graph. After this, increasing temperature does not increase the rate of photosynthesis any further. Some other factor such as shortage of light or carbon dioxide is now limiting the rate of photosynthesis.

Question

f Which factor is most likely to limit the rate of photosynthesis:
(i) at noon on a hot summer day
(ii) at noon in the middle of winter
(iii) at dusk on a summer day?

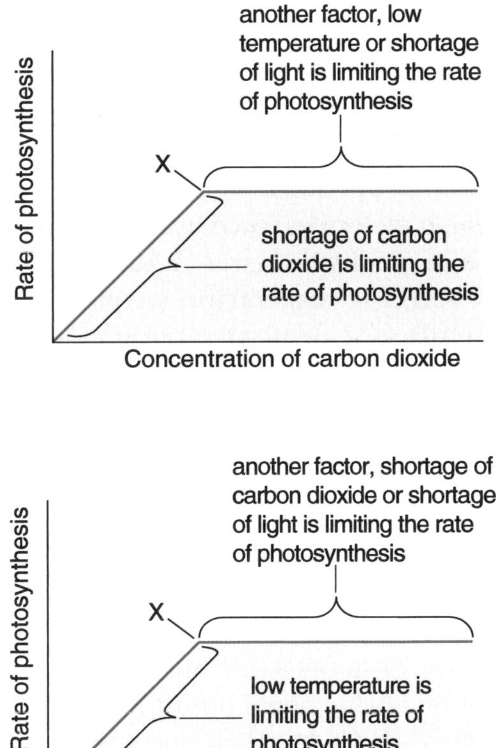

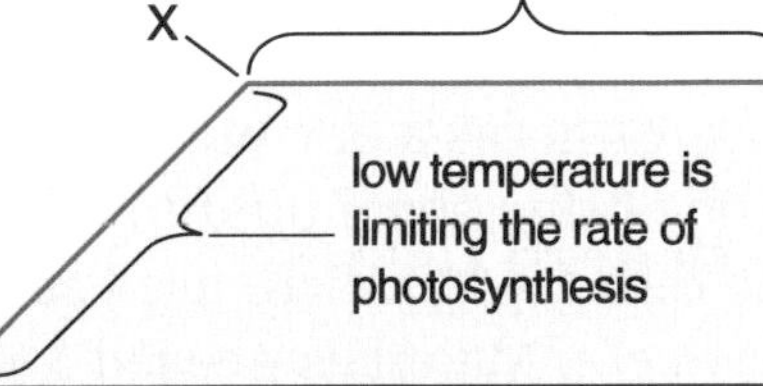

Limiting factors on photosynthesis.

Key points

- Certain factors will stop photosynthesis from occurring even when other needed substances are in plentiful supply.
- These factors include light intensity, carbon dioxide, temperature and light availability.

After photosynthesis

▲ The carbohydrates in chips come from photosynthesis.

Storing carbohydrates

Chips seem to form part of the staple diet of most teenagers. Chips are made from potatoes. In fact, potatoes are called tubers and it is here that lots of starch is stored. Potato plants do not store starch for the benefit of humans; they store starch to survive over winter, and to produce more potato plants the following spring.

Question

a *Starch is insoluble whereas glucose is soluble. Suggest two advantages to the potato plant of storing glucose as starch rather than glucose.*

What do plants use carbohydrates for?

Plants produce carbohydrates for their own use – not for use by animals. Most of the carbohydrates produced by photosynthesis are used by the plant in respiration to provide the energy needed to produce large molecules, such as starch, from smaller ones, such as **glucose**. The rest is used to produce new cells. **Aerobic respiration** (respiration using oxygen) in plants is identical to respiration in animals. The big difference is that plants make their own sugar for respiration whereas animals get theirs by eating food.

Aerobic respiration is an oxidising reaction of glucose to release energy:

glucose + oxygen → carbon dioxide + water + energy

Plants respire 24 hours a day but we can only detect carbon dioxide being given off during the night. Some of the carbohydrates are used as materials to produce new cells. Materials needed for making new cells include:

- cellulose – needed to make the cell walls of the new cells
- proteins – to make proteins the plant needs **nitrate** ions from the soil in addition to carbohydrates
- chlorophyll – to make chlorophyll the plant needs magnesium ions from the soil in addition to carbohydrates.

sugars from leaves are transported into the new potatoes
Remains of potato formed last year. Starch from this potato was used to provide materials and energy for the growth of this year's potato plant.
New potato forming. Here sugars from the leaves are converted into starch for storage over winter.

▲ Potato plant with tubers.

Mineral nutrition

Visit a garden centre and you will see a bewildering variety of products which all claim to make plants grow better. The one thing all these products have in common is that they contain mineral ions. But do plants really need them?

Trials involving over 25 000 plants and 22 different fertilisers carried out over the last two years showed that for most plants there were no significant benefits in adding fertilisers. The exception was cabbages, which are greedy crops that like nitrogen-rich soil.

'Feeding is only worthwhile for a few greedy crops like cabbages and for plants in containers', says Alistair Ayres, Editor of *Gardening Which?*

'But in the majority of cases, it is not worth bothering to feed gardens at all.'

Questions

b *How would* Gardening Which? *carry out the trials described in the extracts?*

c *Do you think that the trials described in the extract gave reliable results? Explain the reasons for your answer.*

d *The evidence from* Gardening Which? *is overwhelming, yet most gardeners continue to buy fertilisers. Suggest an explanation for this.*

Deficiency symptoms

The editor of ***Gardening Which?*** stated that plants grown in containers do need to be supplied with mineral ions if they are to remain healthy.

Question

e *Suggest an explanation for this.*

Tomatoes are often grown in containers. The photograph shows a tomato plant grown in a container in soil which did not contain enough nitrate. Plants need nitrates to make **amino acids** which are then built into protein. Without protein, plants cannot grow. The symptom of nitrate deficiency is stunted growth. Plants need relatively large amounts of nitrate to grow, but they also need small amounts of magnesium. Plants need magnesium to produce chlorophyll. The symptom of magnesium deficiency is yellow leaves.

▲ A plant with nitrate deficiency.

▲ A plant with magnesium deficiency.

Question

f *Why will plants with yellow leaves not photosynthesise?*

Key points

- The carbohydrates made by plants during photosynthesis may be stored as starch or respired.
- Plants need nitrates to produce protein and magnesium to produce chlorophyll.
- Plants deficient in nitrates will have stunted growth, and those deficient in magnesium will have yellow leaves.

Life in the greenhouse

All-year salad

Plants need special conditions to grow well. The conditions outside in fields cannot be easily controlled but in greenhouses it is easier to control conditions for the plants.

▲ Controlled growing conditions.

Question

a *Greenhouses are expensive to build. Can you think of reasons why farmers grow some crops in greenhouses rather than outside in fields?*

In the nineteenth century, most people in this country could only buy tomatoes in summer. Now you can eat tomatoes at any time of the year.

Question

b *Suggest why tomatoes were only available in summer in the nineteenth century.*

▲ Healthy eating at any time of the year.

Crops grow only when plants are photosynthesising. Growers can now get plants to photosynthesise all the year round. They can do this by enhancing light, carbon dioxide concentration and temperature.

Enhancing light

Light is rarely a limiting factor in summer, but some growers supplement natural light with artificial light in winter. The problem with enhancing light is making sure that all the plants receive the extra light. You can do this by moving the light further away from the plants, but this reduces the intensity of the light they receive. Doubling the distance from the light to the plants reduces the light intensity by 75%.

Question

c *Supplementing light is usually only economic when the plants are very small. Suggest a reason for this.*

▲ In greenhouses it can be summer all the year round.

Enhancing carbon dioxide

The very low concentration of carbon dioxide in the atmosphere (about 0.03%) limits the rate of photosynthesis. We cannot improve the carbon dioxide supply to crops growing in fields but we can give plants growing in greenhouses more carbon dioxide.

Growers can increase the amount of carbon dioxide in a greenhouse by burning fossil fuels such as propane. Other methods include using dry ice or simply using carbon dioxide from a gas cylinder. But growers find that they can only increase the rate of photosynthesis up to a set level.

▲ Carbon dioxide enhancement.

Question

d *Getting carbon dioxide from burning propane has one advantage over the other two methods. Suggest what this advantage is.*

Controlling temperature

Keeping a greenhouse at a temperature between 25 and 30 °C in winter requires heating. Most greenhouses are heated in winter by a boiler and radiators – very similar to the central heating system in a house.

Question

e *Look at the graph. What is the best temperature for growing this greenhouse crop?*

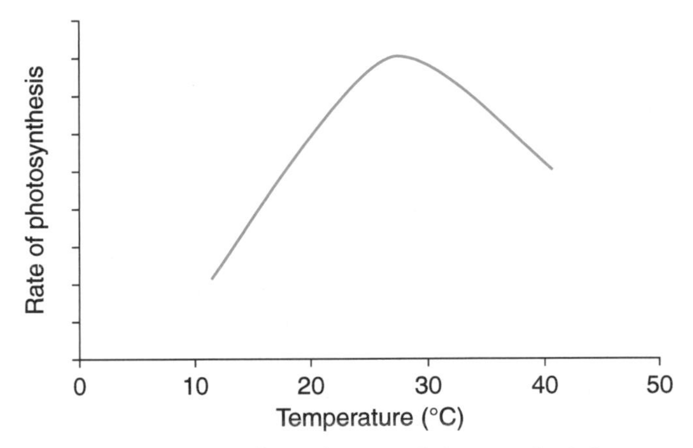

▲ How temperature affects the rate of photosynthesis in a greenhouse crop.

In summer the temperature in the greenhouse will often exceed 26 °C. This is because heat waves from the Sun are trapped in the greenhouse by being continually reflected. This causes the rate of photosynthesis to decrease. There are two solutions to this problem – using blinds and opening vents in the greenhouse roof.

Question

f *How will using blinds affect the rate of photosynthesis? Give the reason for your answer.*

Feeding plants

Plants make their own food, but they need nitrates to produce protein. Growers can control the amount of nitrate and other ions by growing the plants in a solution of mineral salts.

Question

g *Suggest advantages of this system over growing plants in soil, then watering them with a solution of ions.*

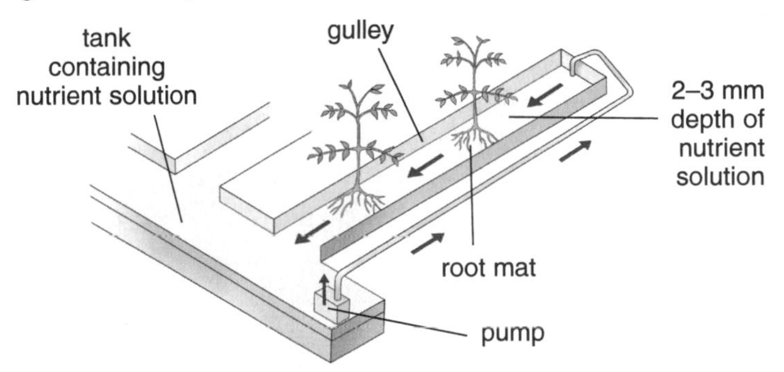

▲ Feeding time for tomato plants.

Key points

- Greenhouses enable the growth of year-round crops.
- Growth only occurs when plants photosynthesise.
- Enhancing limiting factors for photosynthesis can increase yield.
- Whilst plants can make their own food by photosynthesis, they require the addition of nitrates to make proteins.

Getting carbon dioxide to greenhouse crops

The boxes contain information about three ways of supplying carbon dioxide to greenhouse crops. The information in each box is for 4 hectares of greenhouse maintaining a carbon dioxide concentration of 1300 parts per million (ppm).

Propane burners

When propane is burned, carbon dioxide is produced and heat is released. If there is sulfur in the fuel, sulfur dioxide will also be released. About 1.4 kg of water is released for each cubic metre of propane burned. The one-off cost of installing the burners is £28 000 and the daily cost of propane is £58.

Carbon dioxide from flue gases

Natural gas is burned in a boiler and the heat released is used to heat water. This water can be circulated immediately through the greenhouse by pipes or stored for use at night in large tanks. Carbon dioxide is extracted from the flue gases by a condenser. The flue gases contain water vapour. They might also contain carbon monoxide which is a poisonous gas. The one-off cost of the equipment is £44 000 and the natural gas costs £55 per day.

Liquid carbon dioxide

Liquid carbon dioxide is pure carbon dioxide. It is delivered in bulk by tankers and stored in special cylinders. The liquid carbon dioxide is vaporised then delivered to the plants by PVC tubing with a hole punched near each plant. The equipment for storing and vaporising the carbon dioxide is rented for £5 500 per year and the daily cost of the carbon dioxide is £66.

Questions

a *In a table, summarise the advantages and disadvantages of each of the methods of supplying carbon dioxide.*

b *Imagine you are a grower. Which method would you use? Explain the reasons for your answer.*

Net photosynthesis

During the day, plants both photosynthesise and respire. The relationship between gross photosynthesis, net photosynthesis and respiration is given in the equation:

gross photosynthesis = net photosynthesis + respiration

The table shows the rates of gross photosynthesis and net photosynthesis for a cereal crop at different temperatures.

Temperature (°C)	Rate of gross photosynthesis (arbitrary units)	Rate of net photosynthesis (arbitrary units)
12	12	10
19	26	24
26	40	37
34	34	27
41	26	11

Questions

c *Plot a graph of these data. Choose suitable scales for the axes. Label each of the curves.*
d *Describe the effect of temperature on the rate of gross photosynthesis.*
e *Which factor is limiting the rate of gross photosynthesis between 19 °C and 26 °C? Explain the reasons for your answer.*
f *The rate of gross photosynthesis is the same at 19 °C as it is at 41 °C. The cereal crop grows more slowly at 41 °C than at 19 °C. Suggest an explanation for this.*

Growing cucumbers

Cucumbers are now grown mainly in greenhouses. The table shows the yield of cucumbers grown in a well lit greenhouse under different conditions.

Temperature (°C)	Yield of cucumbers (kg/10 plants)	
	0.03% carbon dioxide	0.13% carbon dioxide
12	12	10
19	26	24
26	40	37
34	34	27
41	26	11

Questions

g *In which conditions did the cucumbers give the greatest yield?*
h *Would the grower make most profit by using these conditions? Explain the reasons for your answer.*

Key points

- Enhancing carbon dioxide in greenhouses is expensive; growers need to balance the cost of enhancement against the increase in yield.
- Increasing temperature increases the rate of respiration as well as the rate of photosynthesis; raising the temperature too high may result in a decrease in yield.
- Increasing one factor may have no effect if another factor is limiting.

Energy in ecosystems

Only a tiny percentage (0.023%) of the Sun's energy is used in photosynthesis. It is on this that all life on Earth depends. The material produced by plants is known as biomass. It is called biomass because it has come from processes in living organisms.

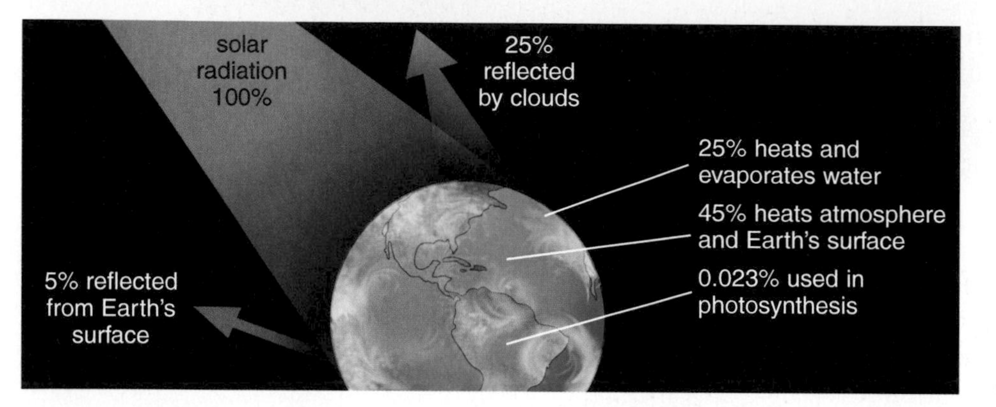

What happens to solar energy reaching Earth.

How much energy is available to humans?

Only a small proportion of this biomass is available to humans. The bar chart shows the biomass produced each year by plants in the different ecosystems of the Earth.

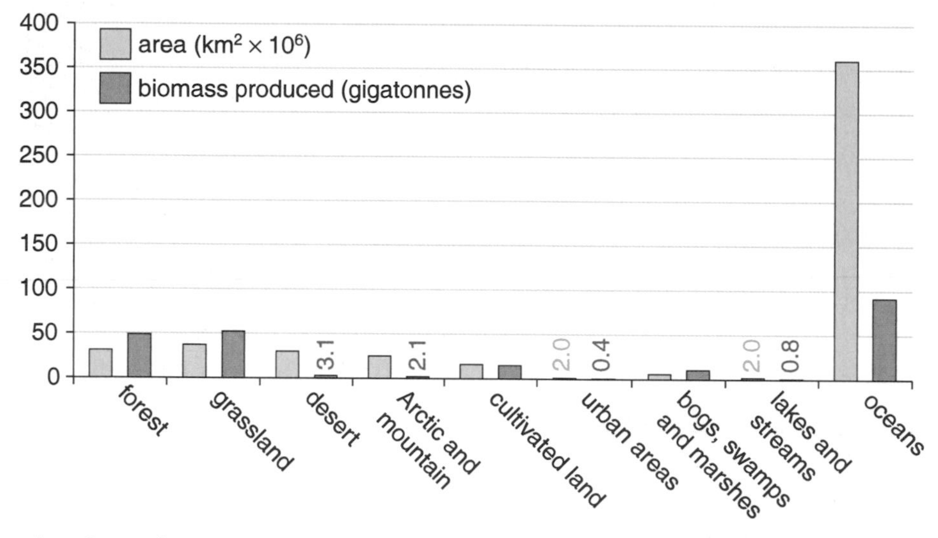

Biomass produced each year by the different ecosystems of the Earth.

The chart shows the areas of the major ecosystems on Earth together with the biomass produced by that ecosystem (1 gigatonne = 10^9 tonnes). We only cultivate a tiny proportion of the Earth's surface. Over 60% of the Earth's surface cannot be cultivated because it is covered by the oceans. We can harvest fish from the ocean, but modern fishing vessels are causing drastic falls in fish populations. Fishing nations have a lot to learn about sustainability. Much of the rest of the Earth's surface cannot be used for cultivation because it is too dry, too cold or too wet. So we have to find more efficient ways of transferring energy from the Sun into food.

How much biomass do humans use?

The pie chart shows the amount of biomass used directly by humans and their domestic animals (cattle, sheep and pigs). The units are gigatonnes (a gigatonne is 1 billion tonnes).

The figures do not include the parts of crop plants that are not eaten, converting natural grassland to grazing land, and clearing of forest land for agriculture and building. When these are added, it is estimated that humans use about 19% of the biomass produced on Earth. And this percentage is rising. This increase is not sustainable, so we must find more efficient ways of transferring biomass.

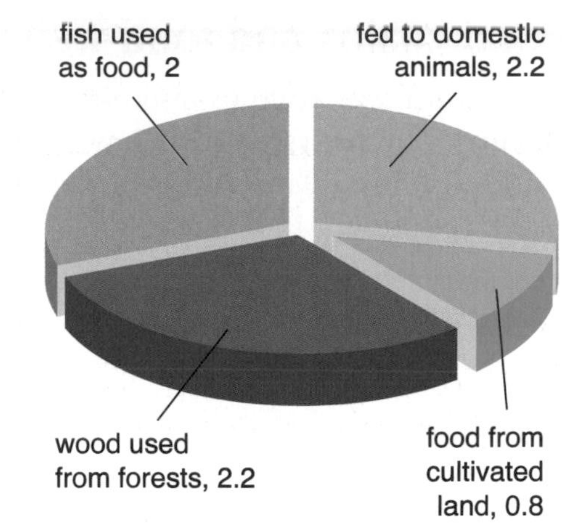

▲ Amount of biomass used (in gigatonnes).

Animals as food

The greater part of the biomass in the food eaten by animals is not converted into food for humans – in effect it is wasted. In some countries, animals are kept in conditions like those in the photograph to increase the efficiency of producing food. Compared with grazing in fields, rearing indoors takes less fodder to produce each kilogram of meat. But are we justified in treating animals like this?

Recycling biomass

▲ Smelly – but necessary!

▲ Are we justified in rearing animals in these conditions?

Farmers spread waste products from their animals onto the land. This makes the crops grow better. The farmer is harnessing natural processes to return mineral ions to the soil. This method is sustainable because it involves recycling. It also reduces our energy demands since the chemical plants producing artificial fertilisers use enormous amounts of energy.

Think about what you will find out in this section

What happens to biomass at each stage in a food chain?	What happens to the waste materials produced by plants and animals?
How can we evaluate the positive and negative effects of rearing animals for food?	How can we recognise that practical solutions to feeding the human population may require compromise?

Ecological pyramids

Food chains and food webs

Many different animals and plants live in the Antarctic oceans. Scientists have worked out many of the feeding relationships in the ocean. Some of these are shown in the diagram.

Question

a *Suggest two ways in which scientists have worked out these feeding relationships.*

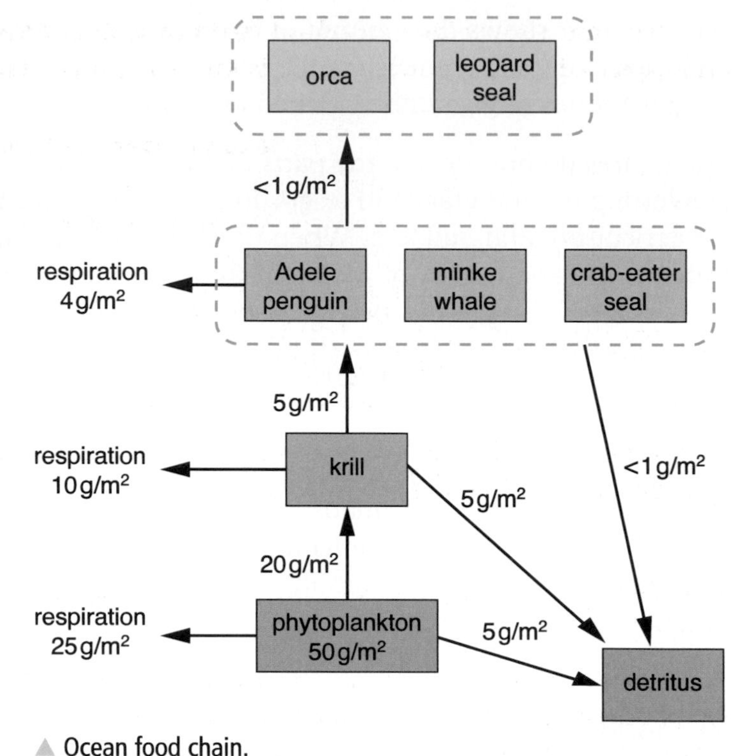

▲ Ocean food chain.

One food chain found in the ocean is:

phytoplankton → krill → Adele penguin → orca

The **phytoplankton** photosynthesise, converting carbon dioxide into carbohydrates using energy in sunlight. Because they produce food, phytoplankton are known as **producers**.

Phytoplankton are eaten by small, shrimp-like animals called krill. Because they eat producers, krill are called **primary consumers**.

Krill are eaten by Adele penguins. Because they eat primary consumers, Adele penguins are called **secondary consumers**.

Adele penguins are eaten by orca. Because they eat secondary consumers, orca are known as **tertiary consumers**.

The arrows in the food chain indicate biomass being transferred from one organism to the next, e.g. from krill to Adele penguins.

Moving biomass

We can measure the biomass of the organisms in a food chain, but this only makes sense if we measure the biomass per unit area of the habitat. So we usually express biomass in grams per square metre (g/m^2).

The diagram shows the approximate transfer of biomass between some of the organisms in the ocean. **Detritus** is the solid waste products produced by organisms and their dead bodies. Organisms which feed on this detritus are known as detritivores.

Questions

b *(i) Suggest how scientists measured the biomass of the krill in 1 m^2 of ocean.*
(ii) Suggest reasons why scientists can only give approximate figures for the transfer of biomass between these organisms.

c *(i) Much of the biomass taken in by animals is lost during respiration. Why do organisms lose biomass when they respire?*
(ii) Calculate the proportion of biomass lost via respiration for the primary consumers and the secondary consumers.
(iii) Apart from size and shape, what is the main difference between the primary consumers and the secondary consumers in the ocean food web?

Pyramids of biomass

The biomass at each level in a food chain or web can be drawn to scale as a pyramid. This type of pyramid is known as a **pyramid of biomass**.

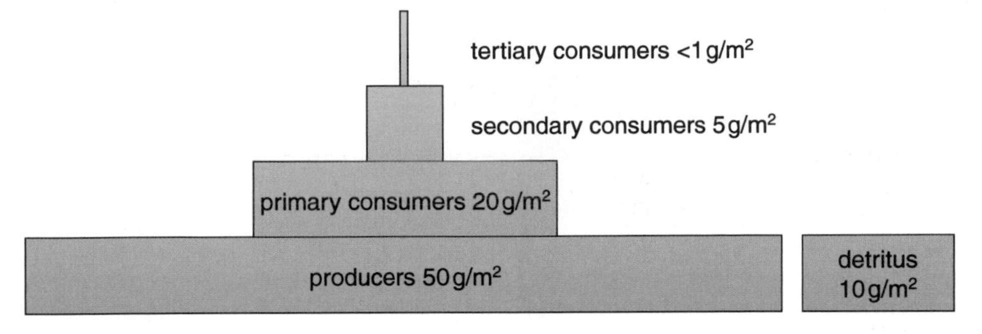

The scale of the drawing is 1 cm = 5 g/m^2. So the width of the base (producers) is 50 g/m^2 ÷ 5 = 10 cm.

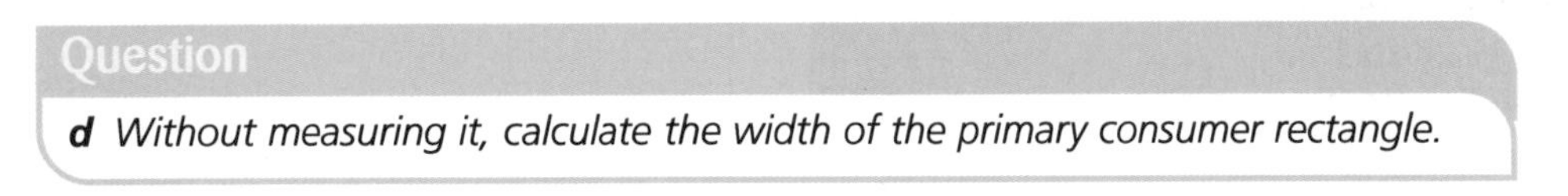

Question

d *Without measuring it, calculate the width of the primary consumer rectangle.*

We always draw the rectangle for the producers at the base, followed by primary consumers, secondary consumers and tertiary consumers. Sometimes this leads to 'pyramids' that are narrower at the bottom than at the top. In GCSE examinations, the pyramids of biomass you will be given will always have a pyramid shape.

The table gives data for the biomass of the organisms living in a marsh.

Organisms	Biomass (g/m^2)
producers	800
primary consumers	40
secondary consumers	10
tertiary consumers	2

Question

e *On graph paper, draw a pyramid of biomass for the data in the table. Don't forget to decide on a suitable scale before you begin your drawing. Label each level in your pyramid.*

Key points

- All plants and animals are part of a food chain.
- Producers are always green plants and always start the food chain.
- Without the Sun nothing would exist as it provides the energy needed by green plants.
- Pyramids of biomass show the actual mass of organisms in grams at each level of the pyramid.

Energy flow

The new oil wells

One hectare of the sugar cane crop shown in the photograph can produce enough fuel to drive a car round the world twice! And this fuel produces practically no pollution.

▲ Sugar cane is full of energy.

Farmers plant small pieces of sugar cane stem. About 9 months later they can harvest up to 190 tonnes of sugar cane per hectare. From these crops, 75 million tonnes of sugar are produced each year. In countries like Brazil some of the sugar cane is fermented to produce ethanol. Many cars there have engines designed to run on pure ethanol which is called gasohol.

The sugar cane crop produces a large mass of living material, and this living material contains a lot of energy. Plants use some of the energy from sunlight to convert carbon dioxide and water into sugars. Photosynthesis has produced the countless millions of tonnes of living matter that exist on Earth. However, plants can only use a small proportion of the energy in sunlight.

At each stage in a food chain, energy is lost to the atmosphere, so that only a small proportion of the Sun's energy absorbed by producers is transferred to the tertiary consumers.

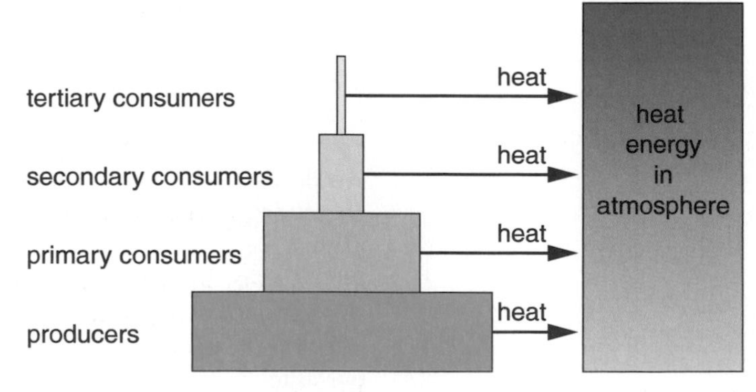

▲ Heat is lost at each stage in a food chain.

Question

a *Scientists measured the amount of sunlight energy that was used by plants in an area of Britain. In 1 year, 7 000 000 kJ of sunlight energy reached each square metre of the area. Of this, only 90 000 kJ was used in photosynthesis. Calculate the proportion of sunlight energy that was used by the plants.*

Measuring energy flow?

Compared with measuring biomass, it is very difficult to measure the amount of energy passing through a food web. Measuring the amount of energy passing through the organisms involves several stages. These include:

- measuring the biomass of each type of organism in $1\,m^2$ of the habitat
- using calorimeters to measure the heat given off when a standard mass of each organism is burned
- measuring the rate of photosynthesis of the producers
- measuring the rate of respiration of all the organisms.

From all these measurements the amount of energy passing through each type of organism per year is calculated. The units for energy flow are kilojoules per square metre per year.

Question

b The first studies on energy flow were carried out on stream habitats. One reason for this choice was that the temperature of a stream remains constant throughout the year.
(i) Which of the measurements described would be affected by changes in temperature?
(ii) How would these measurements be affected by changes in temperature?

Energy flow in a stream

The diagram shows the results for one study on a stream.

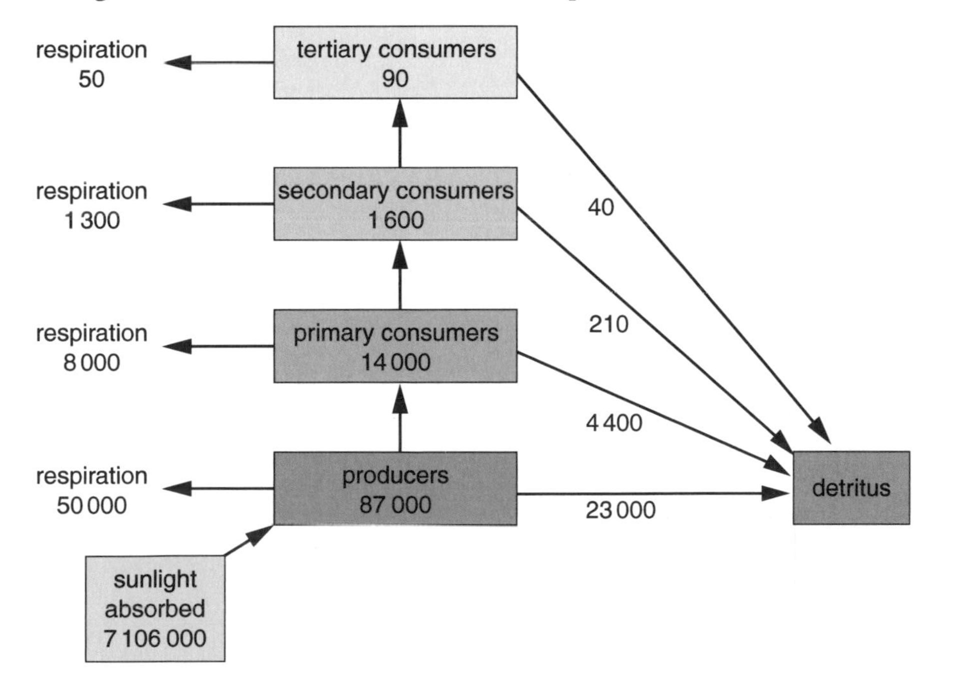

The data for producers shows that of the 7 106 000 kJ per m^2 per year of light energy absorbed, 87 000 kJ per m^2 per year was used in photosynthesis.

Of this 87 000 kJ per m^2 per year:

- 50 000 was used by the plants in respiration
- 14 000 was consumed by primary consumers
- 23 000 became detritus when the producers died.

Question

c (i) Calculate the proportion of the energy absorbed by the producers which was used in photosynthesis.
(ii) Calculate the proportion of energy in the producers which was passed on to the primary consumers.
(iii) Calculate the proportion of the energy in the primary consumers which was passed on to the secondary consumers.
(iv) In which type of organism was the proportion of energy lost via respiration the highest?

Key points

- Energy flows through ecosystems.
- A lot of energy is lost at each level due to respiration.
- Some of the energy is lost as heat.

B2 2.3 Energy diagrams

Sankey diagrams

One of the best ways of describing energy flow is to use Sankey diagrams.

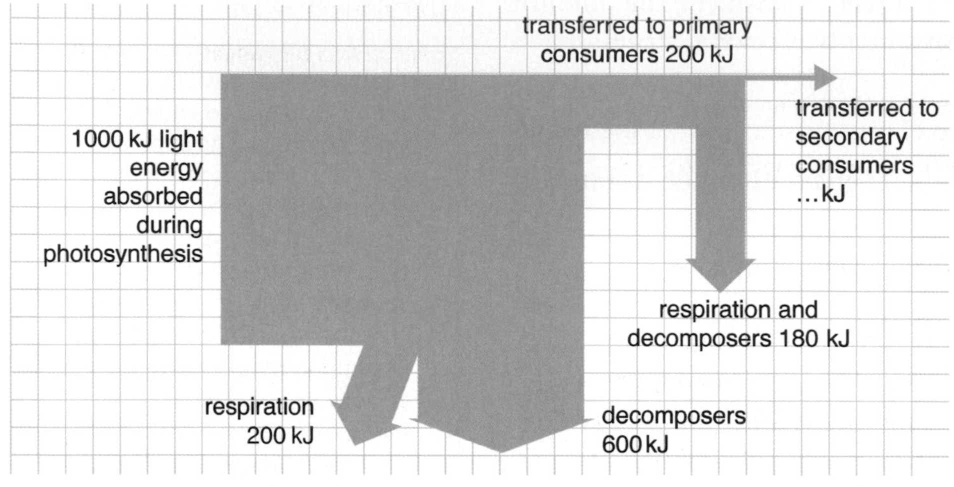

▲ Sankey diagram for a meadow.

▲ How much of the energy in the grass is transferred to the hawk?

The arrows on a Sankey diagram are drawn to scale. The scale for this diagram for a meadow is 1 square = 100 kJ. So the light absorbed by photosynthesis, 1000 kJ, has a width of (1000 kJ ÷ 100 kJ) = 10 squares. Similarly, the arrow for **decomposers** has a width of (600 kJ ÷ 100 kJ) = 6 squares.

Question

a *How much energy was transferred to the secondary consumers?*

Energy flow through grasshoppers

Most primary consumers lose at least 50% of the energy taken in as faeces. This is because plant cell walls are made of cellulose. Very few animals can produce the enzyme cellulase which breaks down cellulose into glucose. So the cellulose passes out, undigested, with the faeces.

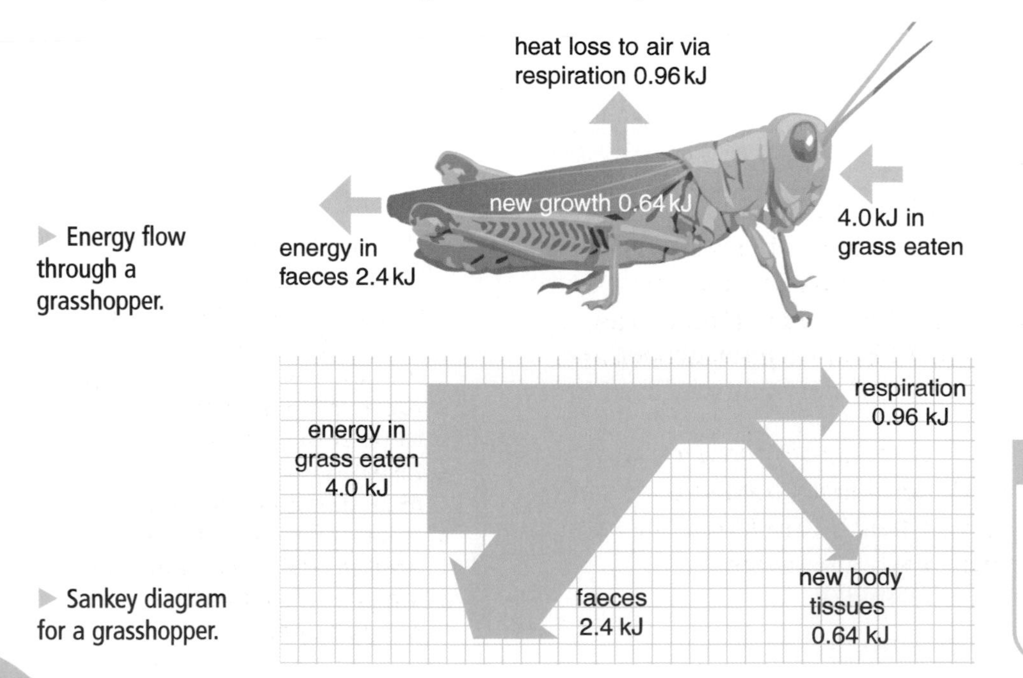

▶ Energy flow through a grasshopper.

▶ Sankey diagram for a grasshopper.

Question

b *Look at the diagram. What proportion of the energy in the grass is:*
(i) lost as heat via respiration
(ii) lost in the faeces?

Energy flow through beetles

The table shows the energy flow through a ground beetle. This beetle eats other insects.

Organisms	Energy (kJ)
Consumed as food	0.17
Lost in faeces	0.03
Heat loss via respiration	0.09
New growth	0.05

Question

c *(i) On graph paper, draw a Sankey diagram for these data. Use a scale of 1 cm = 0.02 kJ.*
(ii) How does the proportion of energy lost via faeces compare with that of the grasshopper?
(iii) Suggest an explanation for this.
(iv) How do these values compare with those for the grasshopper?

Hunting needs energy.

The proportion of energy lost via respiration is much higher for the beetle than for the grasshopper. This is because the grasshopper is surrounded by its food – grass. But the beetle has to search far and wide for its prey – other insects. So, much of the energy from the beetle's food is used by the muscles via respiration.

Energy flow through mammals

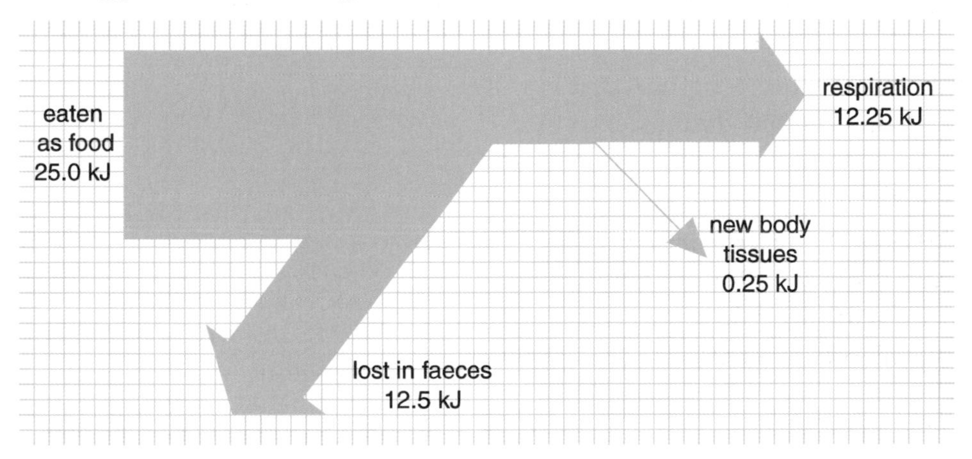

Energy flow through a rabbit.

Rabbits are primary consumers. Like the grasshopper, they eat grass. But the proportion of energy lost via respiration is much higher in the rabbit. This is because the rabbit is a mammal. Mammals maintain a high constant body temperature. You have a body temperature of about 37 °C. This is about twice as high as the average temperature in your classroom. Keeping the body temperature high and constant needs a lot of energy from food.

Birds keep an even higher constant body temperature, about 41 °C. This means that an even higher proportion of their food is used in maintaining a constant body temperature. Most animals have warm blood. But only mammals and birds keep high, constant body temperature.

Key points

- We can show how much energy is transferred through a food chain using energy diagrams.
- A large proportion of energy taken in is lost in faeces.
- Most of the rest is lost via respiration.
- Mammals and birds use a large proportion of their energy intake in maintaining a high constant body temperature.

Food production

Where our protein comes from

What do the photographs have in common? They all show ways of producing protein food for humans. But what people eat depends on where they live and what they can afford. To most people in the world, meat is a luxury they cannot afford. In sections 2.4 and 2.5 let's look at why meat is so expensive compared with plant foods.

Crop plants

Cereal crops usually contain both carbohydrates and protein – that is why your breakfast cereal is so good for you! The pie charts show the compositions of three crops.

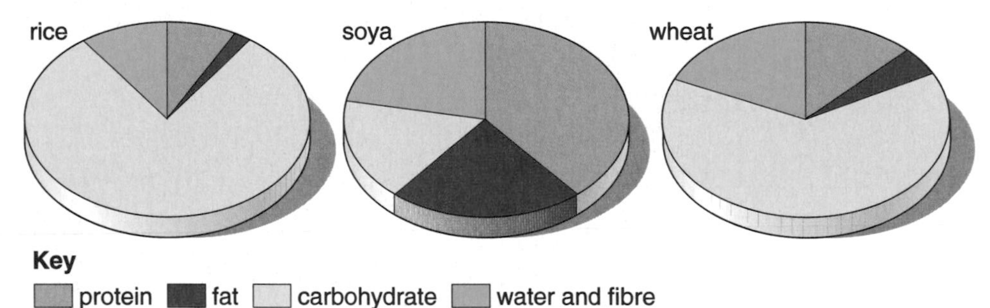

Rice, soya and wheat contain different proportions of protein, fat, carbohydrate, water and fibre.

Question

a *(i) Which crop contains the highest proportion of carbohydrate?*
(ii) Estimate the mass of protein in 10g of soya.

Fish farming

The diagram shows the relative amount of energy at each stage in the food web that produces tuna fish.

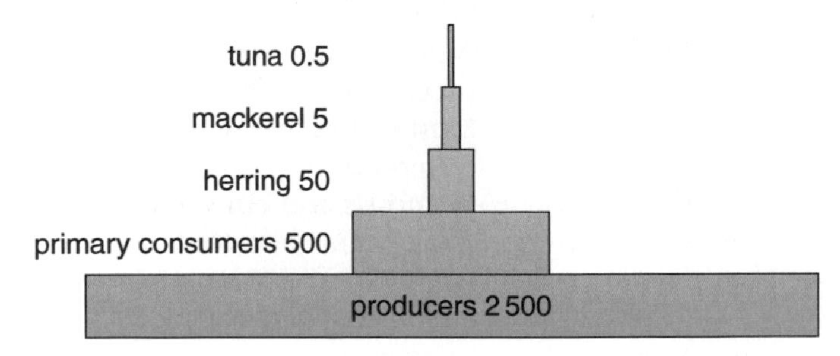

A lot of energy is wasted before we get to eat the tuna.

Question

b *Calculate the proportion of energy in the producers that is transferred to the tuna.*

Catching fish from the sea provides 'free' high-protein food. The main costs involved are for the trawlers and the wages of the fishermen. Unlike farmers who grow crops, fishermen do not have to buy seed, fertilisers and pesticides. They do not have to tend crops for several months. Continual harvesting has led to the collapse of fish stocks in many parts of the ocean. To satisfy the demand for fish, a large fish farming industry has grown up in recent years.

There are over 1000 fish farming businesses in the UK. The main fish species farmed are salmon (139 000 tonnes, mainly in Scotland) and rainbow trout (16 000 tonnes). There is also limited production of other species, such as carp and brown trout.

The yield of fish from a fish pond depends mainly on the diet of the fish. The bar chart shows the different yields obtained from growing fish that are primary consumers or secondary consumers.

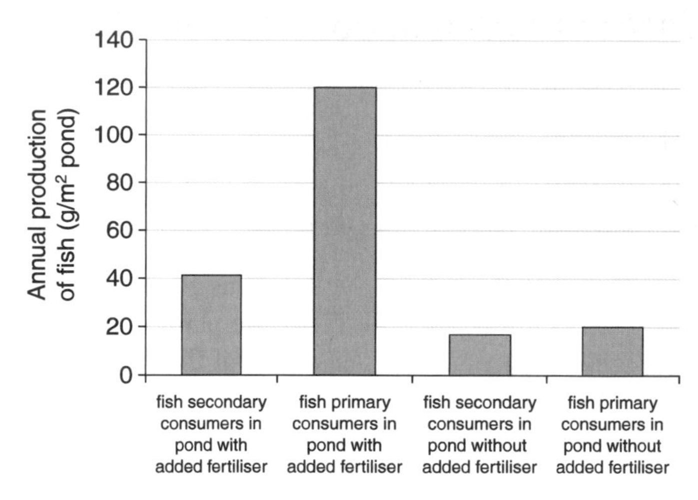

The yield of fish depends upon what the fish eat.

Question

c *(i) Explain why the yield from growing primary consumer fish is greater than that from growing secondary consumer fish.*
(ii) Explain why adding fertiliser increases the yield of primary consumer fish.

Pros and cons of fish farming

In Scotland, salmon are farmed in large net cages in sea lochs. The young salmon are fed on fish meal made mainly from sand eels. It takes 5 tonnes of sand eels to produce 1 tonne of salmon. This is equivalent to the amount of sand eels found in 1 km^2 of sea. Other fish and sea birds also feed on sand eels. The large amount of faeces from the farmed fish sinks to the bottom of the sea.

Salmon farming in Scotland.

Question

d *(i) Give one benefit to consumers of farming salmon.*
(ii) What effects might farming salmon have on populations of other species?
(iii) What effects might farming salmon have on the local environment?

Key points

- Wasted energy in the food chain means that eating meat or fish is not the best or cheapest way for humans to obtain their energy.
- Over-fishing is causing the loss of fish stocks in the oceans.

Removal of waste

Sewage disposal

Until the nineteenth century, human sewage was dumped either on farmland or into rivers. (In many developing countries this still happens.) But as cities grew, rivers became grossly polluted so another solution had to be found. A modern sewage works looks like an oil refinery. It uses the same microorganisms that have been breaking down waste materials since life on Earth began.

High-tech sewage disposal.

Soil microbes

Faeces in the soil are broken down by **microbes**. These include bacteria, **fungi**, and single-celled organisms. Bacteria and fungi digest faeces and other waste products, by producing digestive enzymes. The enzymes pass out of the microbe's body and onto the waste materials, which they break down into soluble compounds. These soluble compounds are then absorbed into the body of the microbe. Soil microbes obtain much of their food from waste materials such as faeces. The breakdown of waste materials by microbes is known as **decay**.

Question

a *All the material in faeces is broken down by microbes. Which enzyme must they produce that is not produced by most animals?*

Sewage works

The residents of a British city will produce thousands of tonnes of faeces in a year. The faeces are flushed down toilets and sent to a sewage works. Here, processes use similar microbes to those found in soil. The sewage is run into large tanks through which air is bubbled. The microbes in the tanks digest 80% of the waste material within a few hours. The oxygen in the air speeds up the activity of the microbes.

In an activated sludge tank, microbes quickly break down waste material.

Question

b *Which process that occurs in the microbes uses oxygen?*

Faeces from cows on a farm are not dumped. They are spread on the fields as a 'natural' fertiliser. The microbes in the soil break down the faeces into simple materials including mineral ions such as nitrates, which plants need to grow.

Question

c *What do plants make out of sugars and nitrates?*

Farmers need to use fertilisers because they harvest crops. These crops have taken nitrates out of the soil. Farmers must replace the nitrates if they want a good crop the following year. They can do this in two ways. They can buy chemical fertilisers or they can use animal faeces. Using animal faeces is known as organic farming because the fertiliser has come from living creatures, not chemical factories.

Question

d *Suggest some advantages and disadvantages of using natural and artificial fertilisers.*

Compost heaps

Microbes break down dead organisms as well as waste materials. In autumn, in a garden, many plants lose their leaves. These leaves contain nutrients that can be recycled by soil microbes. Many gardeners collect all the dead leaves and use them to make **compost**. Compost contains nutrients such as nitrates, so acts as a natural fertiliser when it is spread on the garden.

Question

e *Explain why each of these factors increases the rate of decay:*
(i) warmth (ii) oxygen (iii) moisture.

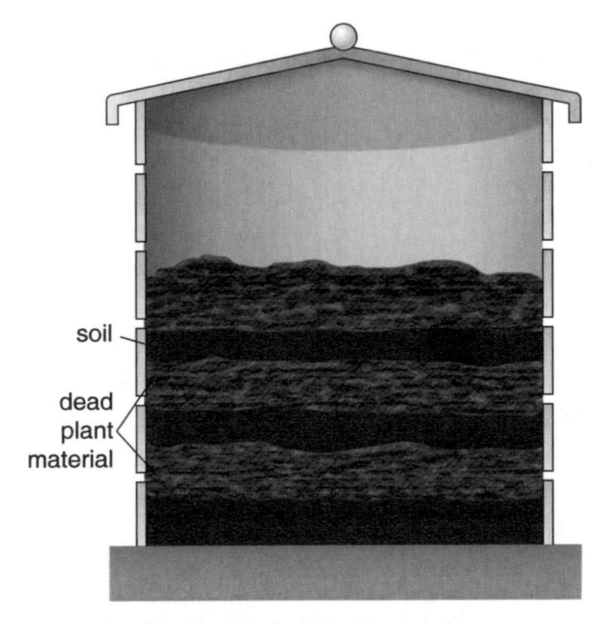

▲ The formation of compost is another example of decay.

The diagram shows one way of making compost. Layers of leaves are put in a large container. A layer of soil is placed between each layer of leaves. The sides of the container are perforated to allow air to circulate through the container. A lid keeps the rain off. The microbes in the soil digest the leaves to form compost. Decay occurs most quickly if conditions are moist, warm and there is a good supply of oxygen.

Disappearing leaves

Squares of leaves, 20 mm across, were placed in mesh bags in the soil. The diagram shows the appearance of the leaves after 3 months.

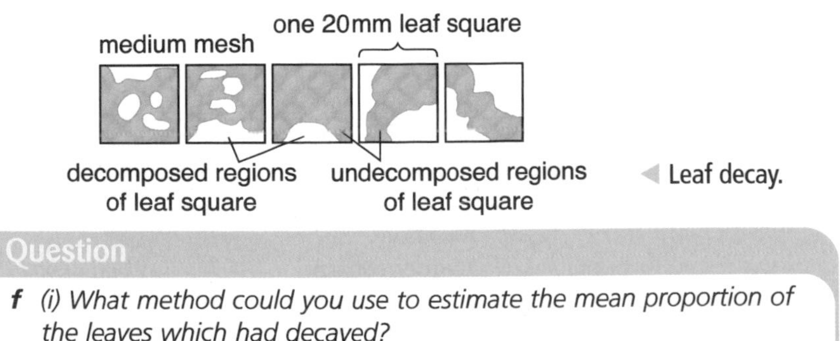

◄ Leaf decay.

Question

f *(i) What method could you use to estimate the mean proportion of the leaves which had decayed?*
(ii) How reliable would be the results of this method of measurement?

Key points

- Waste is broken down by microbes to make it safe.
- As waste decomposes it produces valuable nutrients which can be put back into the soil as fertiliser.
- Compost is a natural fertiliser which recycles waste material.

The ultimate in recycling

Silk is mainly manufactured in China. Silk fibres are made by the silkworm larvae as they turn into pupae. The larvae feed on leaves from mulberry bushes. A lot of waste is produced during this process. This waste can be used to rear fish called carp, which are a valuable food source for the Chinese. So the biomass produced by the mulberry bushes is recycled, as are mineral ions absorbed by the mulberry bushes. The diagram shows mulberry bushes growing around a carp pond.

Scientists have calculated that 36 700 kg of mulberry leaves can be produced per hectare.

Question

a Suggest how scientists could make this measurement.

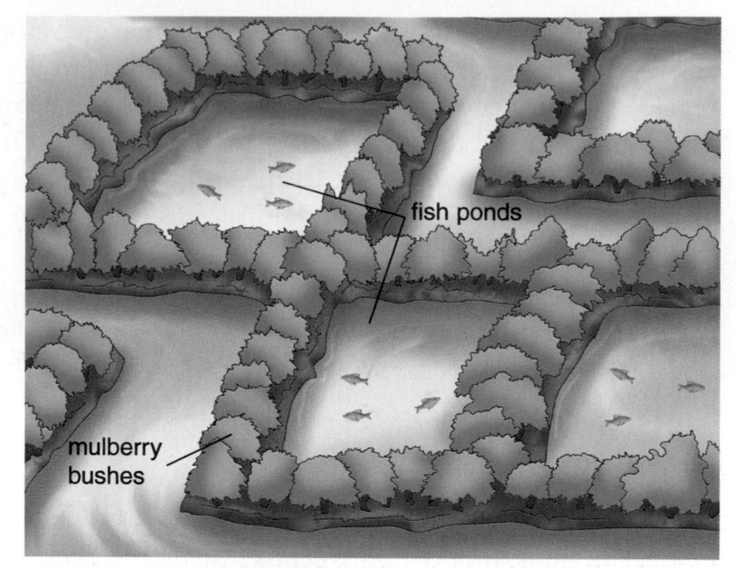

Biological recycling in action.

The silkworm larvae eating 36 700 kg of mulberry leaves produce 2 700 kg of silkworm pupae. From these, 206 kg of silk are produced. However, the waste from the process is placed in the carp ponds. This includes 18 400 kg of faeces from the larvae. From these faeces, 2 290 kg of carp are produced.

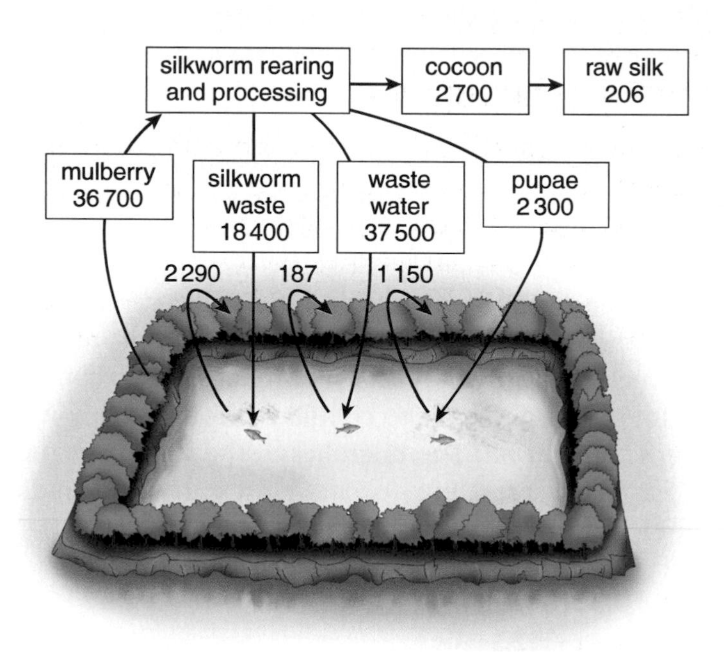

Biomass flow through a mulberry fish pond. The units are kilograms per hectare.

Question

b From the diagram, what mass of carp are produced from the wastes from silkworm farming and processing?

Plants

The mulberry bushes exchange carbon dioxide with the atmosphere. There are two processes involved in this exchange – photosynthesis and respiration. Photosynthesis uses carbon dioxide from the atmosphere. Respiration adds carbon dioxide to the atmosphere.

Question

c Do the mulberry bushes increase the amount of carbon dioxide in the atmosphere over a year, or decrease it? Explain the reason for your answer.

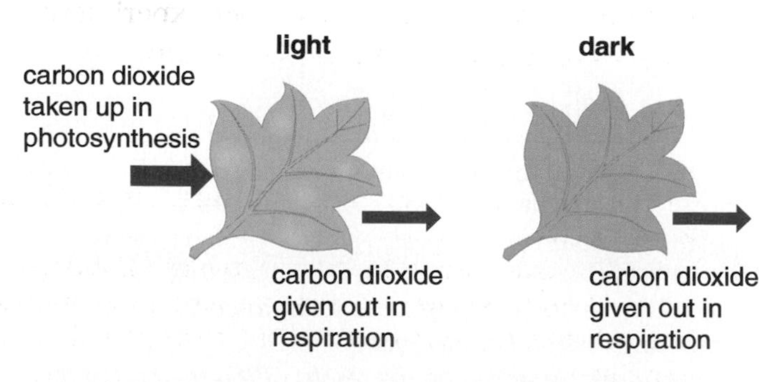

A summary of photosynthesis and respiration in green plants.

Animals

The silkworm larvae are primary consumers. They digest mulberry leaves into sugars, amino acids, **fatty acids** and **glycerol**. All these compounds contain carbon. The carbon compounds are then made into the proteins and fats that form the cells of the larvae. When carp eat parts from the silkworm larvae, the carbon compounds from the larvae are converted into carbon compounds in the carp.

All animals respire. Some of their food is converted into sugars which can be respired. The carbon in the sugars returns to the atmosphere as carbon dioxide.

Question

d *Explain why animals decrease in mass when they respire.*

Decay organisms

When organisms die, small animals and microbes (decay organisms) feed on their bodies. When these organisms respire, some of the carbon compounds from their food (dead organisms) are converted into carbon dioxide.

▲ When living things die, carbon compounds are taken up by decay organisms.

Putting it all together

Now we can group plants, animals and microbes into one cycle. The movement of carbon compounds from the atmosphere into organisms and back again is called the carbon cycle.

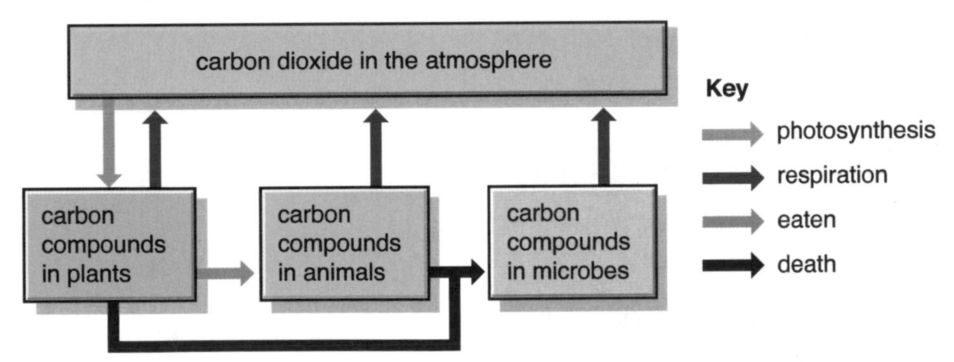

▲ The carbon cycle.

Respiration in soil microbes

The diagram shows the results from an experiment on soil microbes. The flasks were left for 24 hours after setting up.

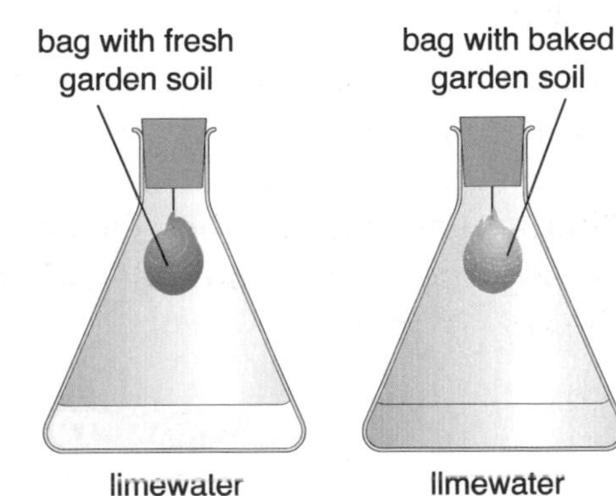

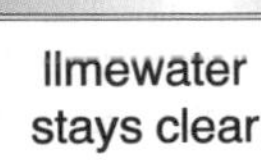

Question

e *(i) List the control variables which should have been applied to the two flasks.*
(ii) Explain the result for each flask.

Key points

- The carbon cycle shows how carbon is recycled from plants and animals by microbes.
- Carbon dioxide is put into the air from respiration.
- Carbon dioxide is removed from the air during photosynthesis.
- When plants and animals die their bodies provide carbon sources for microbes.

Food at any price?

The calf controversy

Veal crates have been banned in the UK since 1990. But in mainland Europe hundreds of thousands of calves are confined in crates measuring just 60 cm wide in order to produce white veal. They are chained by the neck to restrict all movement, making it impossible for them to turn around, or to lie down comfortably. This makes the calves' meat 'tender' since the animals' muscles cannot develop.

Published scientific research indicates that calves confined in crates experience 'chronic stress'. Researchers have also reported that calves confined in crates exhibit abnormal behaviour. This includes head tossing, head shaking, kicking and scratching.

Veal producers severely limit what their animals can eat. The calves are fed an all-liquid milk substitute which is purposely deficient in iron and fibre. This is intended to produce the pale coloured flesh fancied by 'gourmets'. At approximately 16 weeks of age, these weak animals are slaughtered and marketed as 'white' veal.

An expert from Bristol University's division of Food Animal Science has examined veal production techniques. He showed that it wasn't necessary to keep the calves confined in crates to produce the quality of meat desired by the market. All of Europe's veal farmers will have to follow the UK's lead by 2007. However, improvements in livestock living conditions have cost implications for producers.

Do gourmets realise where their food comes from?

Question

a *(i) Suggest why veal crates were banned in the UK before the rest of Europe.*
(ii) Explain why European farmers were given several years' notice of a ban on veal crates.
(iii) Explain why some European farmers still use veal crates.
(iv) Explain why there will be 'cost implications' for making improvements in livestock living conditions.

How far does your food travel?

Brazil
orange juice
5900 miles

New Zealand
apples
11000 miles

Israel
potatoes
2200 miles

Spain
carrots
960 miles

Kenya
mangetout
4500 miles

China
pine nuts
5080 miles

The picture shows the distance travelled by some food items you might put in a shopping basket. The distance travelled by food from where it is grown to the consumer is known as food miles. Food miles have increased rapidly over the last 20 years, as have shopping habits. Together these have had the following effects.

- There has been a 27% increase in shopping for food by car since 1992.
- The number of kilometres travelled by HGV vehicles delivering food has increased by 36% since 1992.
- The amount of food flown into the UK has increased by 140% since 1992.

Question

b *(i) Suggest reasons for each of the increases listed above.*
(ii) Outline the effects on the environment of these increases.
(iii) Suggest ways in which the number of food miles could be reduced.
(iv) Eating locally produced food is not always a solution to the problem. It may take more energy to grow tomatoes in winter in the UK than to grow them in Spain and transport them to the UK. Explain why.

Key points

- Some animals suffer appalling conditions in order for humans to be able to eat 'delicacies' such as veal.
- Improvements in conditions for livestock will have serious economic implications for farmers.
- Since the early 1990s the distance food travels to reach us has increased rapidly.
- Transporting food on this scale has massive effects on the environment.

Enzymes and homeostasis

Many scientific discoveries result from something happening by chance. Scientists who are there when something unexpected happens can make accurate observations and measurements, question why things happen, carry out investigations to find out why things happen, and suggest explanations for what is happening.

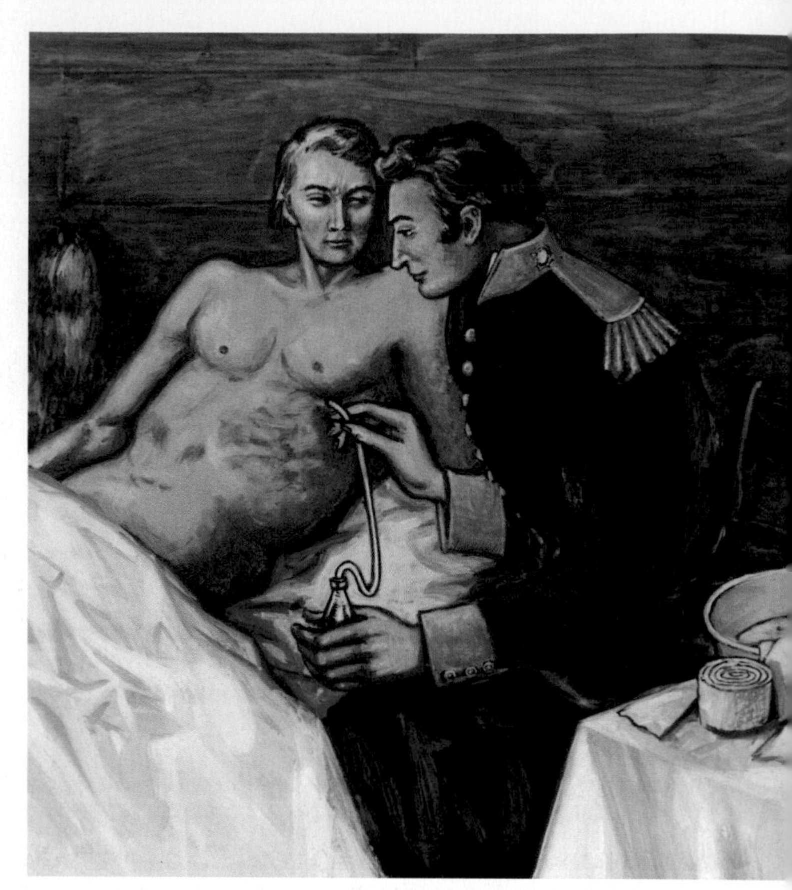

▲ Dr Beaumont carried out experiments on digestion in 1822.

Investigating digestion

Many investigations have added to our understanding of how the digestive system works. One important set of investigations was the result of a remarkable accident. It happened in a remote American village in 1822.

A French Canadian trapper was accidentally shot. Part of his left side was blown off with a shotgun, making a large wound. The trapper was nursed back to health by a doctor in the village, Dr Beaumont. The wound left the trapper with a small hole which exposed the inside of his stomach.

Gathering evidence

Dr Beaumont used the hole into the trapper's stomach to carry out a series of investigations to find out how food is digested.

Investigation 1

The temperature inside the stomach was 38 °C. This was measured by placing a thermometer through the hole. After the trapper had been kept fasting for 17 hours, the doctor extracted some of the fluid from the inside of the stomach. The fluid was put into a test tube and a solid piece of beef was then added. The test tube was kept at 38 °C as this matched the temperature inside the stomach of the trapper.

After 1 hour the beef had started to break up into small pieces and the fluid became cloudy.

After 2 hours the beef appeared as small shreds of fibres.

Investigation 2

Just after taking the sample of fluid from the stomach, a piece of beef attached to a thread was placed through the hole into the stomach.

After 1 hour the beef looked similar to the beef in the first investigation.

After 2 hours the thread was pulled out with no beef attached.

These experiments supported the idea that digestion takes place within the stomach at a specific temperature. In order to prove the theory fully, a control experiment should have been used. For example, a piece of the beef could be placed in a test tube of water at 38 °C, to prove that it was digestive enzymes that were responsible for the breakdown of the beef.

Enzymes

Our understanding of **digestion** is now much more advanced. Scientists have identified the different substances produced by each region of the gut and what each substance does as food passes along. One group of substances involved in digestion is **enzymes**. Enzymes are involved in controlling most of the chemical reactions taking place in the body, including digestion, respiration and making new proteins. Scientists have developed ways of using enzymes to speed up processes that are carried out in industry and in the home.

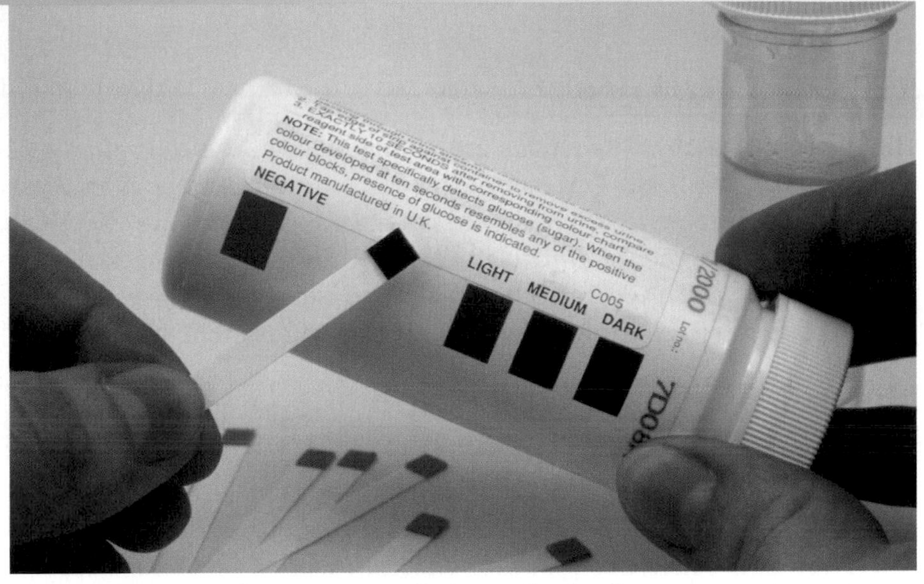

▲ These strips are used to carry out medical tests. The tip of the strip contains an enzyme which reacts with glucose. This makes the strip change colour and is used to measure the amount of glucose in urine samples.

Constant conditions

Enzymes work best under certain conditions. Your body keeps the conditions in the cells at a steady level so enzymes are able to work as efficiently as possible. This is why your body keeps its temperature at 37 °C. Keeping the amount of glucose in your blood at a steady level means that there is always enough fuel for respiration. Maintaining steady conditions inside your body is called **homeostasis**.

▲ By controlling your body temperature you can still be active, even when the temperature around you is very low.

Think about what you will find out in this section

How do enzymes control reactions taking place in digestion and respiration?	Why is it important to control conditions inside the body?
What are the advantages and disadvantages of using enzymes in the home and in industry?	How does the human body maintain a constant body temperature and control the concentration of blood sugar?
How can we investigate enzyme-controlled reactions?	What causes diabetes and how it is treated?
How does the removal of waste products help to keep the conditions inside the body constant?	How can we evaluate the modern methods of treating diabetes?

Properties of enzymes

Biological catalysts

Thousands of chemical reactions are taking place in the cells of animals and plants all the time. The rate of these chemical reactions is increased by the action of enzymes, which are **catalysts**. Catalysts are chemicals that speed up the rate of reactions. As enzymes speed up reactions in living organisms they are called biological catalysts. They catalyse processes such as respiration, photosynthesis and protein synthesis.

Molecules with a special shape

All enzymes are protein molecules made up of long chains of amino acids that are folded to produce a very precise shape. The starting substance of a reaction is called a **substrate**, and the substance it is converted to is called the product. Enzymes work by locking onto substrates. The diagram below shows how this happens. Because of its precise shape each enzyme will only act on one type of substrate – just like a key that fits into a specific lock.

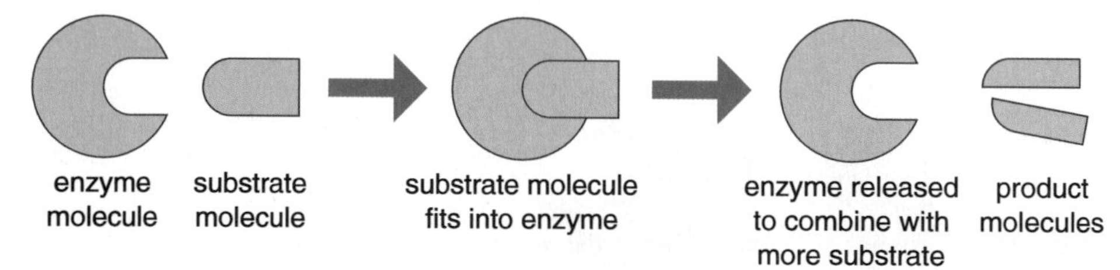
enzyme molecule
substrate molecule
substrate molecule fits into enzyme
enzyme released to combine with more substrate
product molecules

▲ Some enzymes catalyse the breakdown of molecules.

Effect of temperature

Most chemical reactions are speeded up by an increase in temperature. Molecules move around more rapidly as the temperature rises. This causes more collisions to occur between enzymes and substrate molecules, and so increases the rate of reaction.

As the temperature continues to rise the rate of enzyme-controlled reactions falls rapidly. This is because high temperatures destroy the shape of enzymes or **denature** them so they can no longer work. A denatured enzyme has a different shape and so cannot lock onto the shape of the substrate.

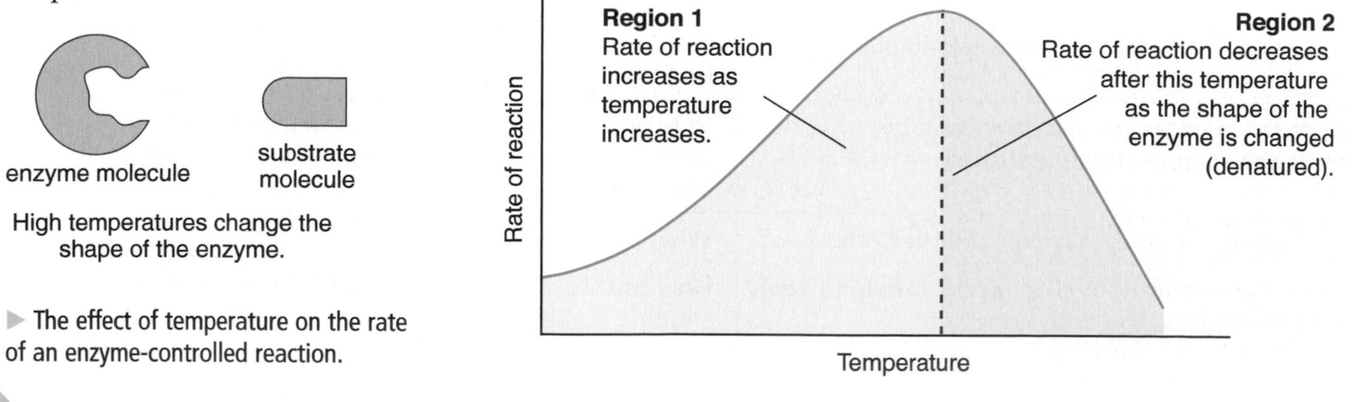
enzyme molecule
substrate molecule
High temperatures change the shape of the enzyme.
Region 1
Rate of reaction increases as temperature increases.
Region 2
Rate of reaction decreases after this temperature as the shape of the enzyme is changed (denatured).
Rate of reaction
Temperature

▶ The effect of temperature on the rate of an enzyme-controlled reaction.

Effect of pH

Different enzymes work best at different pH values. The pH at which an enzyme works best is called the optimum pH. The optimum pH of an enzyme depends on the pH conditions where the enzyme works. For example, intestinal enzymes work best in alkaline conditions so have an optimum pH of 8. Enzymes which work in the stomach have an optimum pH of 2 because the stomach is acidic. Intestinal enzymes will not work at all in very acidic conditions.

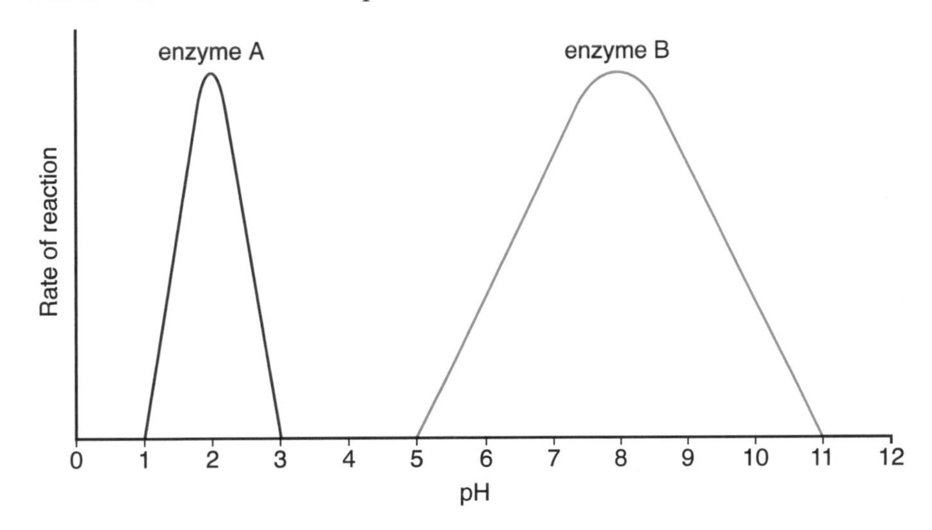

▲ Different enzymes have different optimum pH values.

Question

a *Describe the differences in the way pH affects the enzymes A and B shown on the graph.*

Investigating the rate of reaction

Potato cells contain an enzyme called catalase. This enzyme speeds up the breakdown of hydrogen peroxide. As hydrogen peroxide breaks down, bubbles of oxygen are released forming a froth. A group of students investigated how changing temperature affects the action of catalase. They were provided with potato tissue and hydrogen peroxide. The students decided to measure the rate of reaction by recording the height of froth formed in each test tube.

The results of their investigation are shown in the table.

Temperature (°C)	Height of froth (cm)
15	2.5
25	4.2
35	4.5
45	4.1
55	3.5

Questions

b *Suggest a suitable control for this investigation.*

c *What is the independent variable in this investigation?*

d *One group of students carefully cut the potato into small discs. They used a ruler to make sure the discs were all cut to the same size. Another group of students added the same mass of potato to each test tube.*
(i) Why is it necessary to add the same amount of potato to each tube?
(ii) Which method is the more accurate method – measuring the size or the mass? Give reasons for your answer.

e *Use the results shown in the table to explain the rate of reaction when the temperature increased from:*
(i) 15 to 35 °C (ii) 35 to 55 °C.

Key points

- Catalysts increase the rate of chemical reactions. Biological catalysts are called enzymes.
- Enzymes are protein molecules made of long chains of amino acids. The chains are folded to produce a special shape.
- The shape enables substrates to fit into the enzyme.
- High temperatures destroy or denature this special shape.

Digestion

Getting food into the body

Before your body can use the food you eat, large food molecules must be broken down into molecules which are small enough to be absorbed into your bloodstream. This is the process of digestion. Your diet contains three main types of food – proteins, fats and carbohydrates (mainly starch). Each type of food is made from large, insoluble molecules and is broken down by enzymes into small, soluble molecules during digestion.

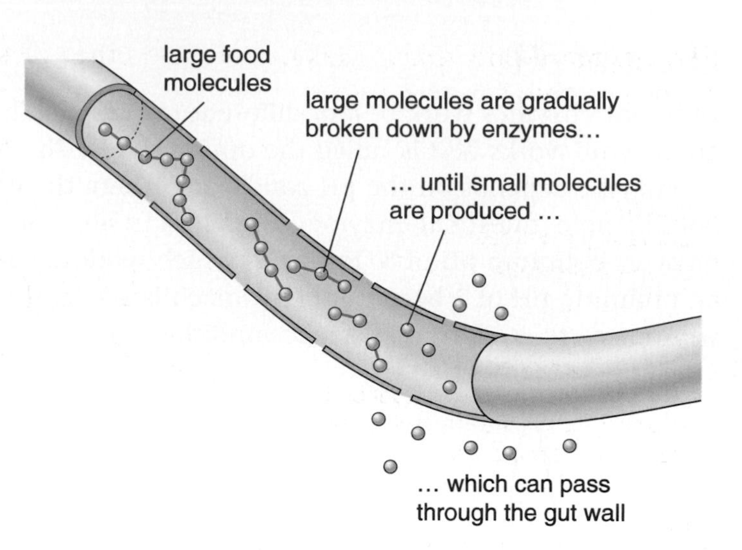
large food molecules
large molecules are gradually broken down by enzymes…
… until small molecules are produced …
… which can pass through the gut wall

Only small molecules can pass through the wall of the gut into the bloodstream.

The right tools for the job

Each type of food needs a particular enzyme to break it down into products that are useful to the body. The diagram shows this. For example, protease enzymes catalyse the breakdown of proteins and will not break down any other type of food. The table shows where each type of enzyme is produced in the body.

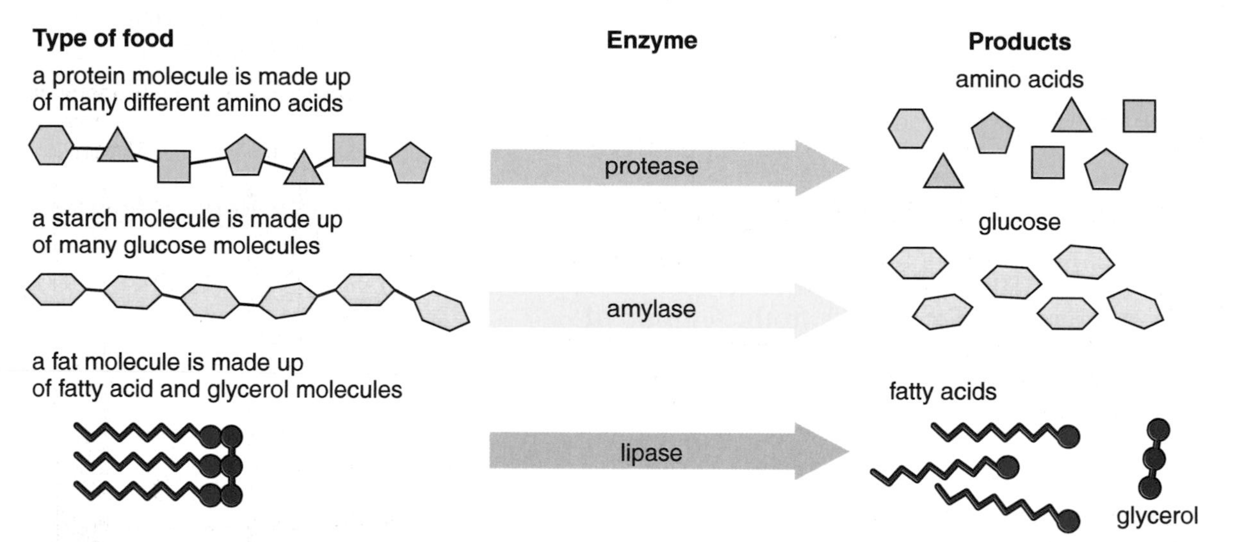
Type of food
Enzyme
Products
a protein molecule is made up of many different amino acids
protease
amino acids
a starch molecule is made up of many glucose molecules
amylase
glucose
a fat molecule is made up of fatty acid and glycerol molecules
lipase
fatty acids
glycerol

Type of enzyme	Where is it produced?
protease enzymes	stomach pancreas small intestine
amylase enzymes	salivary glands pancreas small intestine
lipase enzymes	pancreas small intestine

Digestive enzymes are produced by specialised cells in glands and in tissues lining the gut. The enzymes pass out of the cells into the gut where they act on food molecules. Each part in the digestive system releases different enzymes onto food. Other substances are also released to control the pH in each region of the gut. For example, the stomach produces hydrochloric acid because the enzymes there work best in an acid solution.

The diagram below shows the enzymes and other substances produced as food passes along the gut.

Question

a *Explain the benefit of controlling the pH in each region of the gut.*

mouth
Salivary glands produce **amylase** enzymes.

stomach
produces **protease** enzymes. Hydrochloric acid is also produced. The enzymes in the stomach work best in acid conditions.

liver
produces bile, which is stored in the gall bladder. Bile neutralises the acid produced by the stomach and provides alkaline conditions.

gall bladder

pancreas
produces protease, amylase and **lipase** enzymes.

small intestine
produces protease, amylase and lipase enzymes. Enzymes in the small intestine work best in alkaline conditions.

▲ Different types of food need different enzymes to break them down into smaller molecules.

Investigating the action of protease enzymes

An investigation was carried out using two protease enzymes, X and Y, taken from two different regions of the digestive system. A protein called albumen (egg white) was used. Albumen dissolves to form a cloudy solution. The solution becomes clear after the albumen has been digested. The diagram shows the results of the investigation.

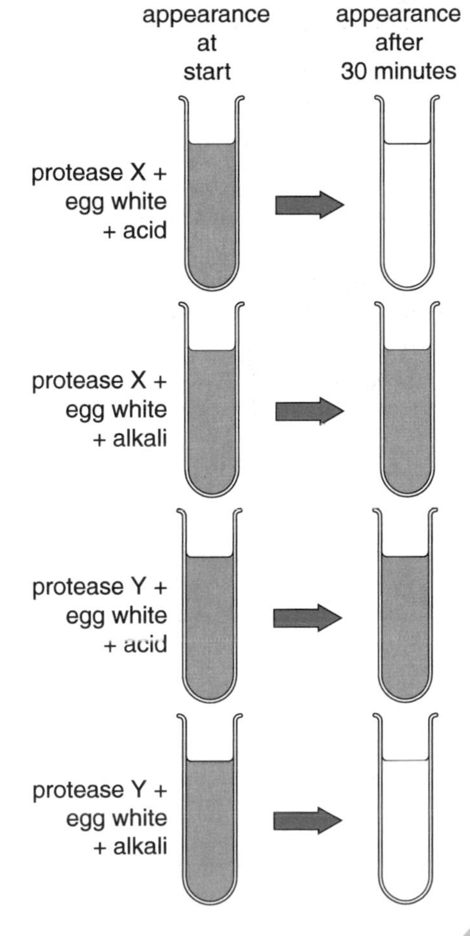

Question

b *Use the results of the investigation to answer the following.*
(i) Explain why some of the solutions become clear and some don't.
(ii) In which conditions does protease X work more rapidly?
(iii) Where in the gut would protease X be found? Explain your answer.
(iv) Name a substance which is released into the gut to provide the optimum conditions for protease Y to work.

Key points

- Digestive enzymes pass out of specialist cells into the gut where they catalyse the breakdown of large food molecules.
- The three types of digestive enzymes are proteases, amylases and lipases.
- The pH in different regions of the gut is controlled so that enzymes work most effectively.

Investigating digestion

The effect of amylase

The enzyme amylase is produced in the salivary glands, pancreas and small intestine.

This enzyme catalyses the breakdown of starch:

$$\text{starch} \xrightarrow{\text{amylase}} \text{glucose}$$

The diagram shows an investigation to find out how temperature affects the rate of starch digestion by amylase.

▶ Droplets of red iodine change to blue-black in starch solution.

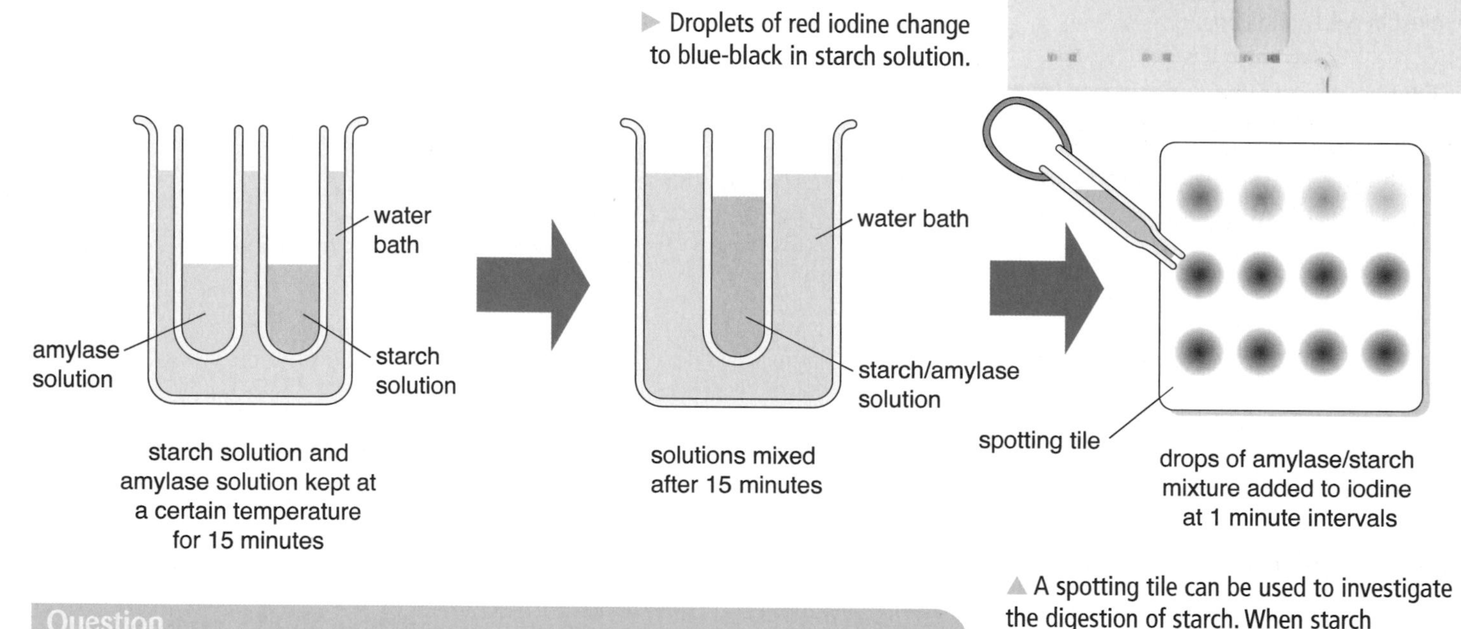
water bath
amylase solution
starch solution
starch solution and amylase solution kept at a certain temperature for 15 minutes
water bath
starch/amylase solution
solutions mixed after 15 minutes
spotting tile
drops of amylase/starch mixture added to iodine at 1 minute intervals

▲ A spotting tile can be used to investigate the digestion of starch. When starch is added to iodine solution the colour changes from deep red to blue–black.

Question

a *Explain why the following procedures were carried out.*
(i) The investigation was repeated using starch solution without amylase.
(ii) The starch solution and the amylase solution were placed separately in the water bath for 15 minutes before they were mixed together.
(iii) The time taken for the blue–black colour to disappear was measured.

The investigation was repeated using a range of temperatures. The results are shown in the table.

Temperature (°C)	Time taken for blue–black colour to disappear (minutes)			
	Reading 1	Reading 2	Reading 3	Mean
5	25	22	28	25
15	15	17	16	16
25	7	9	5	7
35	6	4	5	4
45	2	2	10	2

Questions

b *Explain why reading 3 at 45 °C was ignored.*

c *Use the results to plot a graph to show the effect of changing temperature on the rate of digestion of starch.*

d *Use your graph to predict how long it will take to digest starch at 20 °C.*

Breaking down lipids

Lipids (fats and oils) are broken down by lipase enzymes in the small intestine. Lipase enzymes are produced by the pancreas and small intestine.

$$\text{fats and oils} \xrightarrow{\text{lipase}} \text{fatty acids + glycerol}$$

Food leaving the stomach is very acidic. Lipase and other enzymes in the small intestine work most effectively in alkaline conditions. Bile, which is produced by the liver, is released to neutralise the acid from the stomach and to make conditions alkaline in the small intestine. Bile is stored in the gall bladder before being released.

Investigating the digestion of lipids

The diagram shows an investigation carried out to find out how bile affects the action of lipase on milk. Milk was used because it contains fat. Lipase was added to different mixtures and any change in pH was recorded.

The results are shown in the table.

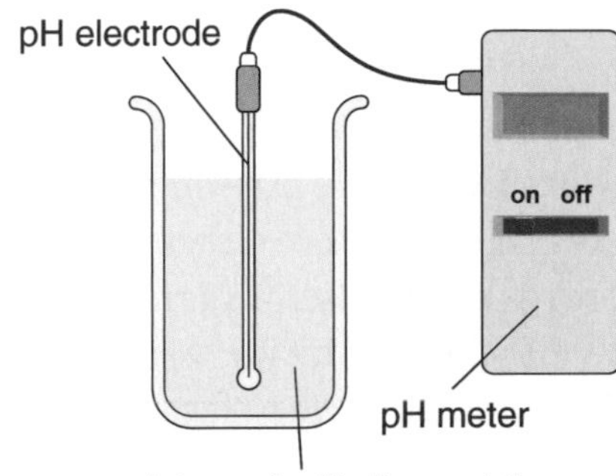

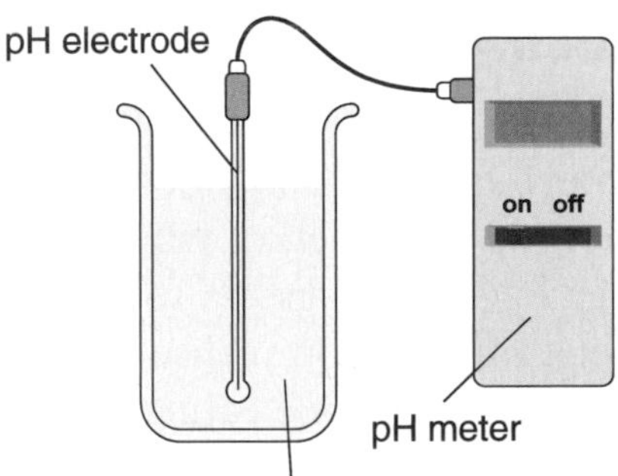

Time (minutes)	pH	
	Lipase + milk + sodium carbonate + bile	Lipase + milk + sodium carbonate
0	8.5	8.5
5	8.0	8.3
10	7.8	8.2
15	7.5	8.1
20	7.4	8.1
25	7.3	8.0

Questions

e *Explain why the pH decreased during the investigation.*

f *A student concluded that lipase works best in alkaline conditions and this was why fat was digested more quickly with lipase and bile. Explain why this is not a complete conclusion.*

g *Another student found out that as well as making conditions alkaline, bile also breaks fat up into tiny droplets, as shown in the diagram. Use this information to explain the results.*

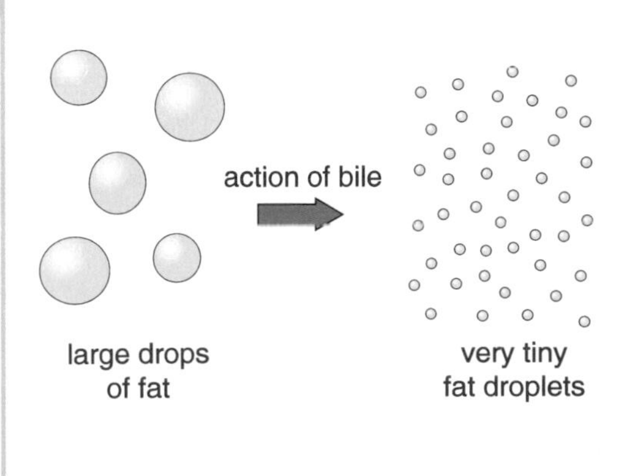

▲ The action of bile on fat.

Key points

- Lipase enzymes break down lipids into fatty acids and glycerol.
- Lipase enzymes work best in alkaline conditions.
- Bile neutralises the acid added to food in the stomach and provides alkaline conditions.

Industrial uses of enzymes

Enzyme technology

For many years, inorganic catalysts (those not containing carbon) have been used in industrial reactions – for example, iron is used in the production of ammonia in the **Haber process**. Recently there have been many developments in the use of enzymes as industrial catalysts – a process known as enzyme technology. Because enzymes are highly efficient catalysts, only a small amount of enzyme is needed to produce a large quantity of product. This is why they are more useful in industrial processes than inorganic catalysts.

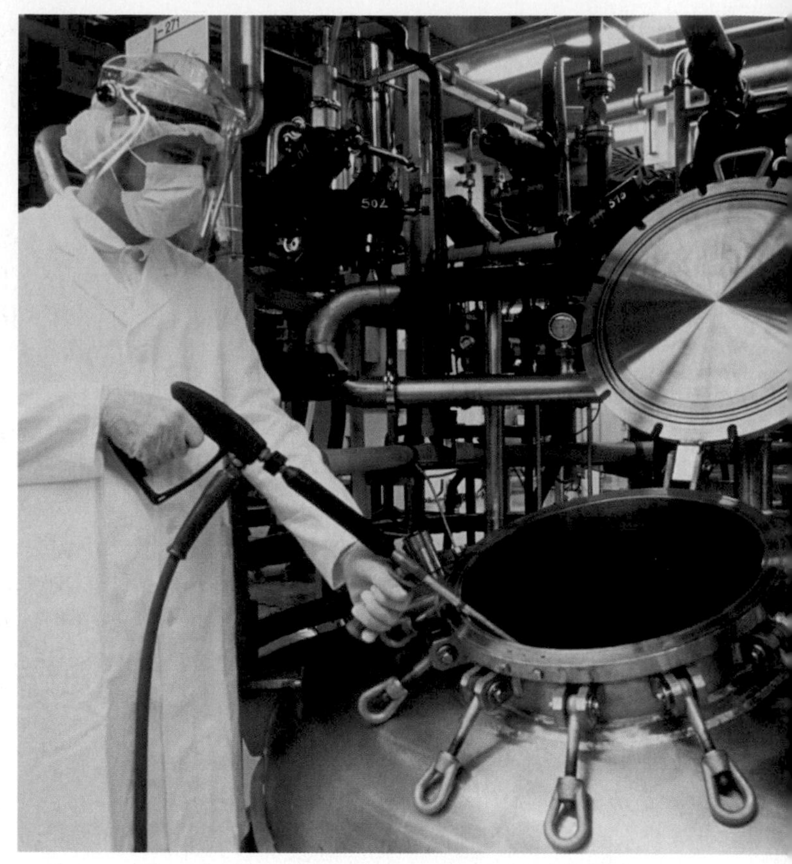

▲ These tanks, called fermenters, contain large quantities of bacteria. The bacteria grow very rapidly inside the fermenters, producing industrial amounts of enzyme.

Enzymes from microorganisms

Most enzymes are obtained from microorganisms, such as bacteria. Many of the enzymes produced by microorganisms are passed out of the cell, which enables scientists to extract the enzyme from the microorganism. Microorganisms can be grown relatively cheaply inside very large fermenters. As they can multiply very fast, microorganisms produce large amounts of enzymes quickly.

Enzymes in the food and drinks industry

Enzymes are now used a lot in the catering industry, from pre-digested baby foods using proteases (produced by bacteria), which makes them softer for babies to eat and absorb, to making the centres of sweets soft. Did you realise that every time you eat a cream egg, you are eating a piece of an industrial process? The hard, creamy centre is wrapped in chocolate and then enzymes are injected into the centre of it to make it runny.

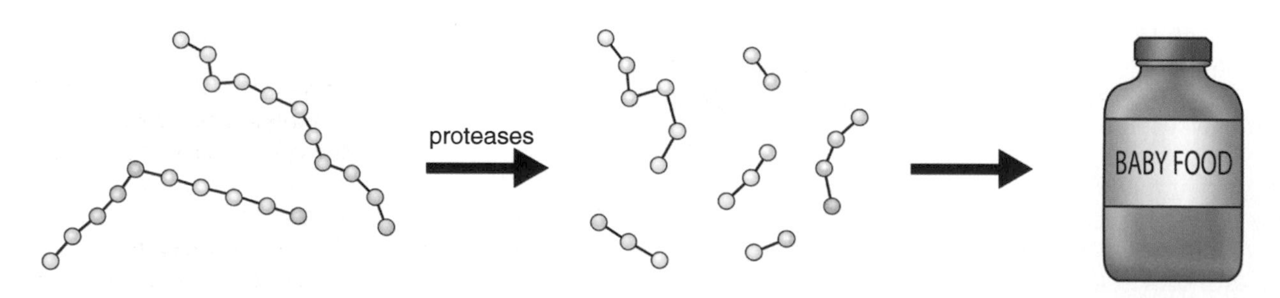

▲ Protease enzymes are used to break down long chain protein molecules into short chains which are easier for babies to digest.

Enzymes are also used in the production of sugar syrups used as sweeteners in the drinks and food industry. The stages in the production of sweeteners and the enzymes involved in each stage are shown on the right.

Fructose and glucose are both sugars with the same energy value. Fructose is a much sweeter sugar than glucose, so less needs to be added to foods and drinks to make them taste sweeter. This is very useful in the production of slimming foods and low-calorie drinks.

Reusing enzymes

When a reaction is complete, the enzyme and product are mixed up with each other. It is very expensive for an industry to keep producing enzymes and to keep separating products from enzymes. To avoid this expense scientists have found ways of fixing enzymes to the surface of small beads. This is called immobilising the enzyme. The diagram shows how isomerase enzymes fixed in this way can be used over and over again. Also, there are no enzyme molecules mixed with the fructose syrup that is produced, as they are all trapped in the beads.

Questions

- ***a*** *Explain why fructose syrup, rather than glucose syrup, is an ingredient of low-calorie drinks.*
- ***b*** *Explain why it is less expensive to use industrial enzymes fixed to beads.*
- ***c*** *The amount of fructose produced is affected by how fast the glucose syrup flows through the reactor. The graph below shows the result of increasing the rate of flow of glucose syrup into the reactor.*
  (i) What rate of flow should scientists use in the reactor? Explain your answer.
  (ii) How much more fructose is produced when the rate of flow is increased from 3 to 4 dm^3/minute?

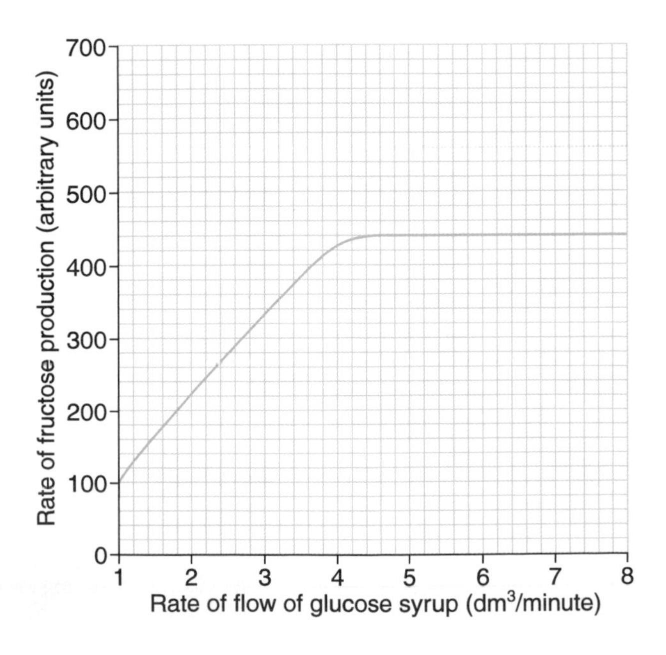

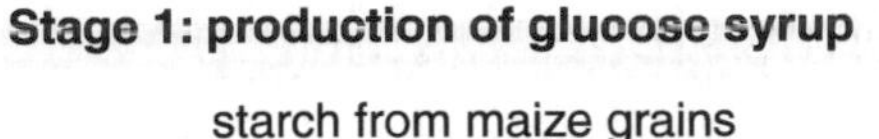

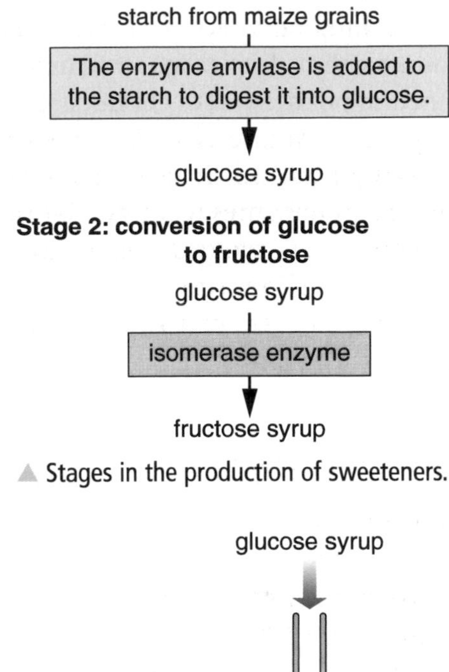

▲ Stages in the production of sweeteners.

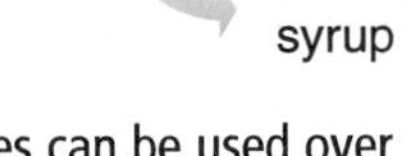

▲ Immobilised enzymes can be used over and over again.

Key points

- Some microorganisms produce enzymes which pass out of the cells. These enzymes have many uses in the home and in industry.
- Uses of enzymes in industry include production of baby foods, sugar syrups and sweeteners.

Enzymes in the home

Biological washing powders

All washing powders contain detergents to dissolve stains so that they can be washed away. Biological washing powders also contain protease and lipase enzymes. Protease enzymes catalyse the breakdown of proteins present in stains such as blood, grass and egg. Lipase enzymes catalyse the breakdown of lipids in stains such as fat, oil and grease. The protein and fat molecules in stains are broken down into smaller, soluble molecules which dissolve easily in water and can be washed away.

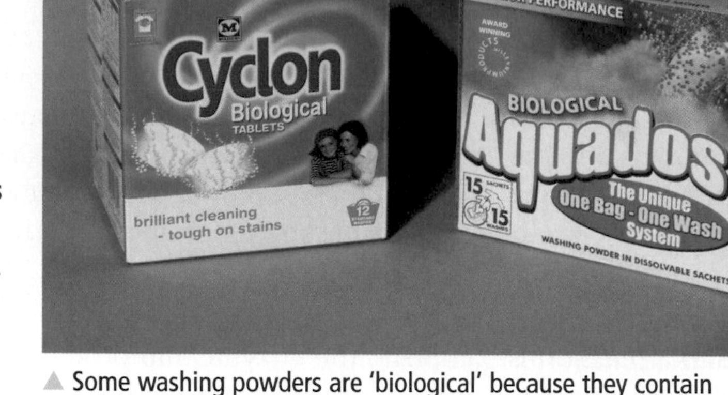

Some washing powders are 'biological' because they contain enzymes from bacteria.

Question

a *What will be formed when the proteins in blood are broken down by the washing powder?*

Enzymes from bacteria

The first biological washing powders that were produced would only work in warm water. However, proteases and lipases have now been produced which work at much higher temperatures. Most of the enzymes in washing powders are obtained from bacteria living in hot springs, which means the bacteria are adapted to live in water above 45 °C. The enzymes obtained from these bacteria will work at moderately high temperatures. This is useful because the detergents in washing powders, which get rid of greasy stains, work best at higher temperatures.

Questions

b *Explain how biological washing powders help to remove greasy stains from clothing.*

c *Explain how finding bacteria which live in hot springs has helped in the development of biological washing powders.*

The enzymes produced by bacteria living in hot springs will work at high temperatures.

Investigating biological washing powders

A group of students carried out an investigation to find out conditions in which biological washing powders work best. The students used photographic film to demonstrate the action of the washing powders. The film contains black grains stuck on by a layer of gelatin. Gelatin is a protein. When the gelatin is broken down by the enzymes in the washing powder the film becomes clear as the black grains come away.

The students prepared a 1% solution of washing powder by dissolving 1 g of powder in 100 cm^3 of water. The diagram shows how the students designed the investigation.

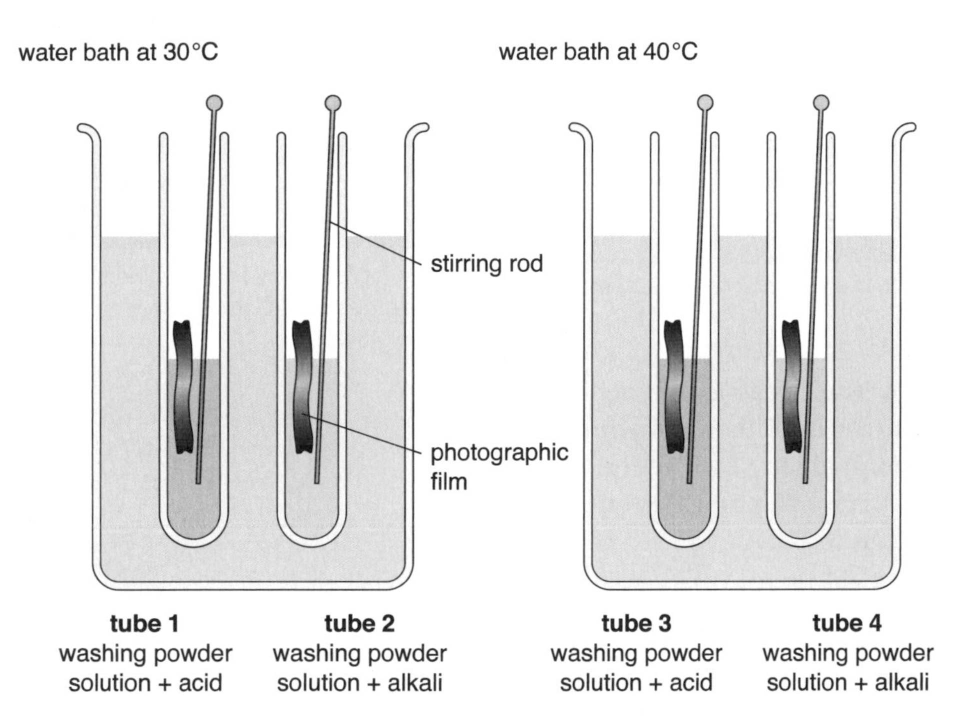

The students timed how long it took for the film to go clear. Their results are shown in the table.

	Tube 1	Tube 2	Tube 3	Tube 4
Temperature (°C)	30	30	40	40
pH	4	8	4	8
Time taken for film to go clear (minutes)	remained black	25	40	10

Questions

d *Name the type of enzyme that caused the film to go clear.*

e *(i) Explain why a 1% solution of washing powder was used in all four test tubes.*
(ii) Explain why a stirring rod was used.

f *Which variable in the investigation had the greatest effect on the rate of reaction? Use the results to explain your answer.*

Key points

- Biological detergents contain protein-digesting and fat-digesting enzymes.
- The enzymes in washing powders are produced from bacteria.

Respiration

Living things need energy

Plants and animals need energy to stay alive. Energy is used by:

- all organisms to build up large molecules from smaller ones (for example, building up proteins from amino acids or making starch from glucose needs energy)
- animals to contract muscles during movement
- mammals and birds to keep their body temperature steady in colder surroundings
- plants to make amino acids from sugars and nitrates.

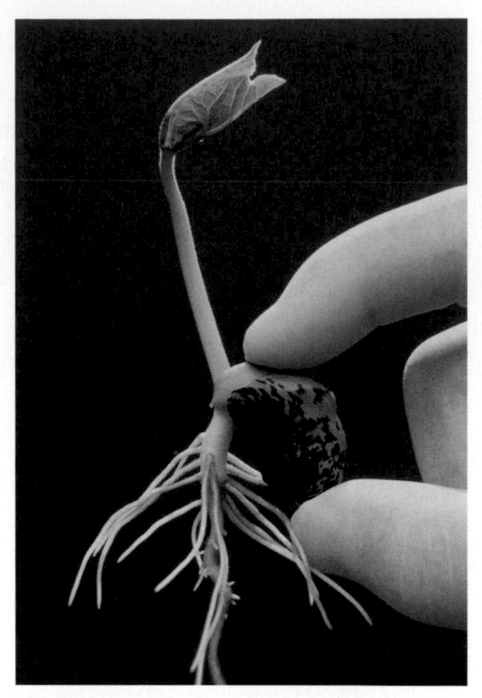

▲ The food store in a seed provides the energy to build large molecules during growth.

Releasing energy

Energy, in the form of glucose, is made available to cells by the breakdown of food molecules. The process of releasing energy from glucose is called respiration.

Question

a *What sorts of foods would be broken down to give the glucose required for respiration?*

The energy in glucose can be released in a single reaction. When this happens heat energy is released as glucose burns. This is **combustion**. During respiration glucose is broken down gradually by a series of reactions, each catalysed by a different enzyme. This releases energy in small amounts so that it can be used by cells.

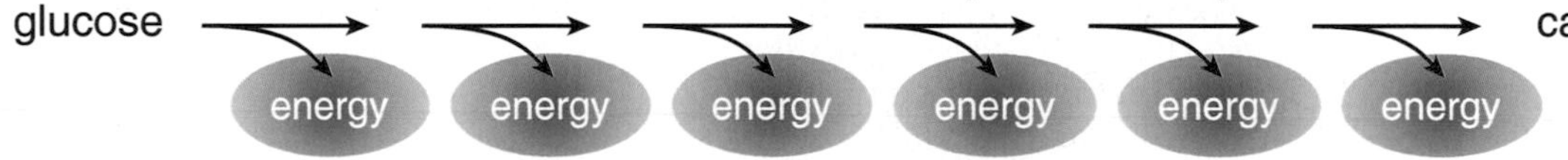

▲ The gradual breakdown of glucose enables cells to use the energy released.

▲ Energy is released as heat when glucose breaks down rapidly during combustion.

▲ Energy can be used by cells when glucose is broken down gradually during respiration.

Using oxygen

Respiration which uses oxygen is called aerobic respiration. During aerobic respiration reactions take place which use glucose and oxygen to release energy. These reactions are summarised by the equation:

glucose + oxygen → carbon dioxide + water + energy

Powering cells

Most of the reactions in aerobic respiration take place inside the mitochondria found in the cytoplasm of cells. Cells such as muscle cells, which need lots of energy, contain large numbers of mitochondria. Because mitochondria release the energy needed by cells they are called the 'powerhouse' of the cell.

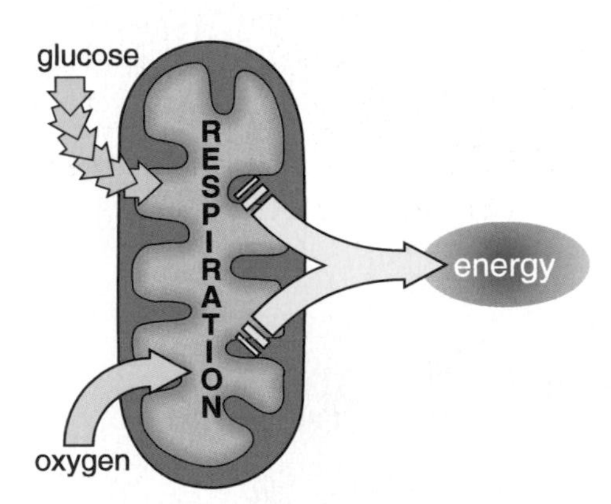

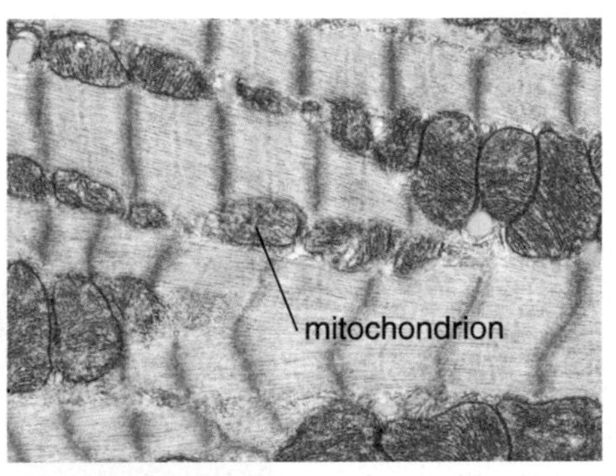

▲ Cells using a lot of energy need a lot of mitochondria. This shows muscle tissue.

Question

b *Nerve tissue contains many mitochondria. Suggest why this type of tissue needs many mitochondria.*

Building large molecules

Cells use energy to synthesise large molecules such as proteins. Proteins are made by joining amino acids together to form long chains. Amino acids produced by the digestion of proteins are absorbed into the blood. They are transported around the body by the blood for use by cells to make the proteins they need. The joining of amino acids is catalysed by enzymes found inside cells. Protein synthesis needs energy, which is provided by respiration.

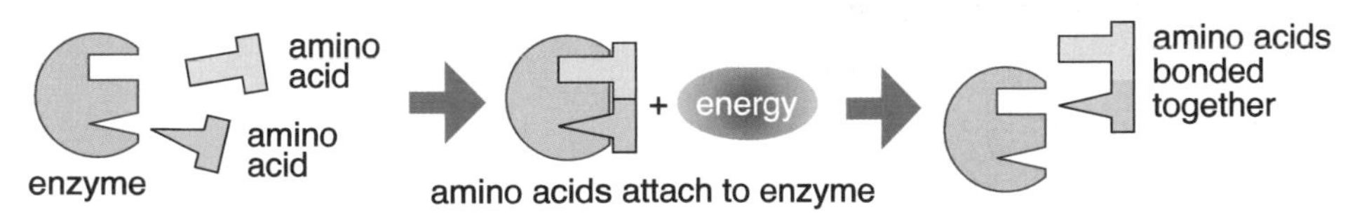

Animals get the amino acids they need from food. Plants make all their own amino acids. Nitrate ions in the soil are absorbed by roots. In the root cells, enzymes convert nitrates and sugars into amino acids using enzymes and energy from respiration.

nitrates + sugars + energy from respiration→ amino acids

Question

c *Food provides the amino acids needed to make body proteins. Describe the steps involved in:*
(i) getting amino acids into the body
(ii) using the amino acids to make proteins.

Key points

- In aerobic respiration oxygen and glucose are used and energy is released.
- Most aerobic respiration takes place inside mitochondria.
- Some of the energy released in respiration is used to make larger molecules from smaller ones, to enable muscles to contract and to keep temperature steady.
- Some of the energy released in respiration in plants is used to make amino acids and proteins.

Homeostasis

Staying the same

Even though the temperature of the air around you might be freezing in winter and very hot in summer your core body temperature stays at 37 °C. Keeping conditions steady inside the body is called homeostasis. For example, your body temperature, the amount of glucose in your blood and the amount of water and ions in your body are all kept at a steady level. These conditions are being adjusted all the time by your body to prevent any big changes.

Homeostasis helps your cells to work as efficiently as possible. The chemical reactions in cells are controlled by enzymes. Enzymes work best in particular conditions, so keeping the conditions at a steady level provides enzymes with the best working environment.

▲ Although there may be very large temperature changes around you, your core body temperature stays very steady.

Getting rid of waste

Chemical reactions in body cells produce waste. This waste includes carbon dioxide and **urea**. The diagram shows the organs that remove waste products from your body.

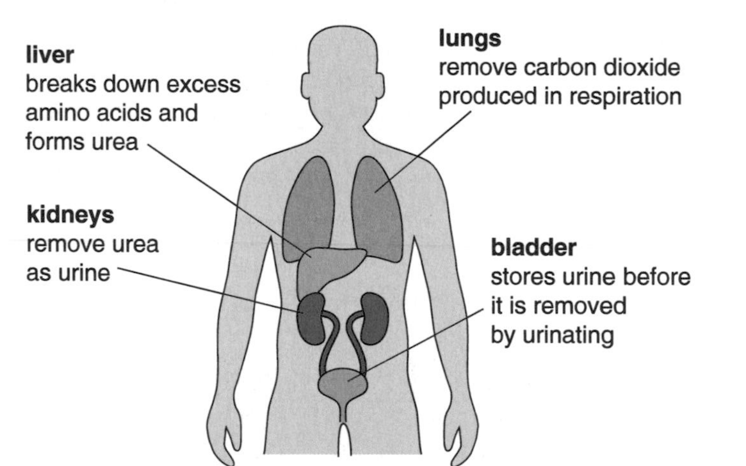

Your lungs play a part in homeostasis by keeping the concentrations of oxygen and carbon dioxide at the best level for respiration. The rate of respiration inside muscle cells increases when you exercise. More waste carbon dioxide will be produced. When this happens you change your breathing to get rid of carbon dioxide more quickly. The graph shows the amount of air taken in and out of the lungs at rest and during exercise.

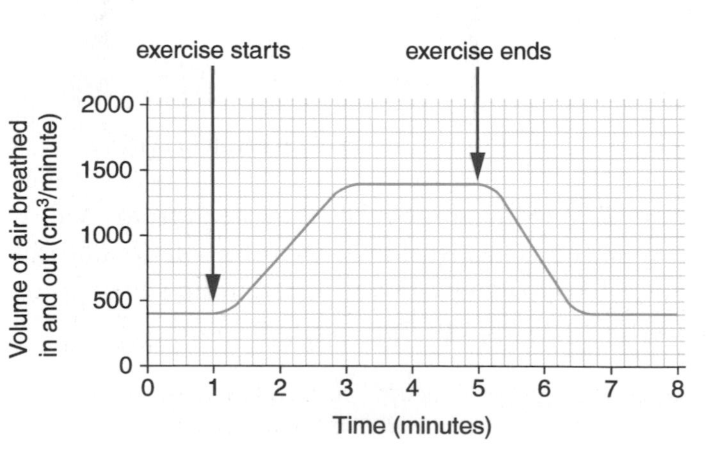

Questions

a *Calculate the increase in the volume of air breathed in and out each minute after 3 minutes of exercise.*
b *The air that is breathed out contains 4% carbon dioxide. Calculate the volume of carbon dioxide breathed out each minute (i) at rest (ii) after 3 minutes of exercise.*

Your body needs amino acids for building new cells. You get these amino acids by digesting the protein in food. Your body is unable to store the excess amino acids it does not use. Instead, these are broken down in the liver to make urea. The urea is taken from the liver to the kidneys where it is converted into urine. Urine is stored in the bladder before being removed from the body.

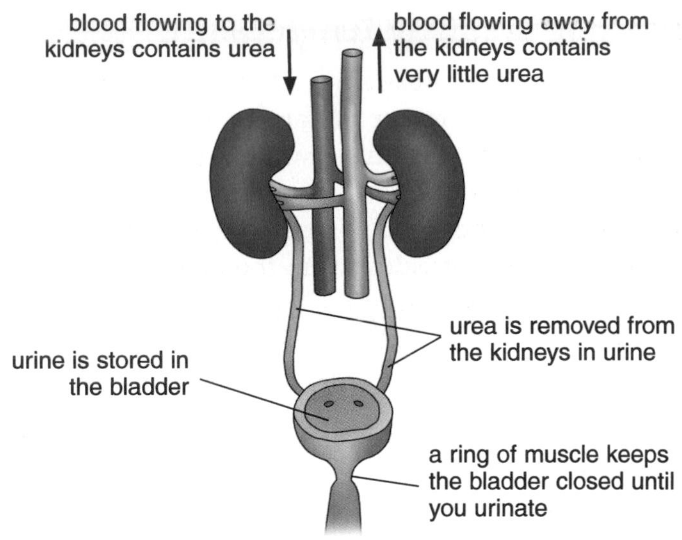

▲ The kidneys convert urea to urine.

Controlling water and ions

Your body cells need a constant amount of water. If the amount of water is not controlled then too much water may move in or out of cells and damage them. The diagram here shows how water is lost and gained by your body each day.

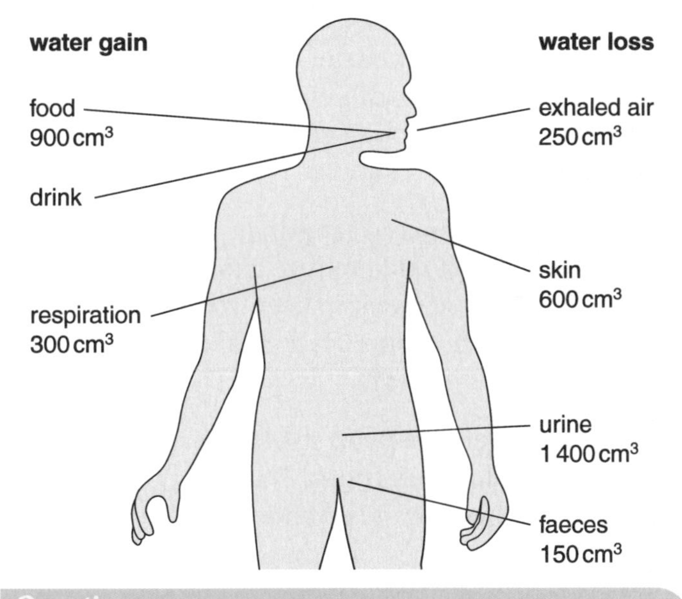

Your body has to balance the amount of water it takes in with the amount it loses. Sweating helps to cool your body down, so on a hot day more water is lost as sweat. When you lose water you become thirsty so you take in more water by drinking more fluids.

When you sweat you lose ions as well as water. Sweat contains about 0.15 g of salt (sodium chloride) per 100 cm^3 water. Sports drinks help to replace both the water and the salt.

▲ When you sweat you lose both water and salt. This is why athletes drink sports drinks – to replace both salt and water.

Question

c *The body balances water losses with water gains. Use the information in the diagram to calculate:*
(i) the volume of water taken in by drinking
(ii) the percentage of water lost via the skin.

Key points

- Waste products include carbon dioxide and urea.
- Carbon dioxide is formed in respiration; urea is formed from excess amino acids.
- Carbon dioxide is removed from the lungs.
- Urea is produced in the liver and removed as urine from the kidneys.
- Internal conditions, including water and ion content, are controlled at a steady level.

Controlling body temperature

Keeping a steady temperature

Your body temperature stays at about 37 °C even though the temperature of your surroundings may change. The photograph shows the temperature of different parts of the body. As a person exercises more heat is produced. To keep body temperature steady, excess heat is lost from the skin.

▲ The red colour shows where most heat is lost from the body.

Staying warm and staying cool

Your skin helps to keep your body temperature constant. Blood vessels supplying blood to the skin capillaries can increase and decrease in width. If the core body temperature is too high, an increase in width causes more blood to flow through the capillaries. This is called **vasodilation**.

On the other hand, if the core temperature is too low a decrease in width of the blood vessels reduces blood flow through capillaries. This is called **vasoconstriction**.

Your skin also contains sweat glands. When you exercise you get hot and produce a lot of sweat. Sweat glands release sweat to keep you cool. As sweat evaporates from the skin surface it takes heat from the body and so cools it down.

When you get very cold you shiver. Shivering is caused by muscle contractions. The energy for contraction comes from respiration which releases some energy as heat.

preventing over-heating

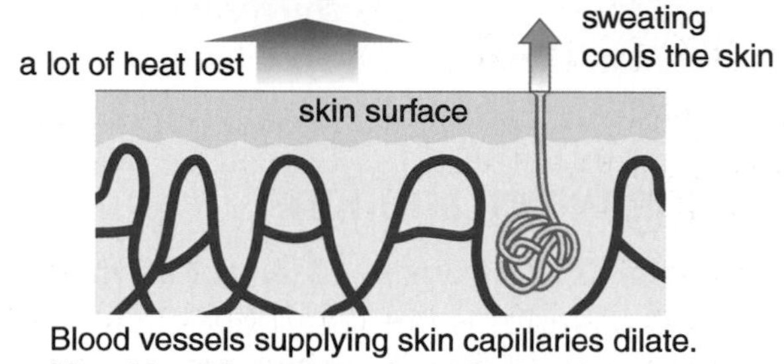

Blood vessels supplying skin capillaries dilate.
More blood flows through capillaries in the skin.

preventing over-cooling

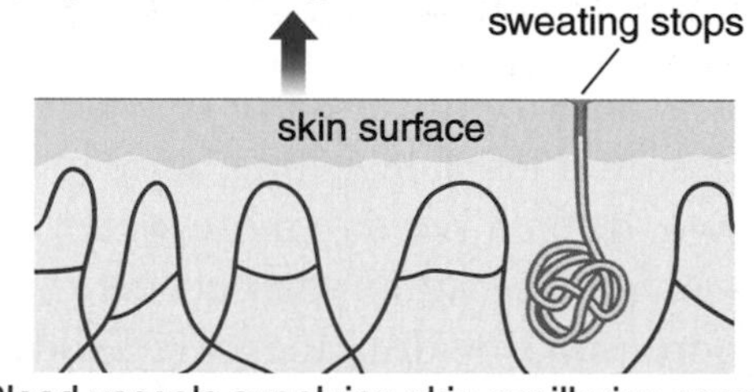

Blood vessels supplying skin capillaries constrict.
Less blood flows through capillaries in the skin.

Question

a *Explain why your skin goes pale when you feel cold.*

Control mechanisms

The water bath you use in the laboratory to maintain the temperature during an investigation shows how conditions are kept steady.

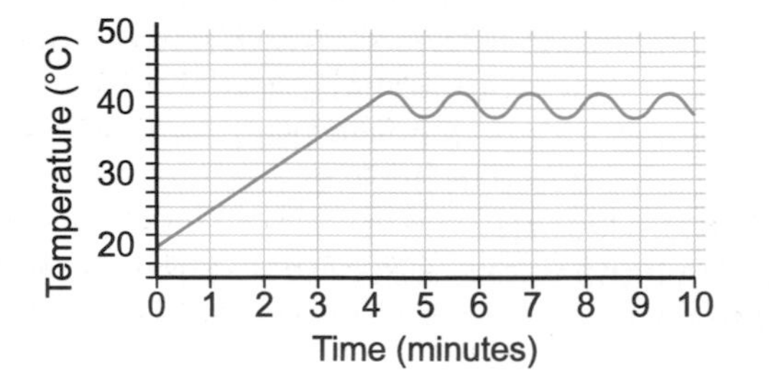

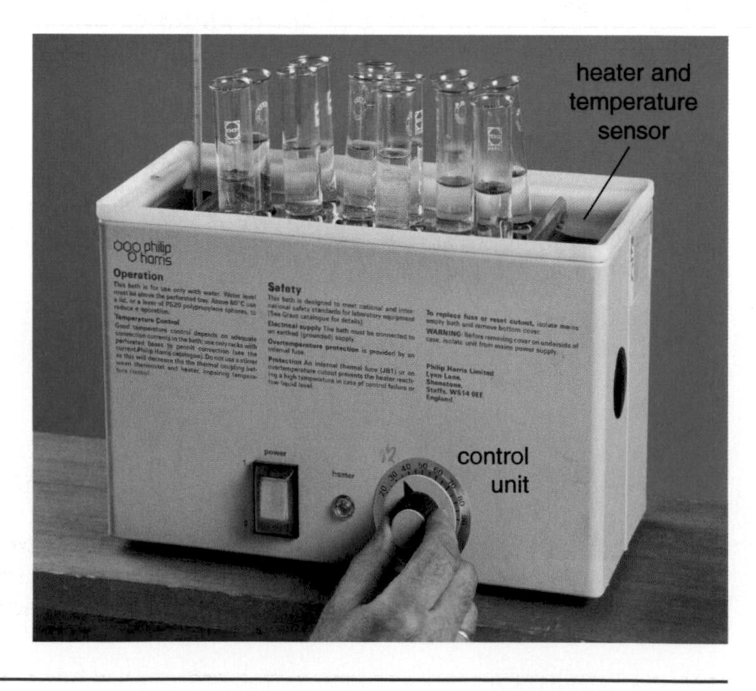

The water bath has three main components to keep the temperature of the water steady. These are the temperature sensor, the control unit and the heating unit. When the temperature of the water in the water bath is too low the heater gets switched on and the water temperature rises. When the temperature gets too high the heater is switched off and the water cools. The graph (page 54) shows the changes in the temperature of water in the water bath. The diagram here shows how these three components work to maintain the temperature in the water bath.

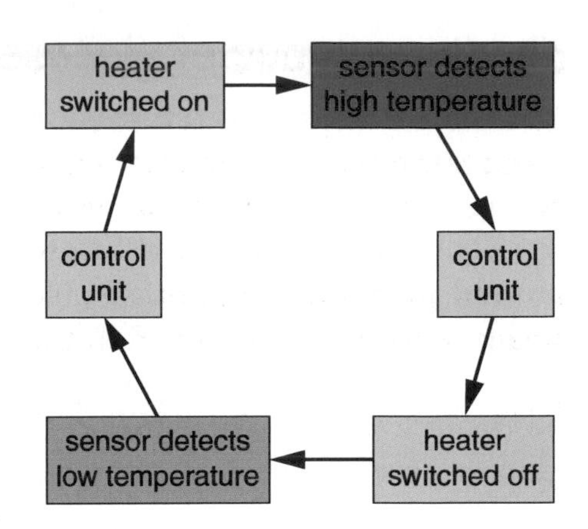

Question

b *At what temperature does the control unit switch the heater (i) on (ii) off?*

Thermoregulation

Maintaining a constant internal temperature is called **thermoregulation**. Body temperature is monitored and controlled by the thermoregulatory centre in the brain. The centre has temperature receptors which detect changes in the temperature of blood flowing through the brain. The centre also receives impulses from temperature receptors found in the skin. The thermoregulatory centre then sends information to blood vessels, muscles and sweat glands to increase or decrease heat loss. The diagram below shows how a constant body temperature is maintained.

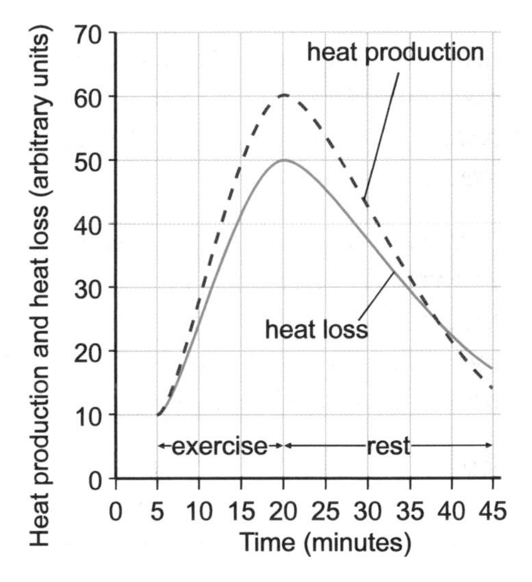

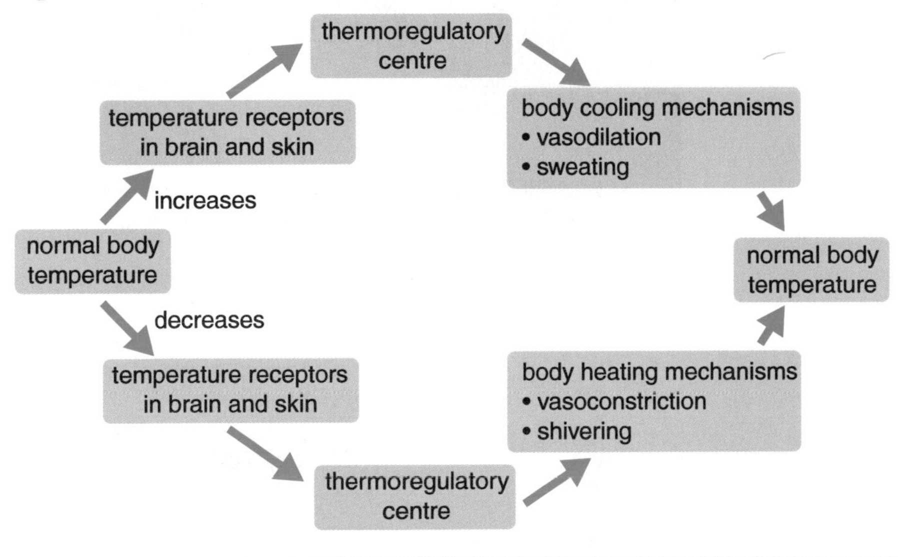

Questions

c *Which part of the body acts in the same way as*
(i) the control unit of a water bath
(ii) the heating unit?

d *The graph (above right) shows heat production and heat loss as a person cycled home from school and then rested.*
(i) What is the maximum difference between heat production and heat loss?
(ii) Explain why heat production increases between 5 and 20 minutes.
(iii) Describe the mechanisms that are used in the body to increase its heat loss.

Key points

- If body temperature is too high, vasodilation and sweating help to cool the body down.
- If body temperature is too low, vasoconstriction and shivering help to increase temperature.
- Body temperature is monitored and controlled by the thermoregulatory centre in the brain.
- Temperature receptors in the brain and in the skin detect changes in body temperature.

Controlling blood sugar

The importance of glucose

The cells in your body are respiring all the time. To do so they need a supply of glucose. Your blood system provides all the cells in your body with a regular supply of glucose. Glucose is transported around your body dissolved in blood plasma. Cells will not be able to respire if the amount of glucose in the blood gets too low. This can damage cells, particularly cells in the brain. A high level of blood glucose also causes water to move out of cells by osmosis. For these reasons it is important that the amount of glucose in the blood is kept steady.

▲ This girl is diabetic and needs regular injections of insulin.

Insulin and blood

The concentration of glucose in the blood is monitored and controlled by the pancreas. The concentration of glucose in the blood rises after a meal. This is because carbohydrates are digested into glucose which is absorbed into the bloodstream.

Cells in the pancreas detect a rise in blood glucose concentration. They respond by releasing a hormone called **insulin** into the bloodstream. Insulin is carried to all body cells. It makes the membranes of cells more permeable to glucose so more glucose is absorbed into cells. This makes the concentration of glucose in the blood come down. In some people, cells in the pancreas do not secrete enough insulin. This condition is called diabetes.

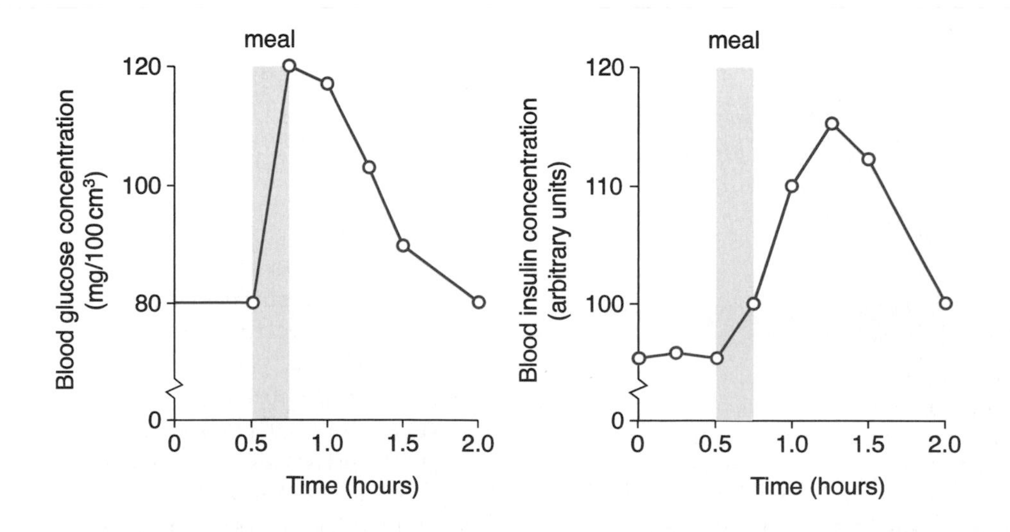

Questions

- ***a*** *Explain why blood glucose concentration increases after eating a meal.*
- ***b*** *Use the information on the graph to explain why the concentration of blood glucose returns to its normal level within a short time.*

Discovering insulin

In 1922, two Canadian scientists, called Banting and Best, announced that they had discovered insulin and successfully used it to treat diabetes. Up to this time, people with diabetes would die at an early age due to severe weight loss.

Much of the work of Banting and Best involved experimenting on dogs. They removed the pancreas from dogs so that the dogs developed diabetes. They also extracted substances from the pancreas to find out if the substances could control blood glucose. Their experiments may seem cruel, but without them they would not have discovered insulin as a treatment for diabetes.

Previous attempts to use pancreas extracts to treat diabetes had failed. Banting thought that this was because the hormone in the extract had been destroyed by digestive enzymes in the pancreas. To avoid this he tried tying the pancreatic duct to prevent enzymes from the pancreas digesting the hormone. He also tried chilling the extract to prevent the enzymes digesting the hormone.

Eventually, Banting and Best injected the extracted fluid from the pancreas of one dog into a diabetic dog. Within 2 hours, the blood sugar in the diabetic dog had fallen, and the dog was soon wagging its tail.

After developing their techniques on many dogs, the scientists were able to make an extract pure enough to try it on a human patient. In May 1922, a 14-year-old boy was successfully treated with an extract that they called insulin.

▲ Nobel Prize winner Frederick Banting with Charles Best (1923).

Questions

c *Insulin is a protein. Explain how enzymes from the pancreas could destroy the insulin hormone in the pancreas extracts.*

d *Explain how the following procedures prevented pancreatic enzymes digesting the hormone:*
(i) tying the pancreatic duct (ii) chilling the extract.

e *Banting and Best carried out their experiments on live dogs. Some people think that carrying out medical research on live animals is ethically wrong.*
(i) Suggest how the scientists would have justified using live animals for their experiments.
(ii) Suggest why some people object to the use of live animals to develop new drugs.

Key points

- The concentration of blood glucose is controlled by insulin produced by the pancreas.
- Insulin increases the uptake of glucose by cells. This lowers the concentration of glucose in the blood.
- Diabetes is a disease caused by the pancreas not producing enough insulin. This causes the blood glucose level to become too high.

Treating diabetes

Living with diabetes

The Type 1 kind of diabetes is a condition that a person has for all of their life. Even so a diabetic can live a long and healthy life by making changes to their lifestyle and working with doctors to find the treatment that suits them best. If diabetes is not managed properly it can cause major health problems.

Treatments for diabetes

There are many different types of treatment for diabetes, and new and better medicines and tests are regularly being developed. Research that monitored the treatment and health of over a thousand diabetic patients over several years showed that a treatment called 'tight control' is one that works well. It involves keeping blood glucose levels under tight control.

To keep glucose levels as close to normal as possible diabetics are advised to:

- watch what they eat and exercise regularly
- have regular injections of insulin
- have regular blood tests to check that their treatment is working.

Insulin devices

Many people who need insulin, inject it using a needle and syringe. Modern needles are very thin so there is minimal discomfort when people inject themselves. Even so, using a needle and syringe can be difficult especially for young children. A number of other devices have been developed to make injecting insulin easier.

Insulin pens are held against the skin and when a plunger is pressed insulin is injected through a small needle. The amount of insulin that is injected from the pen is easily adjusted.

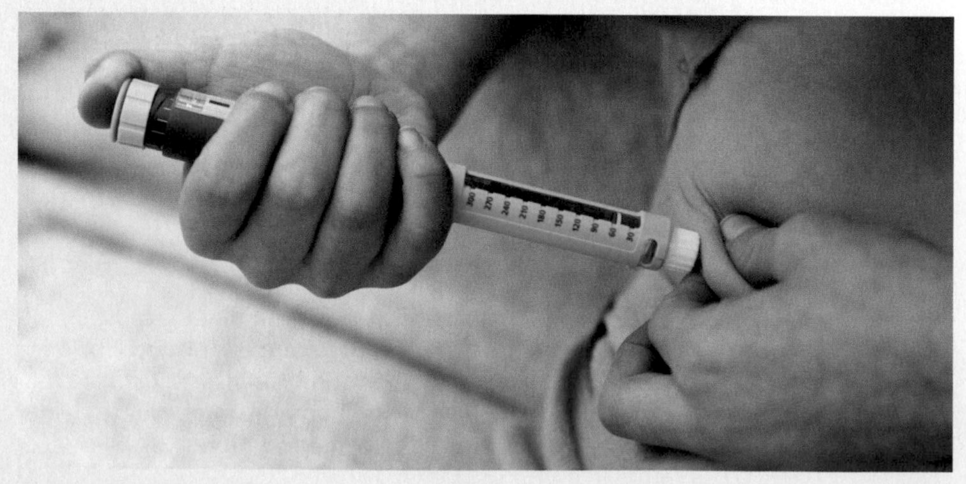

▲ An insulin pen.

Insulin jet injectors don't use a needle. A fine spray of insulin is pushed through the skin under pressure. These devices are expensive and have to be boiled frequently to prevent them causing infection.

Insulin pumps supply a continuous trickle of insulin into the bloodstream. Insulin is pumped through a tube which is attached to the body, usually the stomach, by a special kind of needle. The amount of insulin being pumped can easily be changed. For example, a larger dose of insulin can be given just after a meal. Some people do not like having a tube attached all the time, but soon get used to this. The tube can be taken out for short periods, such as when having a shower. The needles can also become easily infected. This is avoided by changing the needle and tubing every few days.

Questions

a *Which type of insulin device would you recommend? Give reasons why you would recommend this device rather than the others.*

b *As well as injecting insulin, good management of diabetes involves avoiding low glucose attacks (called 'hypos'). Explain why injecting insulin may cause a low glucose attack.*

Testing the level of blood glucose

Regular blood tests show if a diabetic's blood glucose level is dropping too low or climbing too high. Both these extremes are dangerous. Blood testing kits have been developed so that the test can be carried out at home easily and accurately. Modern testing kits and modern insulin devices enable diabetics to have tight control of their blood glucose and to manage their treatment to suit their own personal needs. The information shows how to use a blood testing kit.

Question

c *Describe the features of the glucose testing kit that make it easy to use.*

Key points

- Diabetics can control their glucose levels by careful dieting and consultation with their doctor.
- Insulin can be taken by a variety of methods to help with this control.
- Home test kits to monitor glucose levels can be purchased.

There are kits to help you test your blood glucose. The kits make testing simple. Here's what you do:

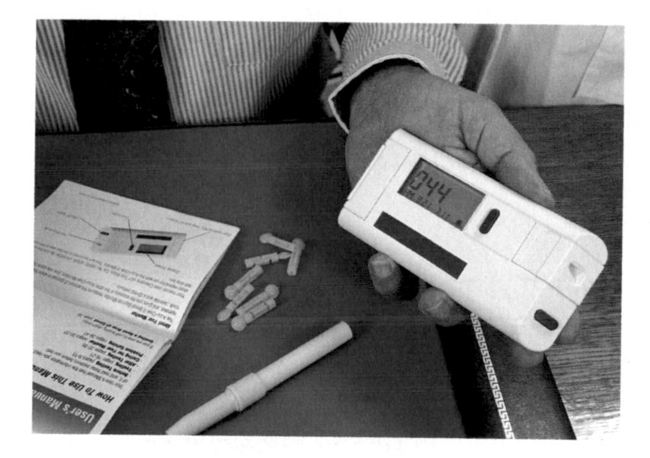

Firstly you have to take some of your blood, usually one or two drops from your finger or your forearm. To get blood for the test, most people use a special needle that springs out when they press a button. You place this needle on your finger. When you press the button, the needle punctures the skin and provides exactly the right amount of blood.

Touch the test strip with blood on your finger. A meter analyses your blood, and tells you how much glucose is in it. The amount is usually shown on a little screen on the meter.

The kit has an internal memory so you don't have to record your results. Just give the meter to your doctor on your next visit.

B2 4 Inheriting disease

The understanding of how genes are inherited enables scientists to predict the likelihood of characteristics appearing in future generations. For example, scientists can use a family's medical history to predict whether a child is likely to be born with an inherited disease. Embryonic cells can now be tested to see if they carry the faulty genes that cause an inherited disease. Such developments provide parents with information to make decisions, such as whether to have children and whether to terminate a pregnancy. This involves making decisions which involve thinking about ethical issues as well as the health and social issues.

Tony's story

When I was a young boy my grandmother died of Huntington's disease. Her illness was terrible as she gradually lost her memory and could not recognise any of her family. Later my uncle was diagnosed with the same disease.

There is a test for Huntington's disease but my mum would not take it. She too developed the disease when she was in her 40s. I thought about having the test to see if I had Huntington's disease, but decided against having it. Instead I chose not to have any children in case I did have it and passed it on to them. But when my wife became pregnant I thought I had to have the test. I was relieved when I found out I did not have the disease. My relief was not just for myself but the thought that I could have passed on a fatal disease to a young child.

▲ This mother says, 'Caring for a child with a genetic disease is very demanding. We love our son very dearly but I'm not sure what we would have done if I had been tested when I was pregnant.'

Huntington's disease

You can't catch **Huntington's disease**. It is a condition passed from parents to their children by a faulty gene. Sadly there is no cure for the disease at present. The disease damages the brain and central nervous system. Unfortunately, the signs of the disease do not usually appear until people are 30–40 years old, by which time they have often already started a family.

Genetic testing

When there is a history of a serious inherited disease in their family, a couple may be offered IVF (*in vitro* fertilisation) treatment and genetic testing. This involves fertilising eggs and sperm from a couple under controlled laboratory conditions. The fertilised egg develops into a tiny ball of embryo cells which can be tested to see if they carry any faulty genes. Only embryos which are known to be free from inherited disease are then implanted into the woman. This process of testing embryos for genetic disorders is called **genetic screening**.

At present only couples with a history of certain inherited diseases, including Huntington's disease, are offered genetic testing and genetic screening so that they do not pass on the faulty gene. In the future these techniques may be used to avoid passing on a wider range of inherited disorders.

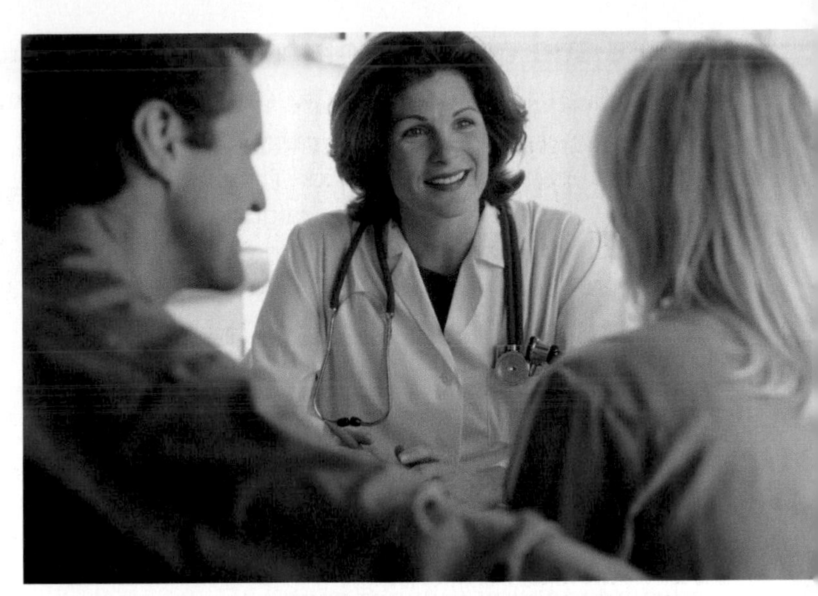

▲ Genetic counsellors provide advice on inherited conditions. This includes talking a couple through all the options they have available to them and what each option involves.

Should embryo screening be more widely available?

11th August 2005

The Human Fertilisation and Embryology Authority (HFEA) announced today that it is to seek the public's views on the appropriateness of using genetic testing to screen out genetic disorders such as inherited breast cancer.

The HFEA want to hear the views of patients, carers, affected families and disability groups as well as medical experts and the wider public.

Think about what you will find out in this section

How are chromosomes and genes passed from parent to offspring?	What are the pros and cons of using stem cells from embryos in research and treatments?
How can we predict and explain the results of genetic crosses?	What are the issues arising from embryo screening?
How are stem cells used to treat conditions such as spinal damage?	How can we explain the pattern of inheritance of genetic disease?

The code for life

Genes and chromosomes

In the nucleus of a cell there are a number of chromosomes. Chromosomes carry the genes that control inherited characteristics. The colour of your eyes, and whether you have straight or curly hair, are examples of characteristics produced by the genes you have inherited.

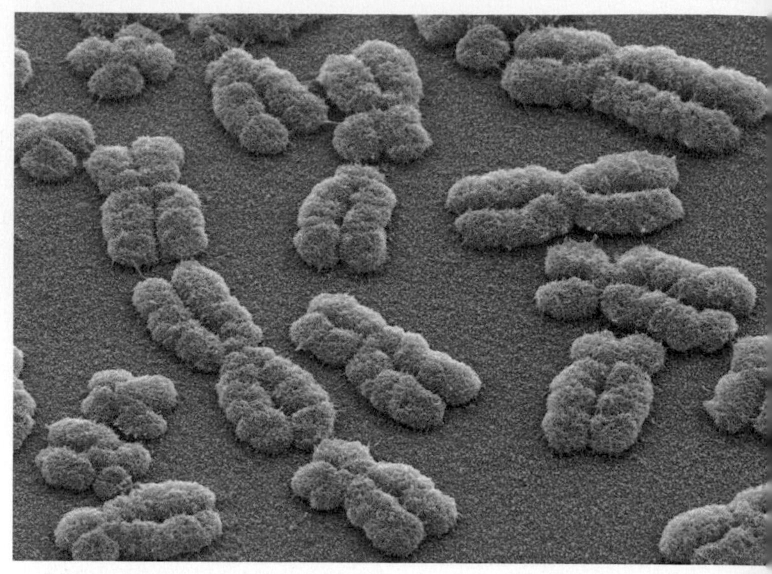

Chromosomes contain the DNA that controls all of an organism's inherited characteristics.

Genetic code

Chromosomes contain a substance called **DNA** (which stands for deoxyribonucleic acid). It is the DNA that contains the coded information for controlling inherited characteristics. The DNA in the fertilised egg that started your life contained all the coded information to make you!

Each gene is a section of a DNA molecule. Each DNA molecule is made of two very long strands. The two strands coil into a double helix like a twisted ladder. The diagram shows the structure of part of a DNA molecule. The strands are made from molecules of the same sugar repeated many times to form a polymer. The two strands are joined together by bases. There are four different bases which can be found in any order.

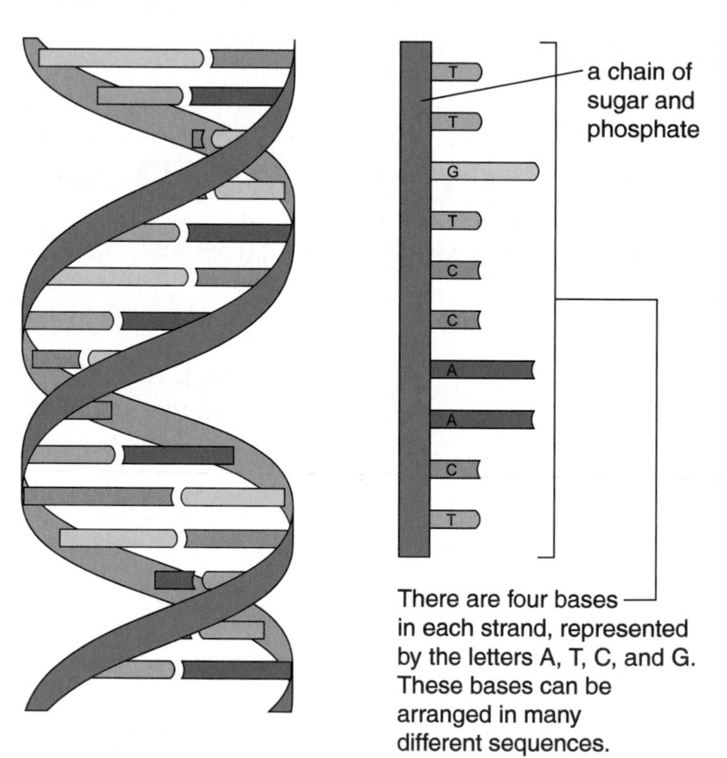
T
T
G
T
C
C
A
A
C
T
a chain of sugar and phosphate
There are four bases in each strand, represented by the letters A, T, C, and G. These bases can be arranged in many different sequences.

Each strand of DNA has bases attached to a long chain.

Question

a *What feature of the DNA structure can act as a genetic code? Use information from the diagram to explain your answer.*

Understanding the code

The code in DNA controls the production of proteins by cells. Each protein is made from amino acids joined together in a specific order. A protein may have over a hundred amino acids arranged in a specific order. If just one amino acid is out of sequence then the protein may not function properly.

The order of the bases in DNA creates a code that controls the order of amino acids in proteins. Each gene codes for a particular sequence of amino acids which make a specific protein. Proteins are the building materials of all living organisms. By controlling the production of proteins, DNA contains the code to control all the inherited characteristics of an organism.

Scientists from all over the world have been involved in working out the code for every protein in the human body. They found that there are only four bases in DNA. The four bases are represented by the letters A, T, C and G. Scientist have also worked out that a group of three bases acts as a code for each amino acid. The diagram (below) shows part of protein X. The code for this sequence of amino acids is also shown.

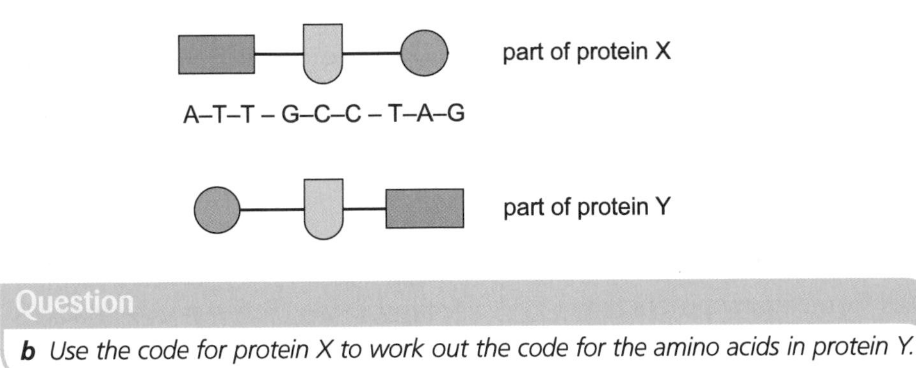

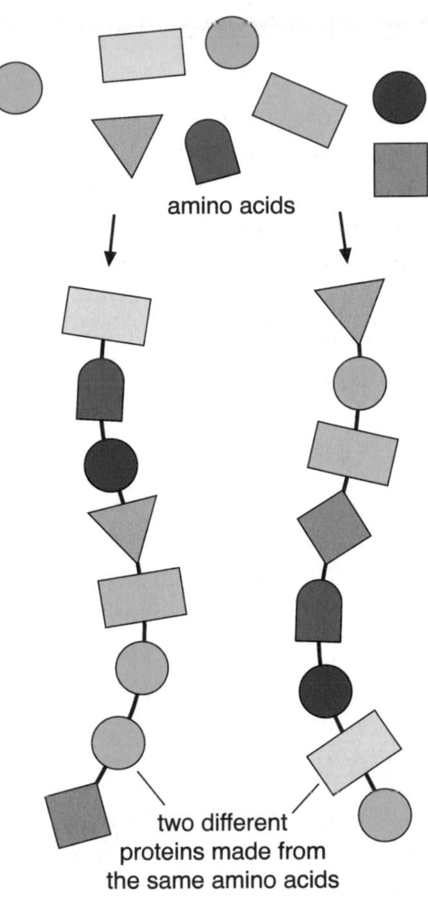

▲ If amino acids are put together in a different order, a different protein is 'spelled out'.

Question

b *Use the code for protein X to work out the code for the amino acids in protein Y.*

DNA fingerprinting

The DNA in your cells is as unique as your fingerprints. Unless you have an identical twin, your DNA is unique to you. Forensic scientists can extract DNA from tiny samples of tissue, such as a drop of blood or a piece of hair. The extracted DNA can then be broken up into sections that look like the bar code found on food packaging. Everyone's DNA bar code is different and produces their genetic fingerprint.

Question

c *A famous pop star was accused of being the father of a woman's child. Genetic fingerprinting was carried out using the DNA from the child, the mother and the accused man. The genetic fingerprints are shown in the diagram.*
(i) Looking at the results do you think that the pop star is the father of the child? Explain your answer.
(ii) Explain why only part of the child's genetic fingerprint matches that of the mother.

mother child pop star

Key points

- A gene is a small section of DNA.
- Each gene codes for a particular sequence of amino acids which make a specific protein.
- The DNA of each person is unique. The unique DNA can be used to identify individuals in DNA fingerprinting.

Passing on chromosomes

Cell division

Your life started from a single cell, formed by the fusion of a sperm and an egg. All the thousands of cells that make up your body developed from this single cell by cell division. During each division, all the chromosomes are copied so that every one of your body cells contains the same chromosomes and genes as the single cell formed at fertilisation. This type of cell division is called **mitosis**. Body cells divide by mitosis to produce new cells that are needed for growth and to replace damaged cells. The diagram shows how new cells are formed by mitosis.

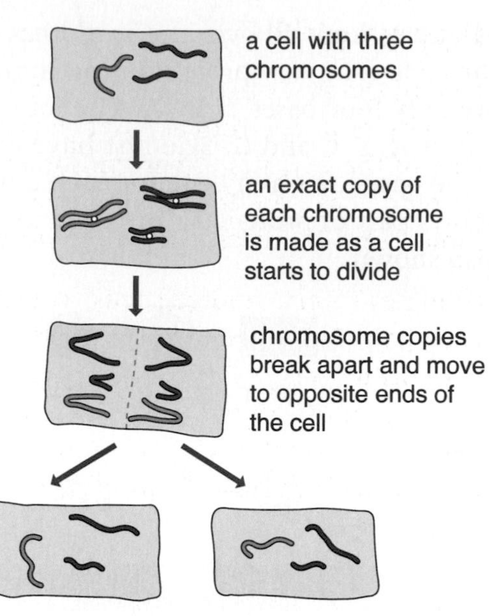
a cell with three chromosomes
an exact copy of each chromosome is made as a cell starts to divide
chromosome copies break apart and move to opposite ends of the cell
two new cells are formed, each containing exactly the same genetic information as the original cell

▲ Cells produced by mitosis are identical to the parent cell.

Halving the number of chromosomes

The chromosomes in body cells are found in pairs. For example, human body cells contain 23 pairs of chromosomes. Gametes (sperm or egg cells) are specialised sex cells formed in reproductive organs by a type of cell division called **meiosis**. During meiosis, the number of chromosomes is halved so that gametes contain only a single set of chromosomes. For example, human egg and sperm cells each contain 23 single chromosomes. The diagram shows how a cell divides to form gametes by meiosis.

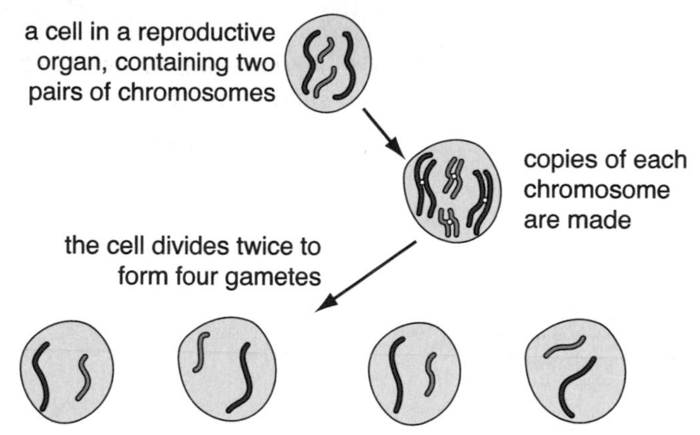
a cell in a reproductive organ, containing two pairs of chromosomes
copies of each chromosome are made
the cell divides twice to form four gametes
each gamete contains one chromosome from each pair

▲ Cells produced by meiosis contain one chromosome from each pair in the parent cell.

Questions

a *The number of chromosomes in a species is always the same. Imagine that gametes carried the total number of chromosomes for the species instead of half the number. How would the number of chromosomes change when gametes fused?*

b *The body cells of cats contain 19 pairs of chromosomes. How many chromosomes will be found in:*
(i) cat liver cells (ii) cat egg cells (iii) cat embryo cells?

Restoring the number of chromosomes

Male and female gametes fuse during fertilisation. The new cell formed will contain chromosomes from both gametes. This means that during sexual reproduction, variation occurs as the resulting mixture of genes provides a unique individual. In humans, the male gamete carries 23 chromosomes from the father and the female gamete carries 23 chromosomes from the mother. When the gametes fuse together, they form a single body cell with 46 chromosomes, arranged as 23 pairs. This body cell then repeatedly divides by mitosis to create a new baby.

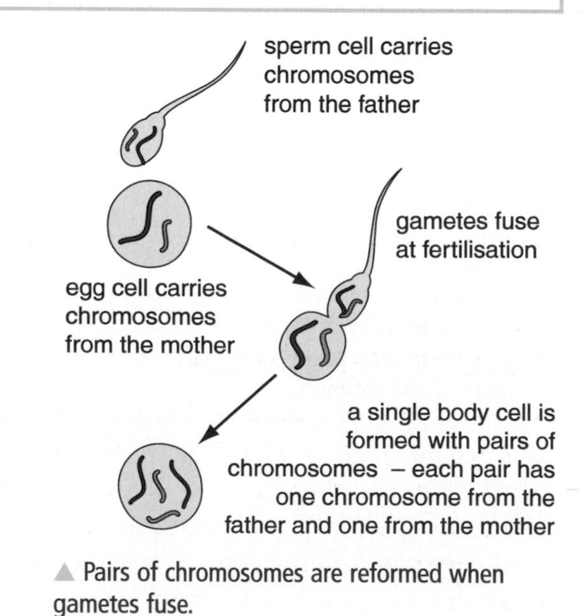
sperm cell carries chromosomes from the father
gametes fuse at fertilisation
egg cell carries chromosomes from the mother
a single body cell is formed with pairs of chromosomes – each pair has one chromosome from the father and one from the mother

▲ Pairs of chromosomes are reformed when gametes fuse.

Boy or girl?

One of the first questions people ask when a baby is born is whether it is a boy or a girl. Your sex depends on the chromosomes you inherit. One of the 23 pairs of chromosomes in your body cells is the sex chromosome pair. In human females, the sex chromosomes are the same (XX). In males, they are different (XY).

Gametes (sperm and egg cells) contain only one chromosome from each pair. This means that gametes will contain only one of the two sex chromosomes. The diagram shows the types of gametes formed in the production of human sperm cells and egg cells. Your sex is determined by the type of sperm cell (X or Y) that fertilised the egg that started your life. So your sex was determined by your father (see diagram below).

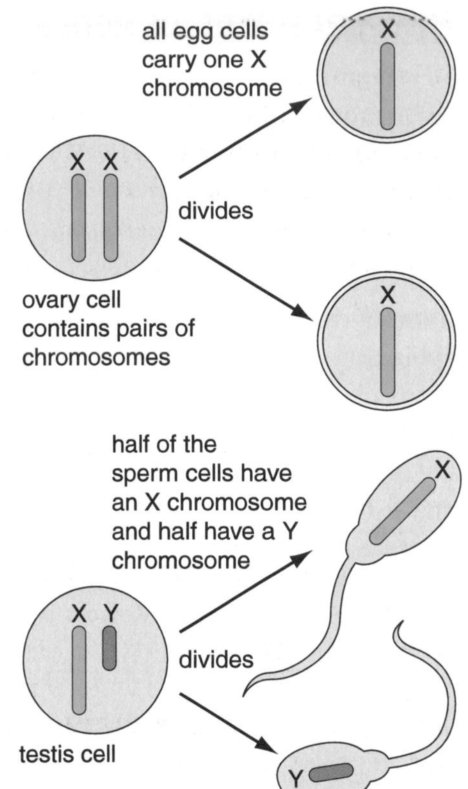

▲ The types of gametes formed in human sperm and egg cells.

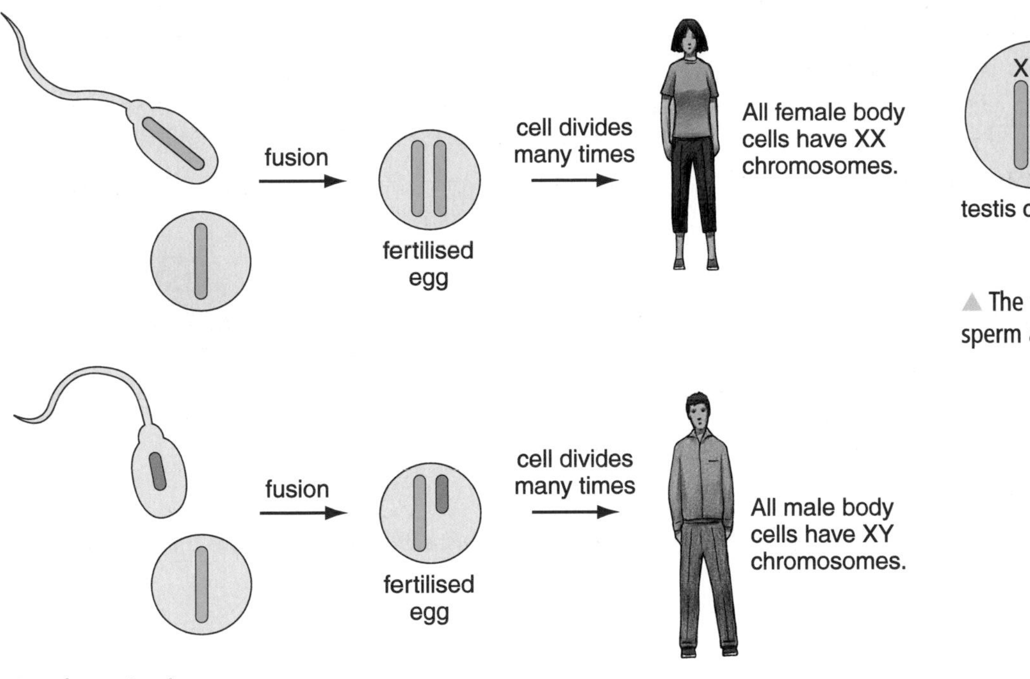

▲ Sex determination.

Question

c *Use the diagram of sex determination to explain why there is a 50% chance of a baby being a girl.*

Asexual reproduction

Meiosis does not occur during asexual reproduction as gametes are not involved. New offspring are formed from a parent by mitosis, producing offspring that are identical to the parent. Because gamete formation and fertilisation do not occur during asexual reproduction, no genetic variation occurs. An example is the 'runner' plantlets produced by strawberry plants.

Key points

- Additional cells are produced by mitosis as organisms grow.
- Cells in reproductive organs divide by meiosis to produce gametes.
- Body cells have two sets of chromosomes. Gametes contain only one set.
- There are 23 pairs of chromosomes in human cells. One of these is the pair of sex chromosomes.

Predicting and explaining inheritance

Different forms of genes

The diagram on the right shows the genes in a pair of chromosomes. Each pair of chromosomes carries genes for the same characteristic in the same place. This means that genes, like chromosomes, are found in pairs.

Genes have different forms called **alleles**. For example, the gene that controls the colour of a rabbit's fur has two alleles. The diagram below shows the alleles present in black and in white rabbits. The genetic make-up of an organism is called its **genotype**. The physical appearance of an organism is called its **phenotype**.

Question

a *What are the possible genotypes of rabbits with (i) white fur (ii) black fur?*

Dominant and recessive alleles

Alleles can be dominant or recessive. Rabbits will always have black fur if the B allele is present on either chromosome. The B allele is an example of a **dominant** allele because the black fur shows up even when the allele is only present on one chromosome.

A rabbit will have white fur only when both alleles are bb, and there is no B allele present. The b allele is an example of a **recessive** allele, because it only shows up if the dominant allele is absent.

An individual is described as being **homozygous** when both alleles are the same (e.g. bb or BB) and **heterozygous** when the alleles are different (e.g. Bb).

Passing on alleles

The diagram on the right shows how the alleles present in parents are passed on to offspring.

Question

b *The first generation rabbits in the diagram grew and developed. Draw a diagram to show the types of gamete that these rabbits will produce.*

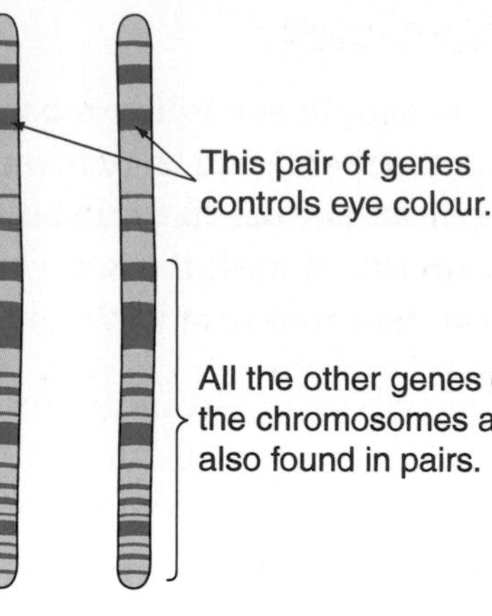
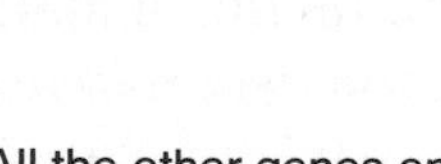

Rabbits are white only when both alleles are bb.

Rabbits with black fur can have either BB or Bb alleles.

White rabbits are homozygous for the recessive allele.

Black rabbits can be homozygous or heterozygous.

Homozygous means 'same alleles'.
Heterozygous means 'different alleles'.

▲ The letters B and b stand for the two alleles that control fur colour.

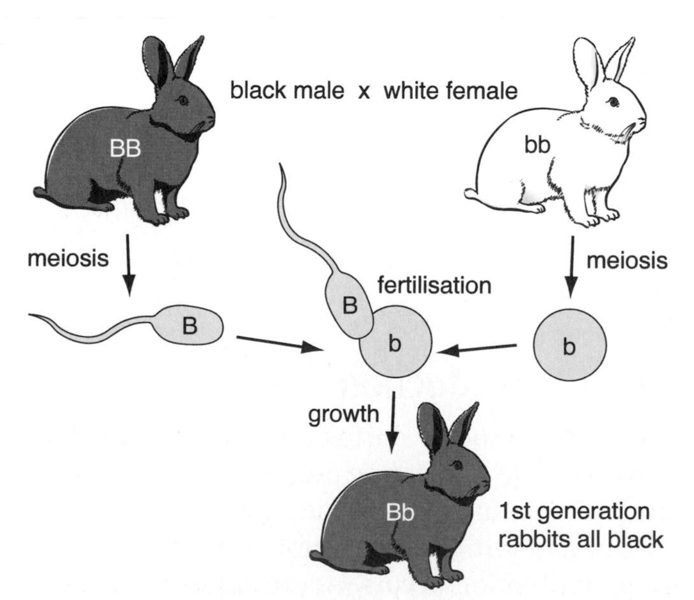

▲ The mating of a homozygous black rabbit and a white rabbit.

Predicting characteristics

Scientists use their knowledge of how alleles are passed on when organisms reproduce, to predict the characteristics of future generations. The diagram on the right shows the results of mating two black rabbits with the genotype Bb. Showing how alleles are passed on when organisms mate is called a **genetic cross**.

Questions

c *What percentage of the rabbits produced from this mating is expected to be white?*

d *When the same rabbits were mated again, a litter of six rabbits was produced with no white rabbits. Explain this result.*

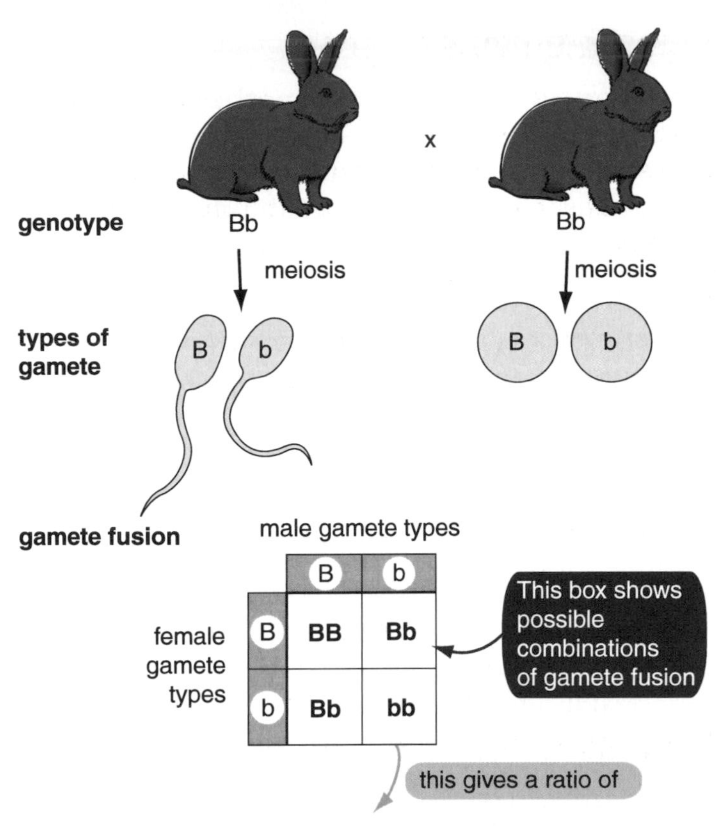

▲ The mating of heterozygous black rabbits.

Genetic diagrams

There is a standard way of explaining the results of a genetic cross. In pea plants a dominant allele R produces red flowers, and a recessive allele r produces white flowers. The diagram below shows what a cross between pea plants with white flowers (homozygous rr) and red flowers (heterozygous Rr) produces. Note how the diagram shows:

- how the alleles are passed on to gametes
- the possible ways that gametes combine
- the expected results.

This type of diagram is called a **genetic diagram**.

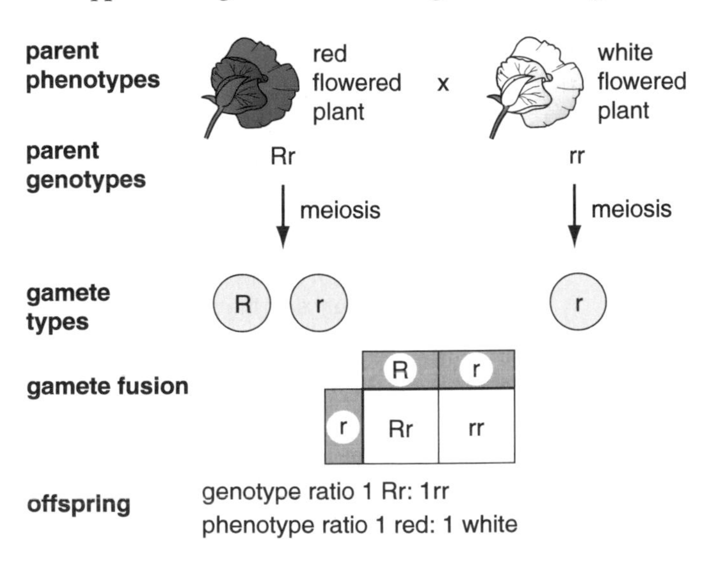

▲ The colour of flowers in pea plants is controlled by a gene. The dominant allele of this gene produces red flowers.

▲ Flower of a pea plant. This pea flower contains the dominant allele R.

Key points

- Different forms of the same gene are called alleles.
- When a cell divides by meiosis each gamete formed contains one of a pair of alleles.
- Gametes fuse randomly at fertilisation forming new pairs of alleles.
- Genetic diagrams are used to explain and predict the outcomes of genetic crosses involving dominant and recessive alleles.

Harmful genes

Inheriting illness

Most illnesses are caused when **pathogens** get into your body. For example flu and colds are diseases caused by virus infections. But not all diseases are caused by pathogens. Some health problems are caused by alleles that are passed on to children from their parents. These are called inherited conditions.

Cystic fibrosis

One child in every 2000 is affected by an inherited condition called **cystic fibrosis**. The photograph shows a girl who has cystic fibrosis. Her body produces a thick, sticky mucus. This mucus can block the air passages to her lungs. The blockages make breathing difficult and increase the risk of infections. Mucus also builds up in the digestive system preventing digestive enzymes being secreted and causing difficulty digesting and absorbing food.

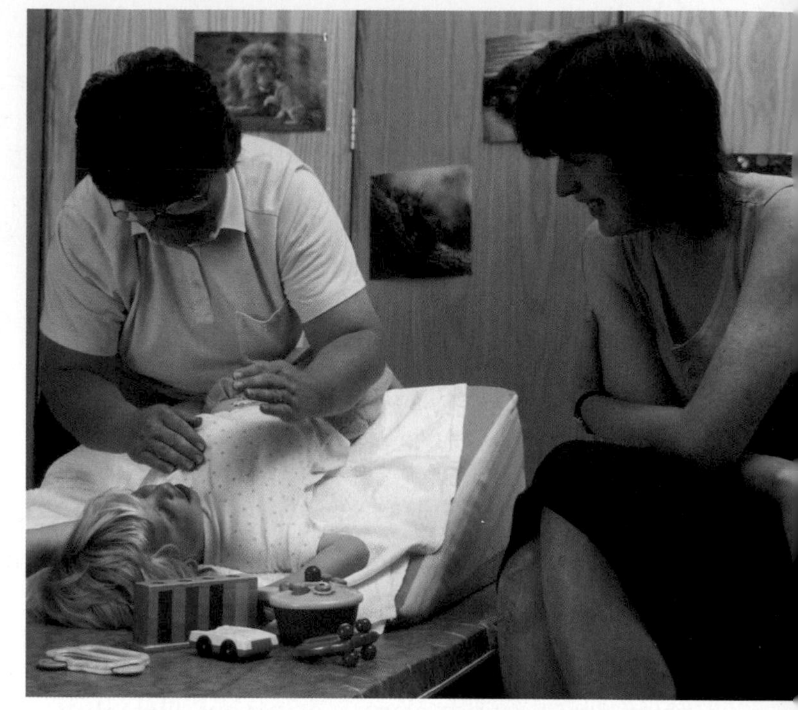

▲ People who have cystic fibrosis need regular physiotherapy to clear their air passages of mucus.

Treating cystic fibrosis

People with cystic fibrosis can be helped with regular physiotherapy and antibiotics. However, as yet there is no cure for cystic fibrosis. Scientists are now working on ways of correcting the faulty gene rather than treating the symptoms of the condition. Correcting the gene in this way is called **gene therapy**.

Gene fix for cystic fibrosis

A DNA spray has shown success in tackling cystic fibrosis.

The gene responsible for cystic fibrosis was discovered in 1989. Since then scientists have been trying to find a way of replacing the faulty gene with a healthy one – gene therapy.

The Cystic Fibrosis Trust says that the disease is the most common life-threatening inherited disorder.

Scientists are hopeful that research in gene therapy will produce a cure for this condition.

Inheriting faulty alleles

Cystic fibrosis is caused by a faulty recessive allele. People can be carriers of cystic fibrosis without showing any of the symptoms of the condition. The diagram of a family tree on the right shows how this allele can be inherited. The faulty recessive allele (shown in black in the diagram) is passed from both healthy parents to produce a child with cystic fibrosis.

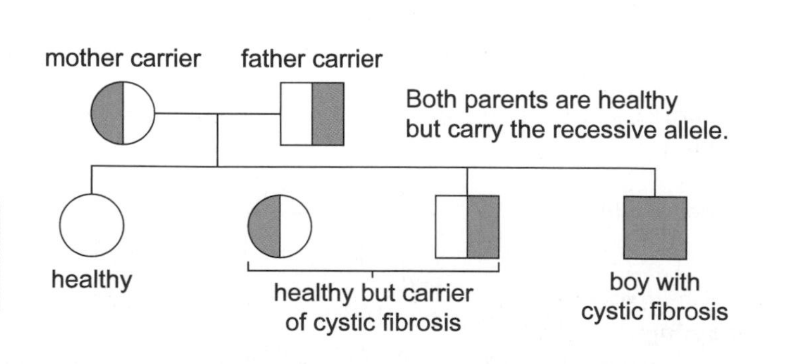

Question

a *A couple are both carriers of cystic fibrosis. What is the chance of them having a child with cystic fibrosis?*

Faulty dominant alleles

Huntington's disease is an inherited condition that damages the brain and other nervous tissues. Sadly, there is no cure. People with Huntington's disease need a lot of care and eventually die from the condition. Signs of the condition do not appear until people are 30–50 years old. This means that they may have children before finding out that they have a condition that can be inherited.

Huntington's disease is caused by a faulty dominant allele, so only one faulty allele needs to be present to cause the condition. It is inherited even if only one parent has the disease.

Interpreting genetic diagrams

The diagrams show the pattern of inheritance of cystic fibrosis and Huntington's disease. Study them carefully and then answer the following questions.

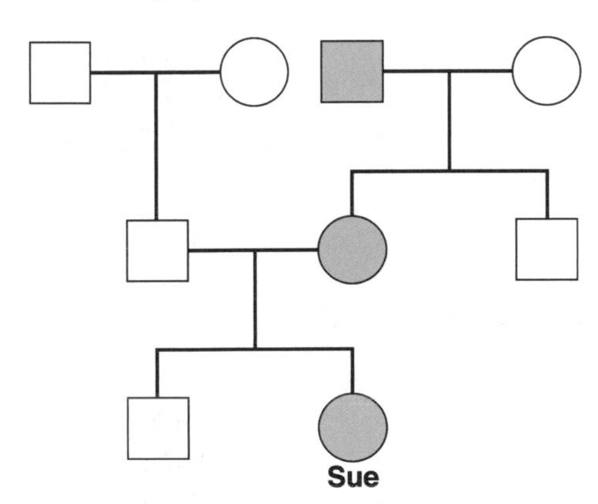

Huntington's disease can be passed on by one parent who has the disease.

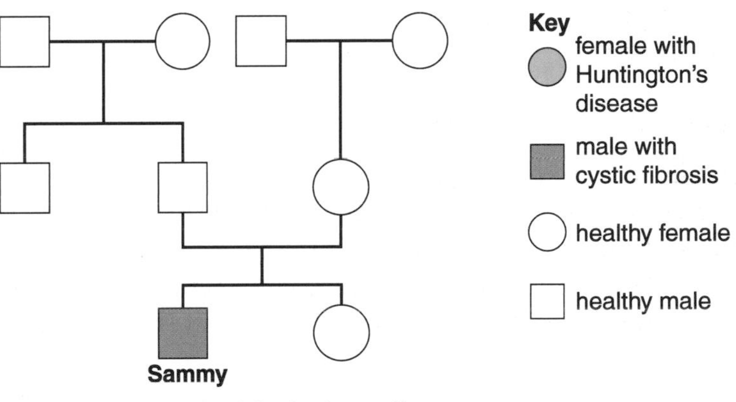

Cystic fibrosis can be inherited even if neither parent has the disease.

Questions

b *Use the symbols c (cystic fibrosis allele) and C (normal allele).*
(i) What is the genotype of Sammy?
(ii) Draw a genetic diagram to show how Sammy has inherited cystic fibrosis even though neither of his parents has the condition.

c *(i) Describe the difference in the pattern of inheritance of Huntington's disease and cystic fibrosis.*
(ii) Explain the reason for the difference in the pattern of inheritance of the two conditions.

Key points

- A family tree can be used to interpret the pattern of inheritance of certain diseases.
- Genetic diagrams are used to explain and predict the outcomes of genetic crosses involving alleles that cause inherited disease.
- Cystic fibrosis and Huntington's disease are both inherited disorders.

Mendel's work

Investigating inheritance

In the 1860s an Austrian monk, Gregor Mendel, carried out a series of very important investigations. His work started by his keen observation of garden plants. He noticed that certain characteristics were passed from generation to generation. To find out how characteristics were inherited, Mendel carried out investigations over 7 years.

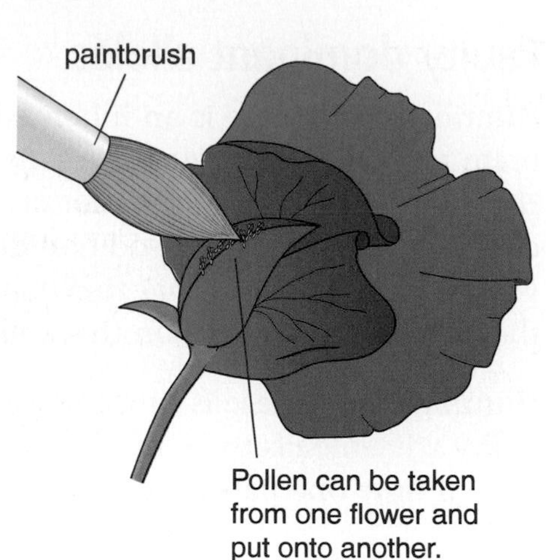

Pollen can be taken from one flower and put onto another.

Controlling plant breeding

In his investigations, Mendel used pea plants with distinctive characteristics, such as flower colour, height and type of seeds. He carefully controlled which plants reproduced by transferring pollen by hand from one selected plant to another. Mendel used hundreds of plants in each of his investigations.

One of Mendel's investigations into how flower colour is inherited in pea plants is shown in the diagram.

Finding an inheritance pattern

parent plants
Mendel crossed red-flowered pea plants with white-flowered plants. He used hundreds of plants.

×

first generation
All the plants had red flowers. Mendel then crossed these first generation plants with each other.

×

second generation
Mendel found that three times more plants had red flowers than white. By counting hundreds of plants, he found ther was a ratio of 3 red : 1 white.

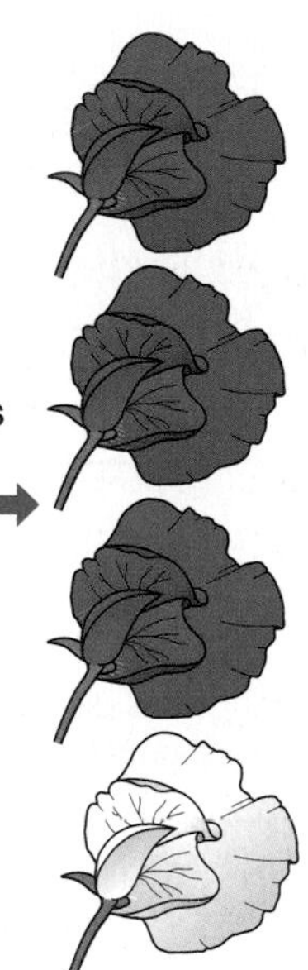

Questions

a *Why was pollen transferred from plant to plant?*
b *Explain why Mendel used a very large number of plants.*

Mendel identified a pattern in the results of his investigations. For example, he saw that some characteristics, such as white flowers, did not appear in the first generation. He also found that characteristics in the flowers in the second generation appeared in a specific ratio, such as 3 red : 1 white.

Mendel explained the pattern of inheritance by making the following conclusions.

- Some characteristics are controlled by a pair of 'inherited factors'. ('Inherited factors' are now called 'alleles'.)
- Only one 'inherited factor' is present in a gamete.
- 'Inherited factors' can be recessive or dominant.

The diagram below shows how these conclusions explain the results of his investigations.

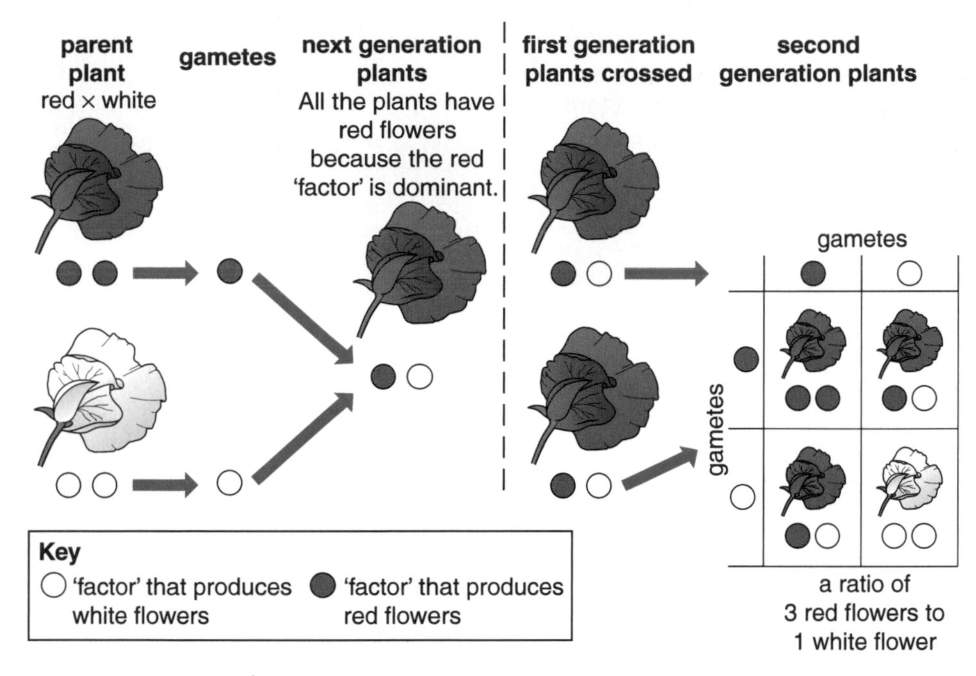

Question

c The results in the second generation were 315 red-flowered plants and 94 white-flowered plants. Mendel expected a 3 : 1 ratio. Explain why the observed results of 315 and 94 can be accepted as a 3 : 1 ratio.

Repeating his investigations

Having found an explanation of how characteristics are inherited, Mendel repeated his investigations using other characteristics of pea plants to confirm his findings. The diagram on the right shows an investigation carried out with plants with round and wrinkled seeds.

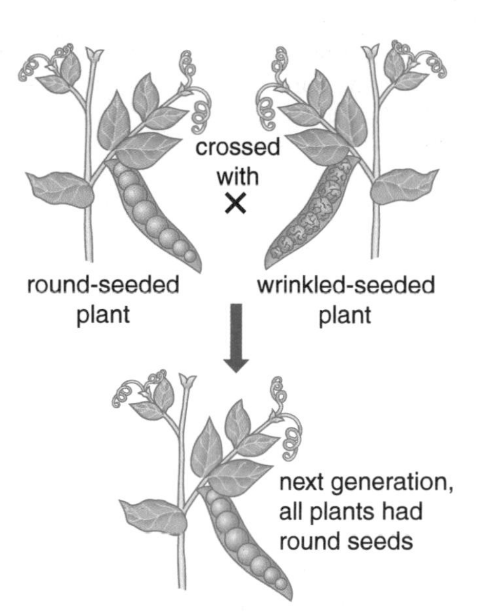

Question

d Draw a genetic diagram to explain the results of Mendel's investigation with round and wrinkled seeds. Use the symbols R and r in your diagram.

A scientist ahead of his time

From the results of his work Mendel developed what he called the 'laws of heredity' which today remain the basic principles of genetics. His work was so far ahead of its time that it was given little attention and was soon forgotten about. Over 30 years later, when scientists knew about chromosomes, they could then confirm Mendel's work and his explanation of 'inherited factors'. These 'inherited factors' are now called alleles.

Key points

- Mendel proposed the idea of separately inherited factors.
- Mendel's discovery and explanation of inheritance was not recognised until after his death.

Designer baby transplant success

July 27th 2004

A boy with a rare, life-threatening blood disorder has successfully been given cells from his perfect match 'designer baby' brother.

Five-year-old Charlie Whitaker needed a stem cell transplant as his only hope of survival. Charlie's baby brother was genetically selected while he was still an embryo. Stem cells were collected from the umbilical cord and used for Charlie's transplant.

Charlie Whitaker

Specialised cells

Your body is made from many different types of cell. Each type of cell, such as muscle cells, nerve cells and skin cells are all specialised to carry out certain functions in your body. Cells which have become specialised are called **differentiated cells** – they have become different so that they can carry out their specialised job. Many types of plant cell retain the ability to develop into different types of specialised cells.

However, animal cells, including human cells, become differentiated at an early stage. Once a cell has become differentiated it cannot develop into another type of cell. For example, when skin cells divide they only produce more skin cells. Cell division in mature cells is used to replace cells of the same type. Some highly specialised cells, such as nerve cells, are unable to continue to divide.

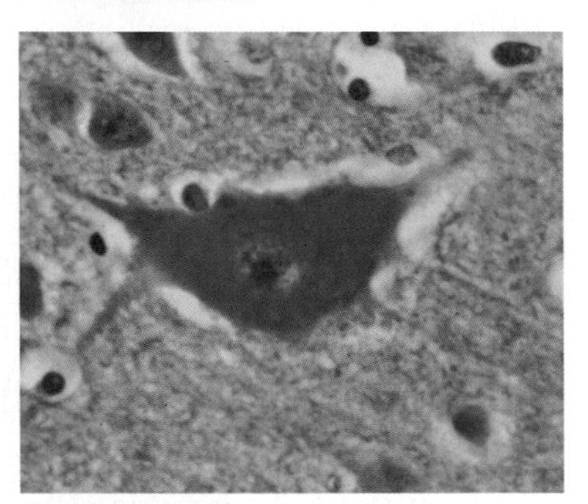

Brain cells are very specialised for transferring impulses. They are so specialised that they can no longer divide.

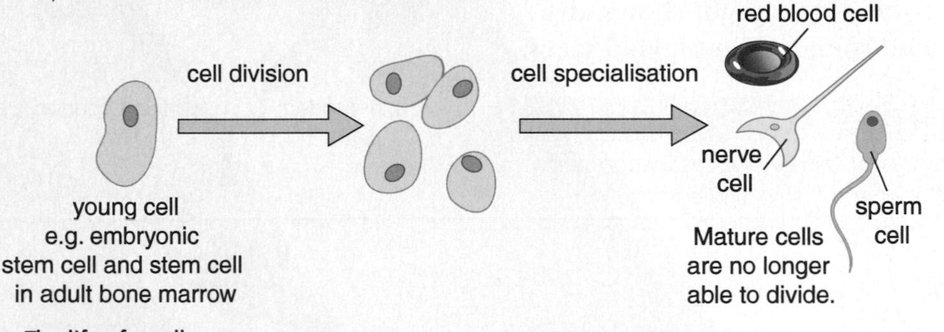

The life of a cell.

Stem cells

Stem cells are cells which have not yet become differentiated. When stem cells divide they can develop into a variety of different cells. Stem cells in the embryo can differentiate into all the cell types in the body – bone, brain, heart, etc. Embryonic stem cells first appear about a week after fertilisation. These are like the 'parent' cells of all the cells of the body.

A few types of cell don't become specialised and continue to be able to divide. For example, stem cells in the bone marrow divide to produce red cells, white cells and platelets in the blood. There are also stem cells in the blood of a baby's umbilical cord.

Question

a *Charlie Whitaker received stem cells from the umbilical cord of his brother, Jamie. Doctors discarded other embryos because they did not provide a tissue match for Charlie. Some people believe that 'designing' a child as a tissue donor is undesirable. Do you think people should be allowed to choose to have their children genetically selected? Explain your views stating the issues involved clearly.*

Using stem cells

Embryonic stem cells have enormous potential in medicine to replace damaged or diseased cells. By using stem cells, scientists are hopeful that a large number of very harmful diseases can be cured. For example, Parkinson's is a disease caused by damage to groups of brain cells. Scientists hope to replace damaged brain tissue by transplanting stem cells from an embryo into the affected part of the brain.

Stem cells could also be used to replace cells damaged by injury. Damage to the spinal cord in an accident can cause permanent paralysis. If stem cells could replace the damaged nerve tissue the paralysis could be reversed.

Stem cells from an embryo, bone marrow and umbilical cord can only be transplanted if they have the right tissue match. One way a tissue match can be achieved is to clone embryo cells. The diagram shows how cloning can produce stem cells to match the tissues of a person needing replacement cells.

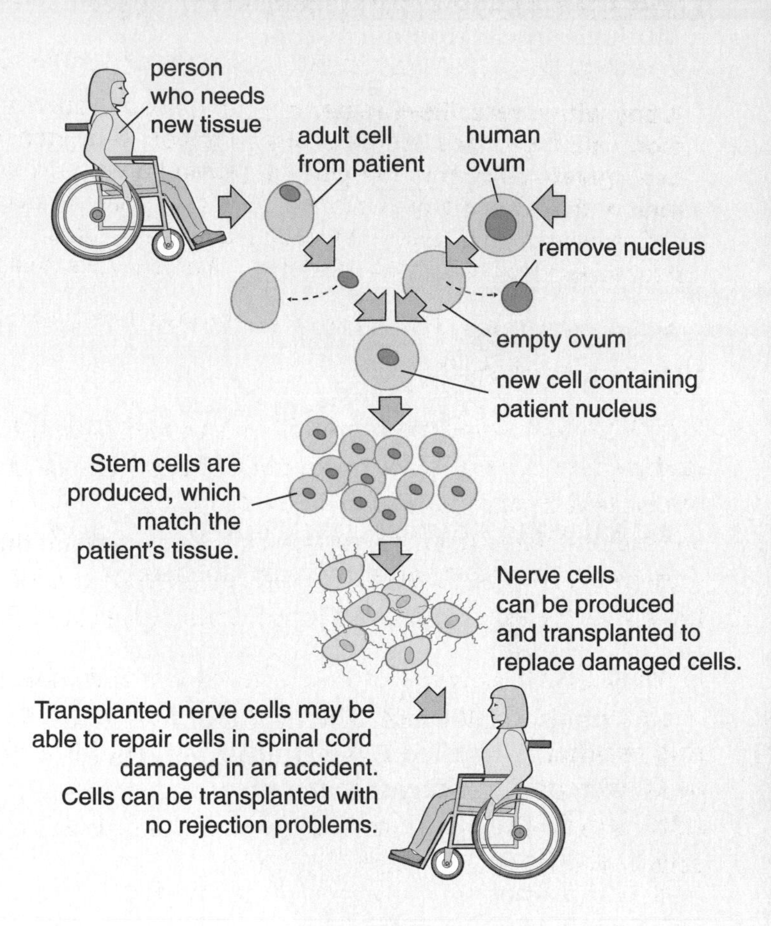
person who needs new tissue
adult cell from patient
human ovum
remove nucleus
empty ovum
new cell containing patient nucleus
Stem cells are produced, which match the patient's tissue.
Nerve cells can be produced and transplanted to replace damaged cells.
Transplanted nerve cells may be able to repair cells in spinal cord damaged in an accident. Cells can be transplanted with no rejection problems.

▲ Christopher Reeve, the actor who played Superman, was paralysed in an accident. He spent a lot of time persuading politicians to change laws to allow research into the use of embryonic stem cells.

Question

b *Explain why using stem cells from an umbilical cord raises fewer ethical problems than using embryonic stem cells.*

Key points

- Cells differentiate to produce specialised cells that carry out a certain job in the body. Differentiated cells lose the ability to divide.
- Stem cells from human embryos and adult bone marrow can develop to produce many different types of cell.
- Treatment with these cells may help conditions such as paralysis.
- We can make informed decisions about social and ethical issues relating to stem cells and embryo screening.

End of unit questions

1 The drawing shows part of a plant cell as seen through an electron microscope.

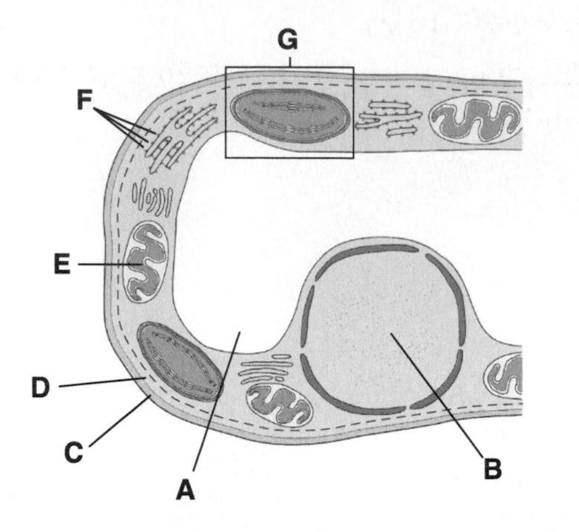

a Name the structures labelled **A–G**. *(7 marks)*

b Give the function of the part labelled:

i **E**

ii **F**

iii **G**. *(3 marks)*

c 1 µm is 1/1000 mm. The length of five of the structure labelled **E** were measured. These were the five measurements:

5.1 µm
5.5 µm
5.8 µm
5.4 µm
5.7 µm

Calculate the mean length of structure **E**. *(2 marks)*

2 The diagram shows the structure of the type of muscle which brings about movements in our limbs.

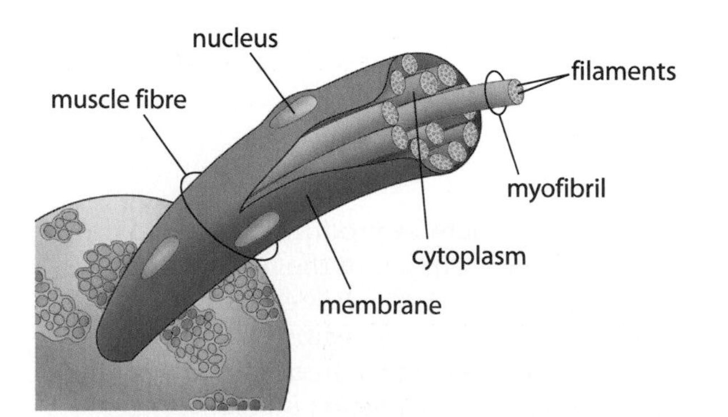

a Give **three** differences between a muscle fibre and a typical animal cell. *(3 marks)*

b There are large numbers of mitochondria in the muscle fibre. Explain the advantage of this to the muscle fibre. *(2 marks)*

c **i** Suggest the function of the filaments. *(1 mark)*

ii Suggest the advantages of having many filaments in a fibre. *(1 mark)*

3 **a** Explain what is meant by diffusion. *(2 marks)*

b The diagram shows four ways in which molecules may move into and out of a cell. The dots show the concentration of molecules.

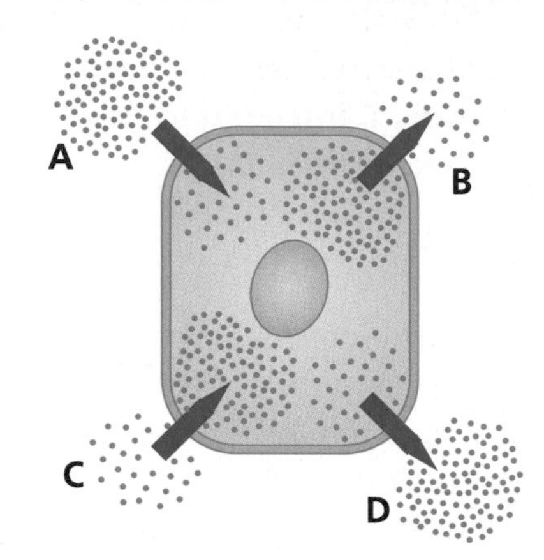

Which arrow, **A**, **B**, **C** or **D**, represents the movement of:

i carbon dioxide during photosynthesis?

ii carbon dioxide during respiration? *(2 marks)*

4 Jody set up an experiment. She used discs of potato, each cut to the same size. In batches of five, she dried the discs on paper towel then weighed them. She put one batch of discs into each of five beakers as shown below.

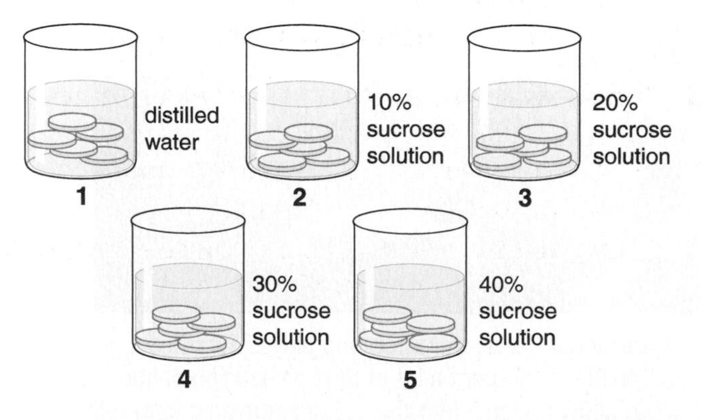

After two hours, she dried the discs then reweighed them.

a **i** What was the independent variable in Jody's experiment? *(1 mark)* ii What was the dependant variable in Jody's experiment? *(1 mark)*

iii Give **two** ways in which Jody tried to improve the reliability of her experiment. *(2 marks)*

b The table shows Jody's results.

i Copy and complete the table. Two calculations have been done for you. *(3 marks)*

	Beaker 1	Beaker 2	Beaker 3	Beaker 4	Beaker 5
Initial mass (g)	9.9	10.5	10.0	10.1	10.4
Final mass (g)	13.0	12.2	9.1	8.0	7.4
Change in mass (g)	+ 3.1				– 3.0
Percentage change in mass	+ 31.3				–28.8

ii Explain why Jody calculated percentage change in mass rather than just change in mass. *(2 marks)*

iii What type of graph should Jody draw to display her results? Explain the reason for your choice. *(2 marks)*

iv Draw a graph of the results. *(3 marks)*

c Give one way in which Jody could have improved the accuracy of her results. *(1 mark)*

d Explain why the discs gained mass in beaker **1**. *(2 marks)*

e i Use your graph to find the concentration of sugar solution in which potato discs would not change in mass. *(1 mark)*

ii Explain why potato discs would not change mass in this solution. *(2 marks)*

5 Match words **A**, **B**, **C** and **D** with the spaces **1–4** in the paragraph about photosynthesis.

A chlorophyll **B** carbon dioxide
C oxygen **D** starch

Light energy is absorbed by a green substance called ____1____ . This energy is used to convert ____2____ and water into glucose. ____3____ is given out as a waste product. Some of the sugar is converted into ____4____ . *(4 marks)*

6 a What is meant by a limiting factor? *(1 mark)*

b Name **three** factors that limit the rate of photosynthesis. *(3 marks)*

c The diagram shows plants in a greenhouse on a hot, sunny day.

Which factor is most likely to limit the rate of photosynthesis in these plants? Explain, as fully as you can, the reason for your answer. *(4 marks)*

d The graph shows the effect of light intensity, carbon dioxide concentration and temperature on the rate of photosynthesis.

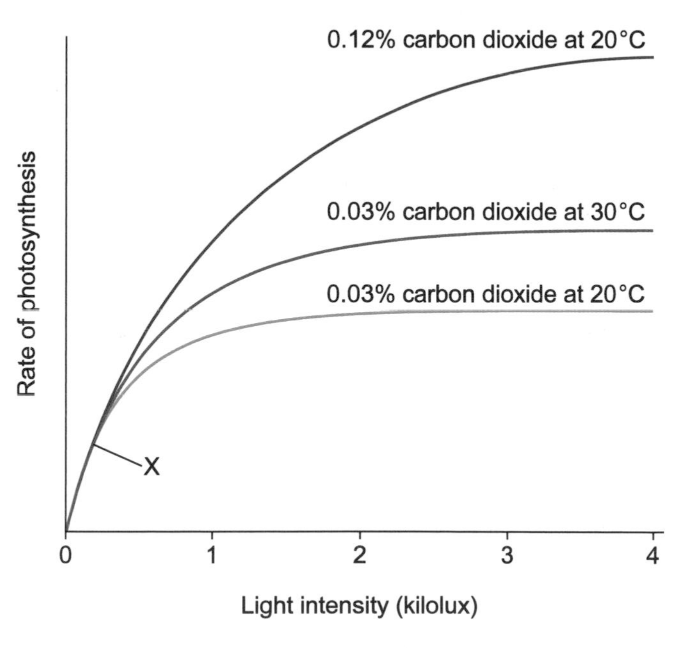

i Which factor is limiting the rate of photosynthesis at **X**? *(1 mark)*

ii In a greenhouse in winter the carbon dioxide concentration is 0.03 %, the temperature is 20 °C and the light intensity is 3 kilolux.
Using the data on the graph, predict whether increasing the carbon dioxide concentration to 0.12 % or the temperature to 30 °C would result in the greater increase in the rate of photosynthesis. Explain your answer as fully as you can. *(3 marks)*

7 a Soil conditions affect the growth of plants. Match each soil condition **1–3** to the effect **A–C** it will have on plants by listing pairs of numbers and letters.

1 soil has a very high concentration of ions
2 soil deficient in nitrate ions
3 soil deficient in magnesium ions

A stunted growth
B leaves yellow
C plants wilt *(3 marks)*

b The graph shows the effect of adding nitrogen fertiliser to the yield of a wheat crop.

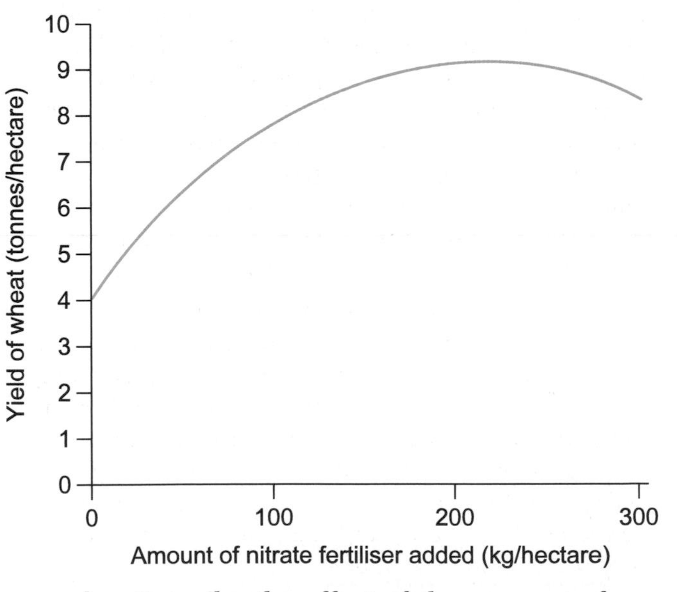

i Describe the effect of the amount of added fertiliser on the yield of the wheat crop. *(3 marks)*

ii Suggest **one** explanation for the yield at 300 kg per hectare of added fertiliser. *(2 marks)*

8 The diagram shows a food web for a wood.

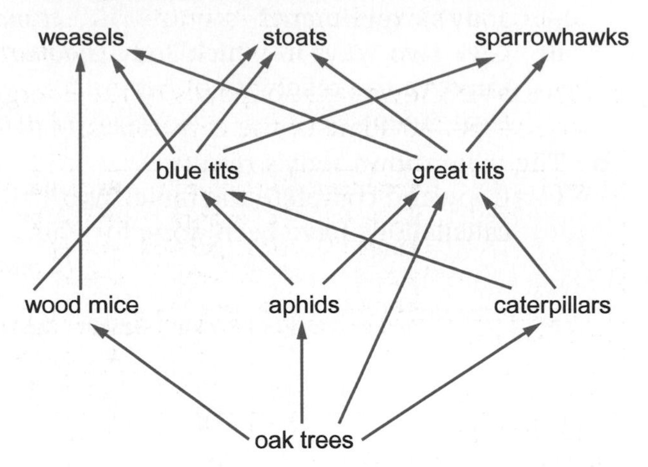

a Name:
i a producer
ii a primary consumer
iii a secondary consumer only
iv a tertiary consumer only. *(4 marks)*

b The table shows the biomass of each type of organism in the above food web.

Type of organism	Biomass (g/m²)
producer	177.0
primary consumer	12.5
secondary consumer	6.6
tertiary consumer	1.0

On graph paper, draw a biomass pyramid (to scale) for the data above. *(3 marks)*

c Explain, as fully as you can, why the biomass of the primary consumers is less than that of the producers. *(3 marks)*

9 A scientist estimated the flow of energy through an area of woodland. Her estimates are shown in the table.

	Energy (kJ per m² per year)
absorbed by trees	4 500 000
used by trees in photosynthesis	45 000
transferred to primary consumers	3000
transferred to secondary consumers	700

a i Calculate the proportion of the energy absorbed by the trees that was used in photosynthesis. *(2 marks)*

ii Explain why the proportion of energy transferred is so low. *(2 marks)*

b Describe what eventually happens to all the energy used by the trees in photosynthesis. *(5 marks)*

c The table shows what happens to the energy transferred to some of the consumers.

Animal	Percentage of energy absorbed into tissues	Percentage of energy incorporated into new tissue	Percentage of energy lost via faeces
small mammal feeding on plants	50	1.5	50
insect feeding on plants	40	16	60
small mammal feeding on animals	80	2	20

i Explain why the proportion of energy incorporated into new tissue by the insect is greater than that for either of the mammals. *(3 marks)*

ii Explain why the proportion of energy absorbed into the tissues is much greater in the mammal that feeds on animals than in the mammal that feeds on plants. *(3 marks)*

10 Waste plant material is often put on compost heaps. Complete the sentences by choosing the correct words from the list.

carbon dioxide cool decay dry
grow moist oxygen respire warm

In compost heaps, waste plant materials ___1___ because they are broken down by microorganisms. This process releases substances that can be used by other plants to ___2___. The waste plant materials are broken down faster when the conditions are ___3___ and ___4___. The microorganisms need ___5___ to ___6___. They break down carbon compounds into ___7___. *(7 marks)*

11 a The table gives the energy output from some agricultural food chains.

Food chain	Energy available to humans from food chain (kJ per hectare of crop)
cereal crop → humans	800 000
cereal crop → pigs → humans	90 000

Explain why the food chain *cereal crop → humans* gives far more available energy than the food chain *cereal crop → pigs → humans.* *(4 marks)*

b Explain, in terms of energy, why many farm animals are kept indoors in temperature controlled conditions. *(4 marks)*

c Read the passage about a pig farm in the USA.

Inside barns, long rows of grunting, snorting pigs fill every available space. Each row contains 100 animals – all pregnant or soon to be. Every animal faces the same direction in a scene of orderliness seldom associated with pigs.

The animals are not lining up by choice; each stands inside a narrow metal crate. The pigs, which can reach 270 kg, will spend much of their three or four years of adult life inside these crates, unable to turn around or even lie down fully because the stalls are just two feet wide. Only when caring for piglets will the sows live outside them for long, and then in different metal crates only slightly wider so they can recline to nurse. This farm outside Chicago is by all accounts a model of pork industry efficiency, cleanliness and productivity, and the metal 'gestation crates' are nothing unusual in the nation's highly industrialised pork business.

But critics of this kind of intensive pig farming – people ranging from animal welfare activists to academic researchers and some big pork buyers – have been raising increasingly pointed and sometimes emotional objections to the crates. Some call the practice inherently cruel, some call it offensive because the confinement produces abnormal behaviours in relatively intelligent animals, and some worry it could endanger the pork industry if consumers begin to focus on it.

Mainstream pork buyers are beginning to take note of the farm animal welfare issue. McDonald's – with its finely tuned understanding of consumers, especially the young – has assembled a task force of outside experts on animal welfare and production specialists to study whether pork suppliers should be required to find alternatives to sow crates.

Pig farmers believe they are being unfairly accused of ignoring the well-being of their animals. Industry officials say the individual stalls contribute to animal welfare by ensuring that all sows are fed, and by preventing otherwise frequent fighting among animals. What's more, they say, crates keep prices low by allowing many sows to be raised, watched and controlled in relatively small areas, and they improve food safety by keeping animals cleaner.

- **i** Give **three** advantages to farmers of keeping sows in gestation crates. *(3 marks)*
- **ii** Give **three** groups of people who are opposed to gestation crates. *(3 marks)*
- **iii** Give **two** reasons why some people are opposed to the use of gestation crates. *(2 marks)*
- **iv** Why might firms such as McDonalds put pressure on farmers to ban gestation crates? *(2 marks)*

12 An investigation was carried out to find out the effect of bile on the action of lipase. Four test tubes were set up as follows.

Test tube	Contents
1	milk, lipase, pH indicator, bile
2	milk, lipase, pH indicator,
3	milk, boiled lipase, pH indicator, bile
4	milk, lipase, pH indicator, boiled bile

Bile is alkaline. The pH indicator is yellow when the pH is 7 or less and red when the pH is over 7. The time taken for the indicator to change colour is shown in the table.

Test tube	Time taken for the pH indicator to change colour (minutes)
1	15
2	38
3	No change
4	15

- **a** What colour was the indicator in test tube 1 at the start of the investigation? Explain your answer. *(2 marks)*
- **b** Explain why the action of lipase caused the indicator to change colour. *(3 marks)*
- **c** Explain why there was no colour change in test tube 3. *(2 marks)*
- **d** What do the results from test tubes 1 and 2 tell you about the effect of bile on the reaction? *(1 mark)*
- **e** One student concluded that bile contains enzymes that digest fats. Which of the results shows that this conclusion is incorrect? Explain your answer. *(2 marks)*

13 Scientists working for a washing powder manufacturer carried out tests on a new protease enzyme that removes protein stains, such as egg and blood. They wanted to find out if the protease was suitable to use in washing powders.

They placed equal sized cubes of egg white into test tubes. The test tubes were placed into water baths and kept at different temperatures ranging from 0 °C to 60 °C. The same volume of the enzyme was added to each tube. The scientists recorded the time taken for the egg white to be digested. The following graph shows their results.

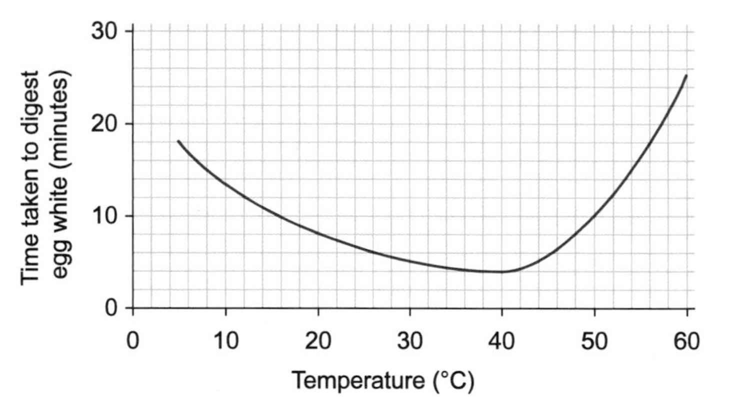

- **a** How long did it take to digest the egg white at 20 °C? *(1 mark)*
- **b** Explain why the scientists used equal sized cubes of egg white. *(1 mark)*
- **c** Explain the change in the rate of digestion
  - **i** between 5 °C and 40 °C *(1 mark)*
  - **ii** between 40 °C and 60 °C. *(1 mark)*
- **d** Is this new protease suitable for use in washing powders? Use the results of this investigation to explain your answer. *(2 marks)*

14 The rate of respiration of an organism can be investigated using a respirometer. A respirometer measures the amount of oxygen used in respiration.

The diagram shows a respirometer containing germinating seeds. The amount of oxygen that is used during respiration is measured by the movement of the coloured liquid in the capillary tube.

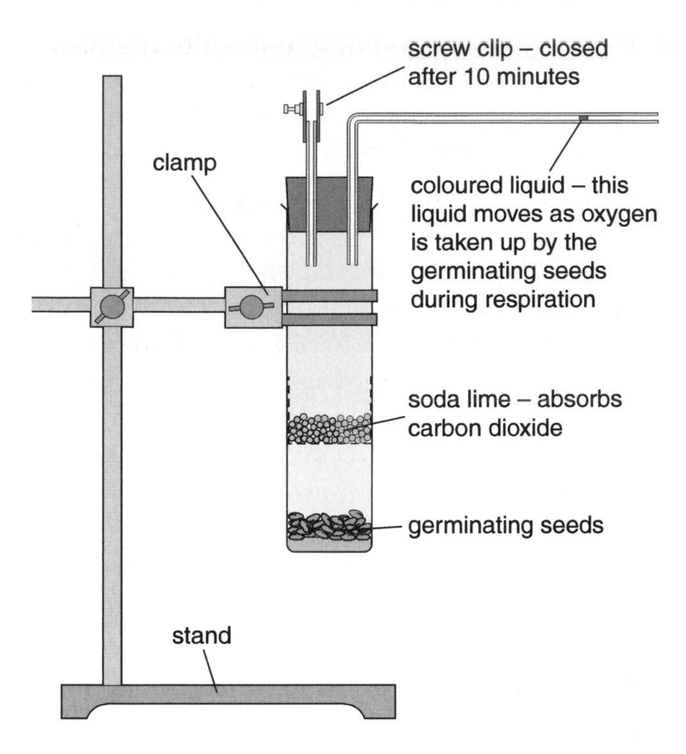

The respirometer was used to investigate the effect of changing temperature on the rate of respiration in germinating seeds. The respirometer was placed in a water bath at 20 °C with the clip open. After 10 minutes the clip was closed and the position of the liquid in the capillary tube was recorded every 5 minutes. The investigation was then repeated with the water bath set at 30 °C. The results are shown in the graph.

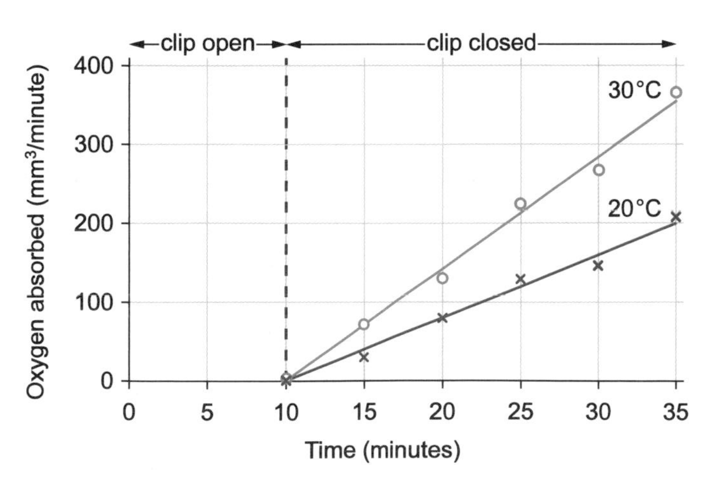

a Explain why the respirometer was left in the water bath for 10 minutes before closing the clip. *(1 mark)*

b Explain why the liquid moved after the clip was closed. *(2 marks)*

c Use the graph to calculate how much oxygen was used per minute at:

i 20 °C

ii 30 °C.

Show your working. *(4 marks)*

d Respiration is controlled by enzymes. Use the results to explain the effect of temperature on the rate of respiration in the germinating seeds. *(1 mark)*

15 a A boy eats food containing 14 000 kJ of energy each day. His body uses 80% of this energy to maintain core body temperature.

i Name the process which releases energy from food. *(1 mark)*

ii Calculate the amount of energy that the boy would use each day to maintain his core body temperature. *(2 marks)*

b Describe how structures in the skin help to cool the body on a hot day. *(5 marks)*

c The diagram shows the mean gain and loss of water in a day.

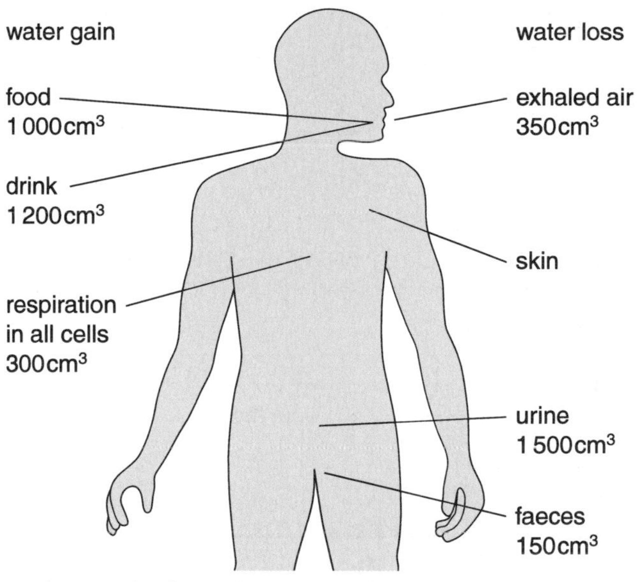

i Calculate the mean daily loss of water from the skin. *(2 marks)*

ii How would a person keep the amount of water at a steady level in their body even when they lose a lot of water on a hot day? *(2 marks)*

16 a Describe how the human body monitors changes in body temperature. *(3 marks)*

b When heat production is low and the external temperature is low, the core body temperature stays constant. Describe two ways that the body prevents core body temperature decreasing. *(2 marks)*

17 The glucose tolerance test is used in hospitals to test for diabetes. A patient being tested must not eat for several hours before the test. At the start of the test the patient is given a glucose drink containing 50 g of glucose dissolved in 150 cm^3 water. The concentration of glucose in the patient's blood is then measured over the next 2–3 hours.

The graph shows the results of a healthy person, a person with severe diabetes and a patient who is suspected of having diabetes.

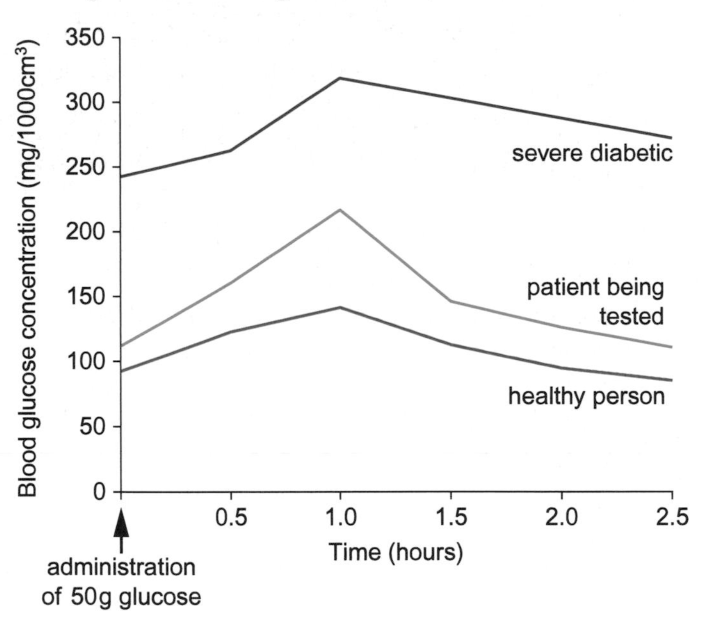

a Explain why patients must not eat for several hours before the test. *(1 mark)*

b Explain why the blood glucose level in the healthy person:

i rises between 0–1 hour *(1 mark)*

ii falls between 1–2.5 hours. *(2 marks)*

c What do the results show about insulin production in the severe diabetic? Explain the evidence for your answer. *(2 marks)*

d What do the results of the patient being tested show? Use the results to support your answer. *(2 marks)*

18 *Streptocarpus* is a common type of house plant with colourful flowers. Plant growers can produce new plants by taking leaf cuttings. This is shown in the drawings.

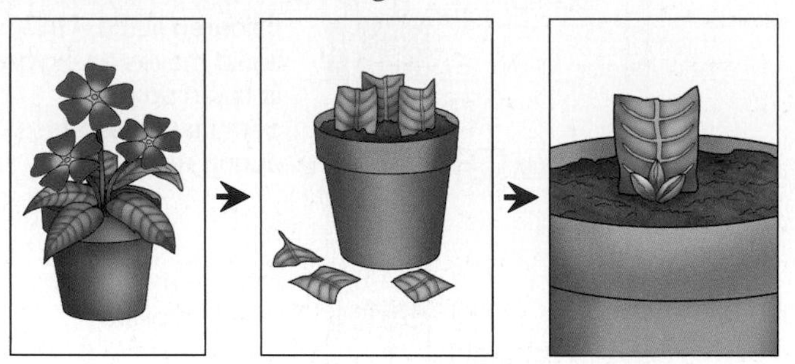

a Name the type of reproduction involved in producing plants from leaf cuttings. *(1 mark)*

b Explain why all the cuttings taken from a single leaf grow into plants which are similar in appearance. *(2 marks)*

AQA 2002

c Explain why plants grown from seeds often have different characteristics. *(3 marks)*

19 The diagram shows the result of cell division. The parent cell contains two pairs of chromosomes.

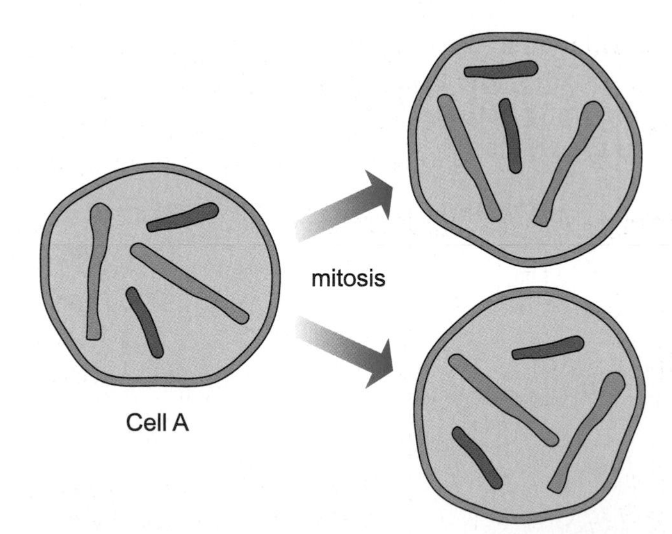

a Describe **two** reasons evident in the diagrams which show that this division is by mitosis. *(2 marks)*

b Describe why mitosis is important in living organisms. *(1 mark)*

c Draw a cell that could be formed by cell A dividing by meiosis. *(2 marks)*

20 The diagram shows fertilisation and development involving different types of gametes.

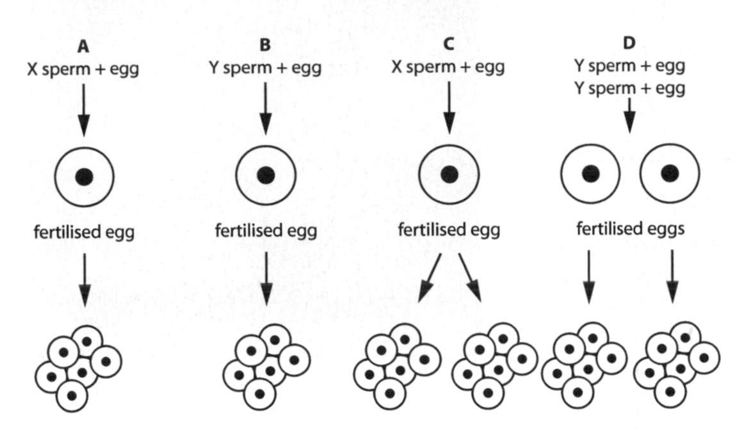

a Which types of fertilisation and development (**A**, **B**, **C** or **D**) would produce:

  i one male baby *(1 mark)*

  ii female twins? *(1 mark)*

b Name the type of cell division involved in:

  i producing sperm cells *(1 mark)*

  ii producing embryo cells from fertilised eggs. *(1 mark)*

c Explain why both groups of embryo cells produced in fertilisation **C** contain identical genes. *(2 marks)*

d In fertilisation **D** the mother released two eggs, both of which were fertilised. Explain why the twins produced were not identical. *(3 marks)*

21 The diagram shows the inheritance of a disorder called PKU in two families. A person with PKU develops a very high concentration of a particular amino acid in their blood. This can eventually lead to severe brain damage. PKU is caused by a recessive allele. A person who is a carrier shows no ill effects.

Use the letter D for the dominant allele and d for the recessive allele.

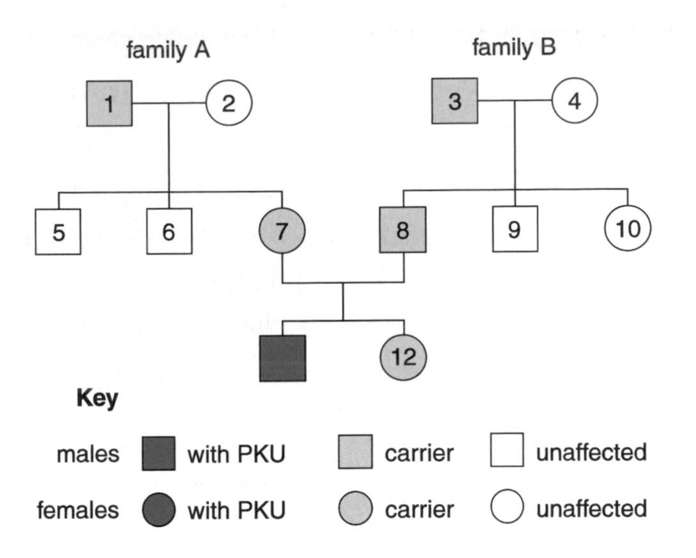

a What is the genotype of person **1**? *(1 mark)*

b What is meant by the term 'carrier'? *(2 marks)*

c Draw a genetic diagram to explain how person **11** inherited PKU when neither of his parents had the disease. *(4 marks)*

d Person **7** is pregnant with her third child. What is the chance that this child will have PKU? *(2 marks)*

22 A mother has undergone a new medical procedure as she gave birth. Stem cells were collected from her baby's umbilical cord and stored for possible use in the future. The mother said that using stem cells in this way could save her child's life in the future. She said it was like having an insurance policy for her child.

'Stem cells are undifferentiated cells that act like the master cells of the body,' her doctor said. 'These cells can be used to treat heart problems and various forms of cancer.'

Stem cells can also be obtained from embryos produced by IVF. Using stem cells from embryos raises more ethical issues than using stem cells from the umbilical cord.

a Explain what is meant by 'stem cells are undifferentiated cells'. *(2 marks)*

b Why are stem cells regarded as the body's 'master' cells? *(1 mark)*

c Suggest **three** reasons why using stem cells from the umbilical cord raises fewer ethical problems than using stem cells from an embryo. *(4 marks)*

Exchanging materials

Specialised organs

Substances are constantly moving in and out of the cells of living organisms. Most substances diffuse in and out of cells. Diffusion is efficient only over very small distances so large, multicellular organisms need specialised organs for exchanging and absorbing substances. For example, your lungs are specialised for gas exchange, and the lining of your small intestine is specialised for food absorption. The roots of plants are specialised for absorbing water and mineral ions and the leaves for absorbing carbon dioxide.

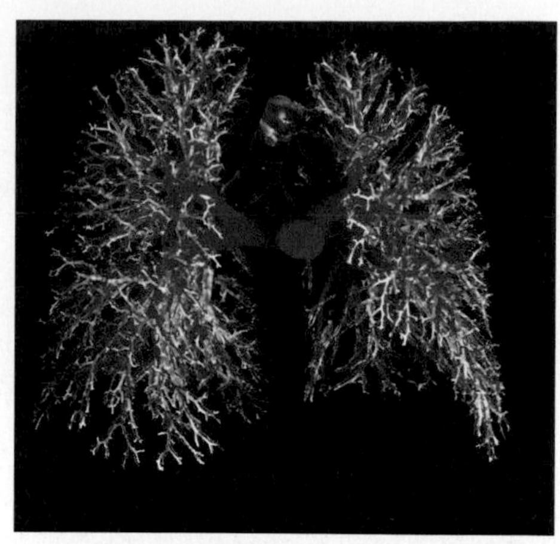

▲ This photograph shows the fine network of blood vessels in the lungs. Each alveolus is covered by a fine network of blood capillaries.

To move these substances around the body, organisms also need a transport system. In the human body, substances are transported by the blood system, pumped around by the action of the heart.

Exchanging gases

The air that enters the lungs is taken into tiny air sacs called **alveoli**. The alveoli are where oxygen diffuses into the blood and carbon dioxide diffuses out of the blood. Each alveolus is only 0.2 mm in diameter and your lungs contain about 700 million of them. The surface area of all of the alveoli is about 80 m^2 – a very large area for gas exchange. Your skin has a surface area of just 2 m^2 which is very small in comparison.

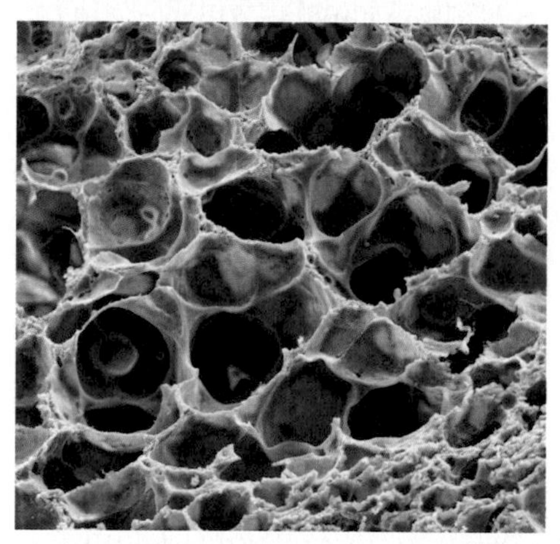

▲ These tiny air sacs are a very efficient surface to absorb all the oxygen your body needs.

Pumping blood

Your heart is a remarkable organ. It beats over 100 000 times a day, around two and a half thousand million times in your lifetime – and never tires like other muscles. If the heart stops beating as a result of a heart attack, other body organs are also damaged. When the heart is damaged, organs do not get the oxygen and other substances they need. Fortunately, doctors can monitor the health of a person's heart using sensitive monitors.

▶ Doctors use sensitive monitors to check the health of a person's heart. A healthy heart pumps blood to all body organs, providing useful substances and removing waste.

Transporting water

All large organisms need a transport system to carry substances around their body. Plants have a transport system too. The tree in the photograph is over 100 metres tall, so substances need to be transported over large distances to reach all its tissues.

Plant roots absorb water and mineral ions. Roots have specialised cells, called **root hairs**, which are long and thin and have a large surface area for absorbing water and mineral ions. Water and ions are transported from the roots, up the stem to the leaves.

▲ Water and mineral ions absorbed by the roots of this Californian redwood need to be transported to the leaves over 100 metres above the ground.

▲ Just behind the growing tips of roots are microscopic hairs. These are specialised for absorbing water and mineral ions.

Think about what you will find out in this section

How substances move by diffusion, osmosis and active transport.	How carbon dioxide enters leaf cells by diffusion.
Diffusion in the alveoli, and active transport and diffusion in the small intestine.	How plants increase the efficiency of absorption by roots and gas exchange in leaves.
How the heart pumps blood around the body in a double circulatory system.	How water and mineral ions are absorbed by root hair cells.
How blood transports food, oxygen and waste products.	How water vapour is lost from leaves by transpiration.

Movement in and out of cells

You will have met diffusion and osmosis in a previous module. Here we introduce another type of movement called **active transport**. Many different types of substance move in and out of the cells of living organisms. The diagram shows some of the substances that are regularly passing in and out of cells. All cells are surrounded by a cell membrane, which forms a barrier. Any substance entering or leaving cells must pass through this barrier. There are three main ways in which substances move in and out of cells: diffusion, osmosis and active transport.

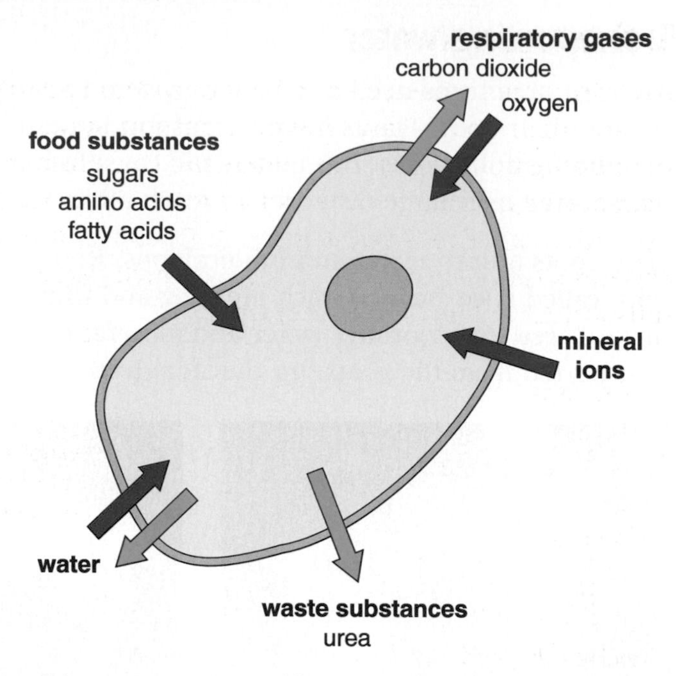

▲Substances entering and leaving a cell.

Substances on the move

Molecules of gases such as oxygen and carbon dioxide are moving about all the time. The molecules of a solution, such as glucose dissolved in water, are also moving in all directions. When molecules move, they spread themselves out evenly. This causes them to move from a higher concentration to a lower concentration until the concentrations become the same. This process is called diffusion. As they move from a higher to a lower concentration, molecules diffuse down a **concentration gradient**. Most substances move into and out of cells by diffusion.

Movement of water molecules

Water moves in and out of cells by a process called osmosis. Osmosis is the movement of water molecules from a higher concentration of water molecules to a lower concentration of water molecules through a **partially permeable membrane**.

Effect of osmosis

As water enters a cell it makes the cell swell up like a blown-up balloon, causing the cytoplasm to push against the cell membrane. When water leaves a cell it deflates because it contains less fluid. The diagram shows the effect of placing red blood cells in different concentrations of solution.

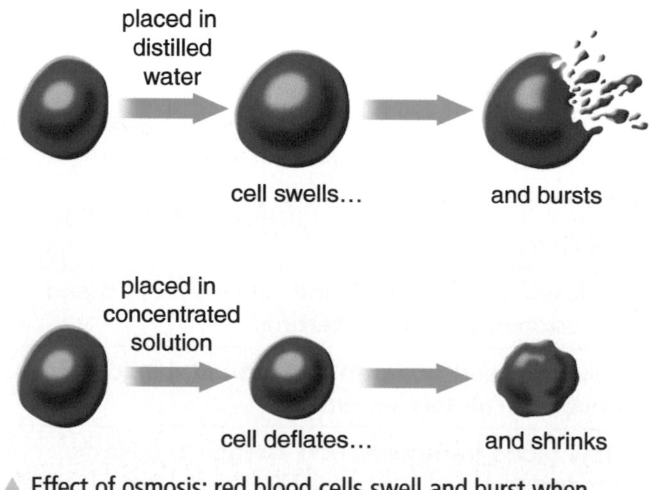

▲ Effect of osmosis: red blood cells swell and burst when placed in water.

Questions

a *Explain why red blood cells shrink when they are placed in a concentrated sugar solution, and burst when they are placed in distilled water.*

b *Explain why a plant cell placed in a dilute sugar solution does not burst.*

Active transport

Sometimes molecules need to be moved from an area of lower concentration to an area of higher concentration. This means moving them against a concentration gradient. This requires energy, which is supplied by respiration. Moving molecules using energy from respiration is called active transport. Respiration is carried out in tiny structures called mitochondria. Cells which use energy for active transport have lots of mitochondria.

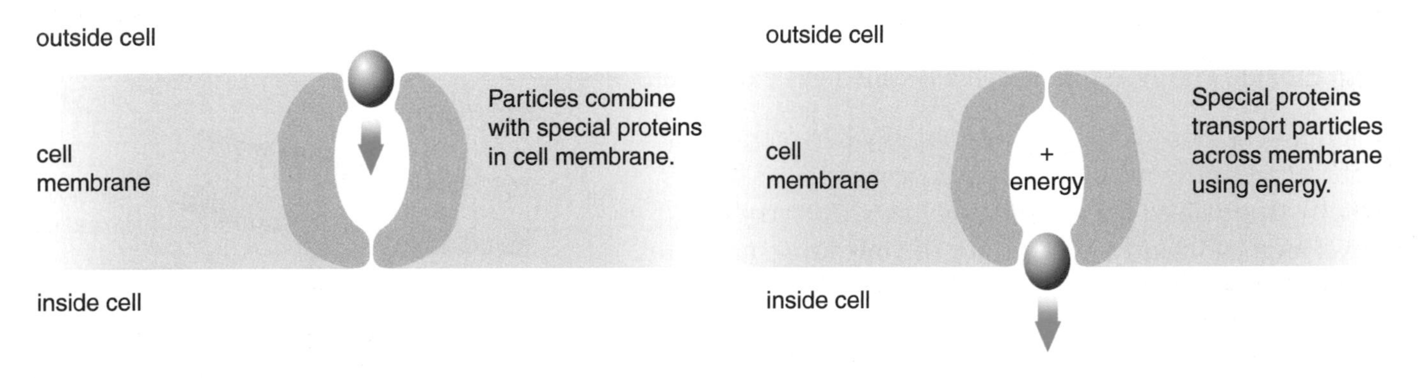

▲ Active transport.

Plants need nitrate ions to make proteins for growth. Nitrate ions are absorbed by active transport from the soil. There is a higher concentration of nitrate ions inside root cells than in the soil surrounding the root. Plants use energy to actively transport nitrate ions across the cell membrane into root cells against a concentration gradient.

Questions

c *Copy and complete the table to compare the ways in which substances pass in and out of cells.*

d *Potassium ions are actively transported into nerve cells. Explain why nerve cells contain large numbers of mitochondria.*

Diffusion	Osmosis	Active transport
Random movement		Selective movement
	From higher to lower concentration	From lower to higher concentration
Along a concentration gradient	Along a concentration gradient	
No energy is needed from respiration		Energy is needed from respiration

Key points

- Dissolved substances move by diffusion, osmosis and active transport.
- When substances move against a concentration gradient by active transport, energy is needed.

Organs involved in exchange

The lungs and the digestive system are involved in exchanging substances. The lungs exchange the gases oxygen and carbon dioxide. In the gut, the small intestine is responsible for absorption of food into the bloodstream.

Exchanging gases

As you breathe you take in the oxygen you need for respiration, and you also get rid of waste carbon dioxide. In your lungs you exchange the oxygen you need for the carbon dioxide you don't need. This is called gas exchange.

Oxygen and carbon dioxide diffuse rapidly between the air in your lungs and your blood. This is because your lungs are highly specialised for exchanging gases. To carry out gas exchange efficiently, the surface of your lungs needs to have the following features:

- Thin walls so that gases diffuse across only a short distance.
- A good blood supply to transport oxygen and carbon dioxide to and from body tissues.
- A large surface area for diffusion.

Getting oxygen into the body

The diagram shows the path taken by air to the lungs. Your breathing system takes air in and out of your body. This provides a regular supply of air containing oxygen, and removes air containing carbon dioxide.

The airways of the breathing system end in very small air sacs called alveoli. The walls of the alveoli are where gas exchange happens, and they provide an extremely large surface area for gas exchange. The total surface area of all the alveoli is about 80 m^2. You can see from the diagram that the alveolar walls are only one cell thick and each alveolus is surrounded by blood capillaries. The air in each alveolus is very close to the blood flowing in capillaries. This means that oxygen diffuses only a short distance from the air in the alveolus to reach the bloodstream.

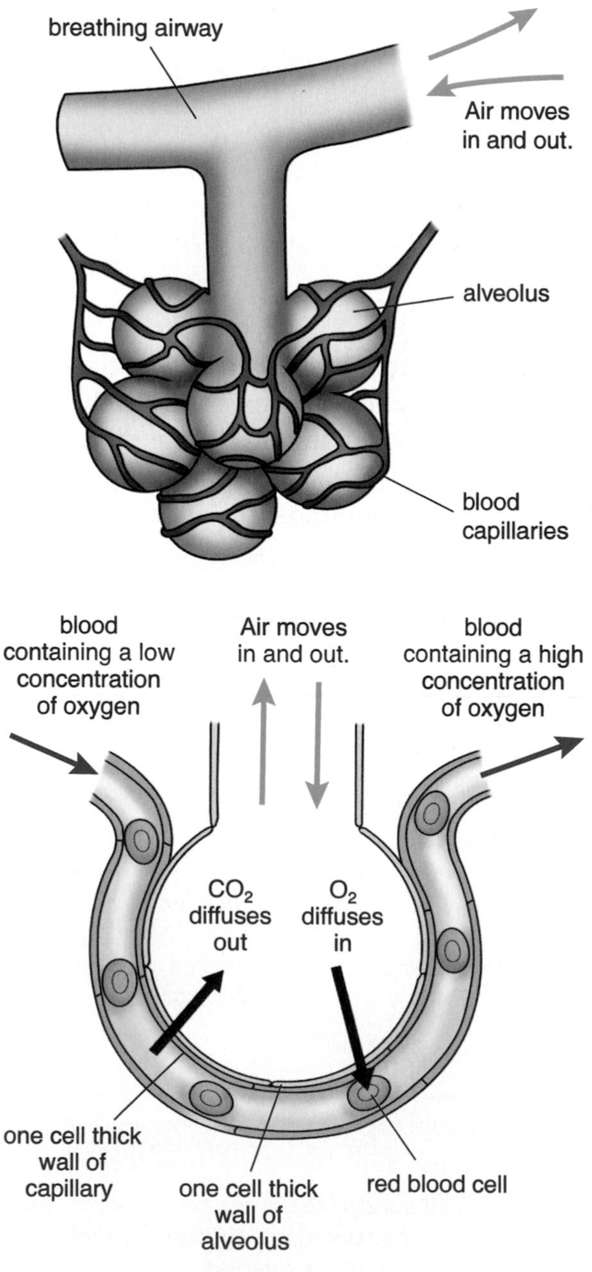

▲ Gas exchange in alveoli.

Question

a *What features of the alveoli provide:*
 (i) *a short diffusion pathway?*
 (ii) *a large surface area?*

Getting food into the body

The wall of your small intestine is very efficient at absorbing food. It is a specialised tissue for absorption. The small intestine can absorb food efficiently for the following reasons.

- Its inner surface contains many tiny folds called **villi**. The large number of villi produce a very large surface area for absorption.
- Each villus contains many blood capillaries to transport absorbed food from the small intestine to the rest of the body.
- Each villus is very thin, so that food molecules diffuse over only a short distance to reach the bloodstream.
- The diagram (right) shows how these features allow absorption to take place efficiently.

Questions

b *Food molecules diffuse over a short distance to reach the blood. What features of the small intestine provide the short diffusion distance?*

c *Diffusion of a substance is more rapid when there is a large difference in concentration between one area and another. What features of the small intestine maintain a steep concentration gradient of food molecules?*

Active transport of food molecules

As food is absorbed by the villi the concentration of food molecules inside the villi increases. This prevents further food molecules being absorbed by diffusion. Once the concentration of food molecules inside the villus is higher than that outside, absorption takes place by active transport. The diagram shows the structure of cells from the villus wall.

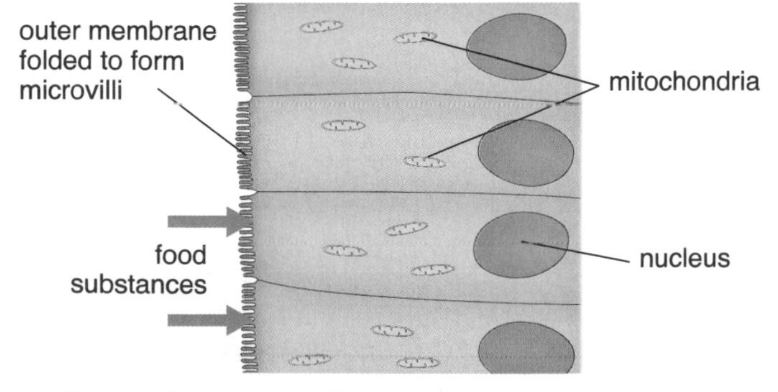

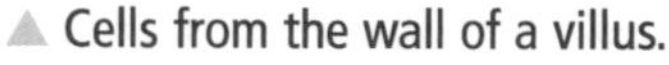
▲ Cells from the wall of a villus.

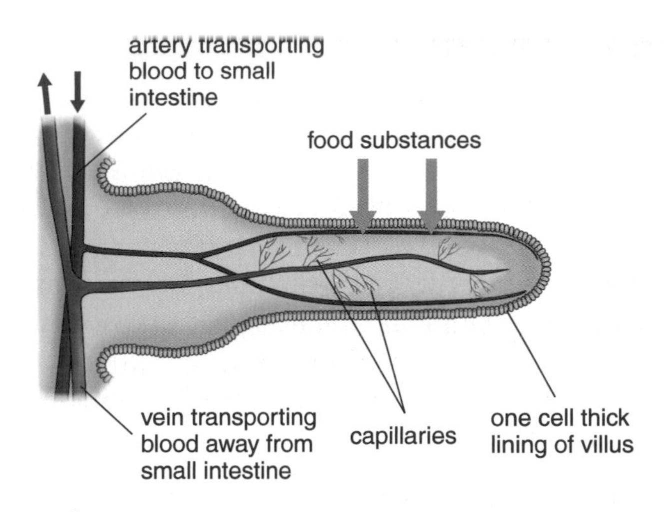

▲ A villus.

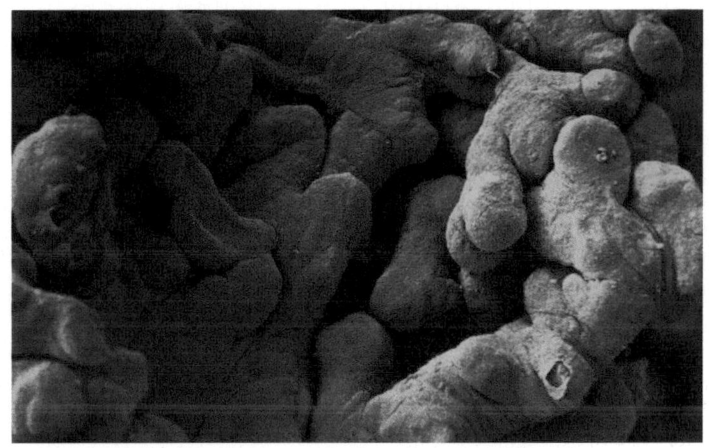

▲ Villi increase the surface area of the small intestine.

Questions

d *Describe one feature of the villus cell which indicates that some food molecules are absorbed by active transport.*

e *Explain the advantage of having lining cells with a highly folded outer cell membrane.*

Key points

- Some organ systems are specialised for exchanging materials.
- In humans, the surface area of the lungs is increased by the alveoli. The surface area of the small intestine is increased by villi.
- Alveoli provide a very large, moist surface with a good blood supply. This enables gases to diffuse readily into and out of the blood.
- Villi provide a large surface area with a good blood supply to absorb the products of digestion.

Pumping blood

Blood flows around your body through a series of blood vessels which make up your circulatory system. Blood is kept flowing around the system by the pumping action of your heart. Your heart beats because the muscles in its walls contract and then relax. When the muscular walls contract, blood is forced out of the heart under high pressure. When the muscular walls relax, the heart fills up with more blood, ready to be pumped out again. Beating at between 60 and 70 beats per minute, your heart is working all the time without pausing to rest.

Supply lines

Blood is pumped around your body through a system of blood vessels. There are three types of blood vessel – arteries, veins and capillaries.

When blood flows out of the heart, it flows through **arteries**. Because it has been pumped from the heart, the blood in arteries is at high pressure. When arteries get to an organ in your body, they branch many times, forming smaller and smaller arteries.

The smallest blood vessels are called **capillaries**. This is where substances pass in and out of the blood. Capillaries are so narrow that only one red blood cell at a time can be squeezed through. They also have very thin walls which are just one cell thick, so that substances can pass in and out easily. Substances needed by the cells in body tissues pass out of the blood through the capillary wall into cells. Substances produced by cells, such as carbon dioxide, pass into the blood through the walls of the capillaries.

The capillaries join up again to form **veins**, which return blood back to the heart. The blood flowing through veins is at much lower pressure.

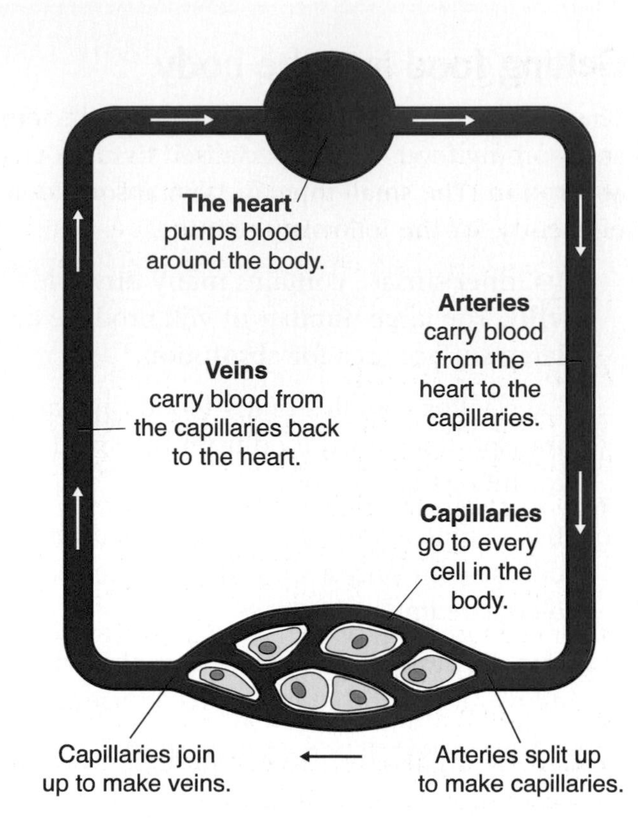
The heart pumps blood around the body.
Arteries carry blood from the heart to the capillaries.
Veins carry blood from the capillaries back to the heart.
Capillaries go to every cell in the body.
Capillaries join up to make veins.
Arteries split up to make capillaries.

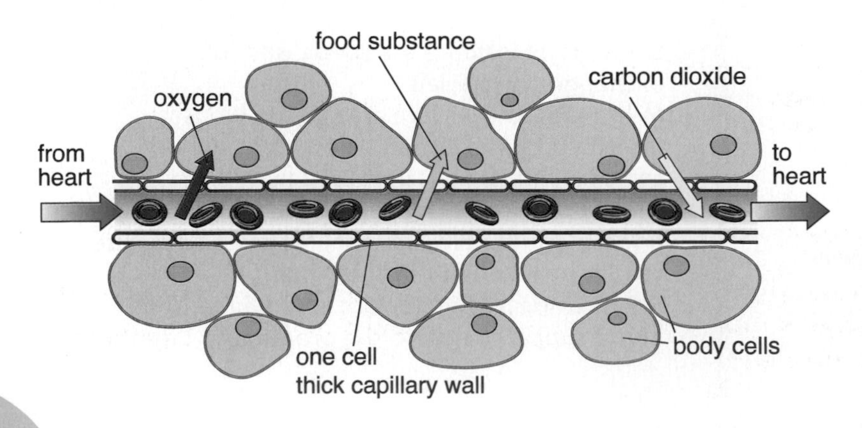
food substance
carbon dioxide
oxygen
from heart
to heart
body cells
one cell thick capillary wall

Question

a *What features of capillaries allow substances to pass in and out easily?*

Taking your pulse

Blood does not flow smoothly through arteries. Every time the heart muscles contract, a surge of blood passes along arteries, causing the artery walls to bulge slightly. When this happens the walls of arteries become stretched. The walls then spring back (recoil) as the heart muscles relax. You can feel arteries stretch and recoil when you feel the pulse in your wrist.

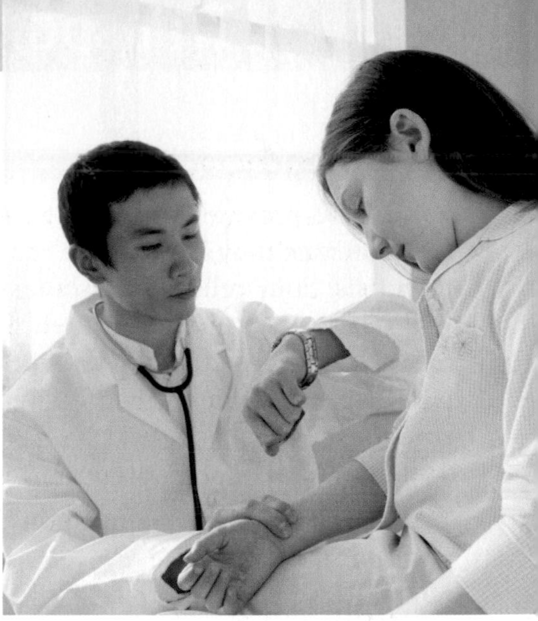

▲ You can feel an artery stretching and recoiling as blood flows along. This is your pulse.

A double pump

The numbers on the diagram show the path of blood as it flows through the heart and around the body.

1. Blood from the right-hand side of the heart is pumped to the lungs.
2. In the lung tissue oxygen diffuses into the blood and carbon dioxide diffuses out. The blood becomes **oxygenated**.
3. Oxygenated blood from the lungs then returns to the left-hand side of the heart.
4. Oxygenated blood is pumped around the rest of the body.
5. In respiring tissues, oxygen diffuses from the blood into body cells, and carbon dioxide diffuses from body cells into the blood. The blood becomes **deoxygenated**.
6. Deoxygenated blood is returned to the right-hand side of the heart.

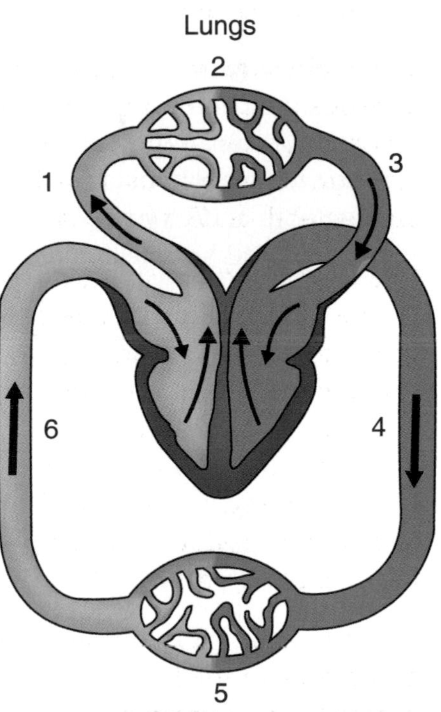

This is called a **double circulatory system** because blood travels through the heart twice as it flows around the body. By having a double circulation, oxygenated blood is separated from deoxygenated blood.

As blood flows around the body, substances can pass in and out of the blood only through capillaries.

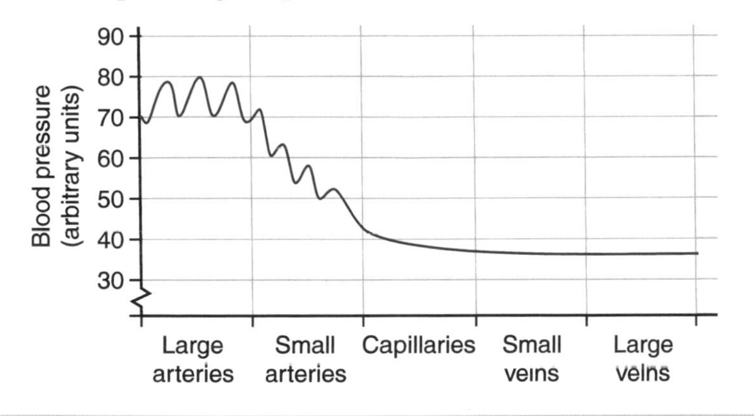

Questions

The graph shows the changes in blood pressure as blood flows through different blood vessels.

b *Explain why the blood pressure in arteries goes up and down.*

c *In which type of blood vessels does the blood pressure fall the most?*

d *What is the pressure of blood as it flows back into the heart?*

Key points

- The heart pumps blood around the body.
- Blood flows from the heart to other body organs in arteries and returns to the heart through veins.
- Substances pass in and out of the blood through the walls of capillaries.

Transporting materials

Your blood

Your blood provides all the cells of your body with the materials they need as well as removing waste materials. Your cells would soon stop working without a good supply of blood. This is why blood has to be replaced quickly if someone loses a lot of blood in an accident.

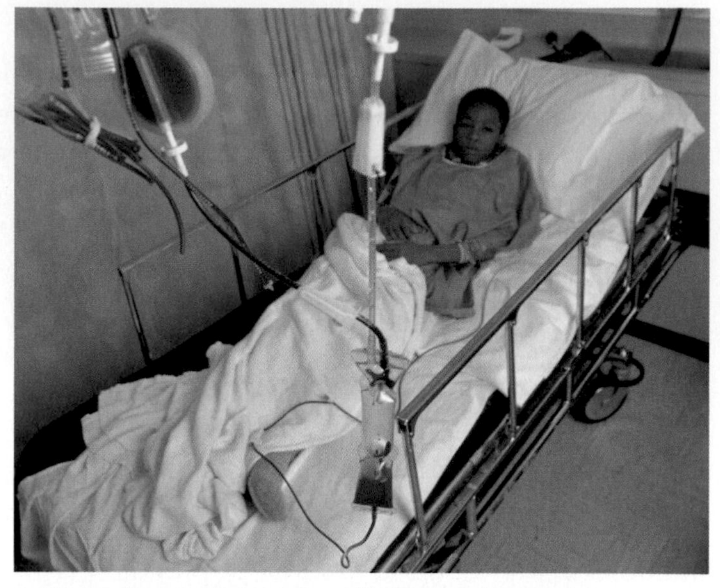

▲ All your body cells need a good blood supply to keep them working properly – if someone loses a lot of blood in an accident it has to be replaced quickly.

Transporting oxygen

Your blood contains an enormous number of **red blood cells**. These are very specialised cells which transport oxygen to all the cells of your body. Each red blood cell contains a protein called **haemoglobin**. Red cells are so specialised that they have no nucleus and consist only of a cell membrane packed with haemoglobin.

As blood flows through the alveoli in the lungs, haemoglobin combines with oxygen to form oxyhaemoglobin. When blood flows through respiring tissues, oxyhaemoglobin splits up into haemoglobin and oxygen. The oxygen that is released is used by cells for respiration.

In the lungs:
haemoglobin + oxygen → oxyhaemoglobin

In respiring tissues:
oxyhaemoglobin → haemoglobin + oxygen

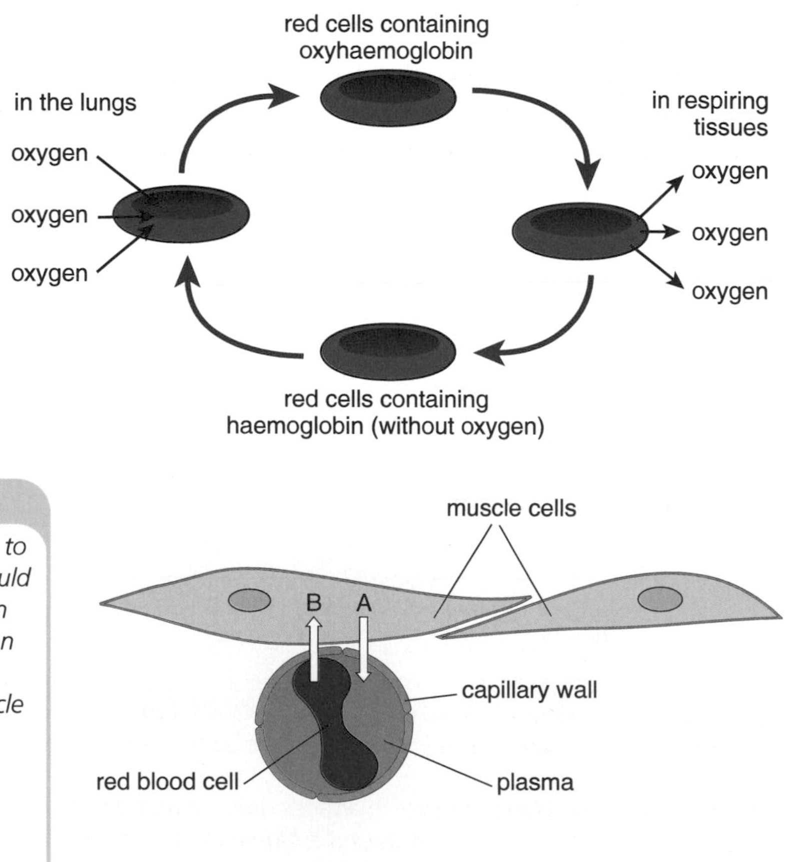

Questions

a *A blood test was developed before the 2000 Olympics to detect the presence of a drug called EPO. This drug could be used to increase the number of red blood cells in an athlete's body. Explain why athletes taking this drug can increase their level of performance.*

b *The diagram shows a capillary supplying blood to muscle cells. Name a substance that diffuses in the direction shown by (i) arrow A, and (ii) arrow B.*

c *Only one red blood cell at a time can pass along a capillary. Explain how this increases the efficiency of diffusion of oxygen into body cells.*

Transporting food and waste materials

Plasma is the liquid part of blood. Plasma contains mainly water with a number of dissolved substances, including the following.

Carbon dioxide is produced as a waste product of respiration. Carbon dioxide is transported in plasma from respiring cells to the alveoli in the lungs. As blood flows around the alveoli, carbon dioxide is removed from the blood by diffusion and then breathed out.

Soluble sugars, amino acids, fatty acids and glycerol are produced as products of digestion in the small intestine. These soluble products are absorbed from the small intestine and transported in plasma to other body organs.

Urea is a waste product formed from excess amino acids. Urea is made in the liver and transported in blood plasma to the kidneys where it is removed from the body in urine.

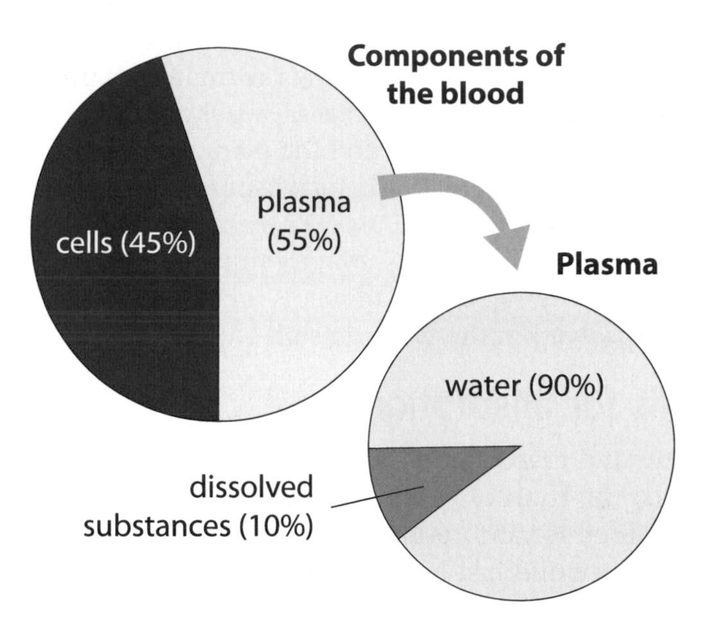

▲ Blood plasma contains mainly water with a number of dissolved substances.

Changing blood

The substances present in your blood change as the blood circulates around your body. This is because different substances pass in and out of the blood as it flows through each organ in your body.

Questions

The diagram shows the blood vessels connected to the main organs of your body.

d *Give the letter of the blood vessel which contains the highest concentration of (i) oxygen, (ii) glucose, and (iii) carbon dioxide.*

e *Give the letter of the blood vessel with the highest blood pressure. Explain your answer.*

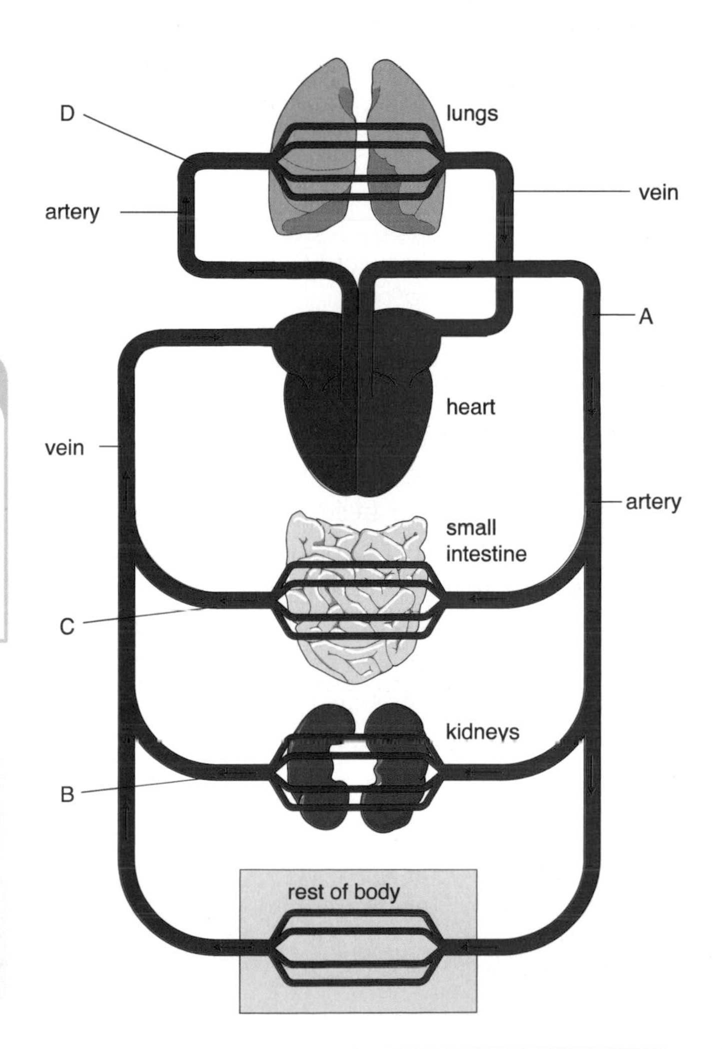

Key points

- Blood contains red blood cells which transport oxygen around the body.
- Haemoglobin combines with oxygen in the lungs to form oxyhaemoglobin.
- Oxyhaemoglobin splits up in respiring tissues to release oxygen.
- Plasma transports carbon dioxide, urea and the products of digestion.

Water and minerals

Green plants make their own food by photosynthesis. As well as using raw materials in photosynthesis, large amounts of water are used to transport materials around the plant. Plants also need certain **minerals** such as nitrates and magnesium. Nitrates are used by plants to make amino acids and proteins. **Magnesium** is used to make chlorophyll. These minerals are obtained as ions dissolved in the water in soil.

▲ Root tip with root hairs.

Cells for absorption

Water and mineral ions are absorbed from the soil by the roots of plants. The parts of the root specialised for absorption are the **root hairs**. Root hairs are found just behind the growing tip of roots. Each root hair is a tube-like extension of a cell. By growing between soil particles, each root hair is surrounded by water containing dissolved ions. This means that water and ions diffuse over only a very small distance to reach the root hair cell. By having lots of root hairs, the surface area for absorption is greatly increased.

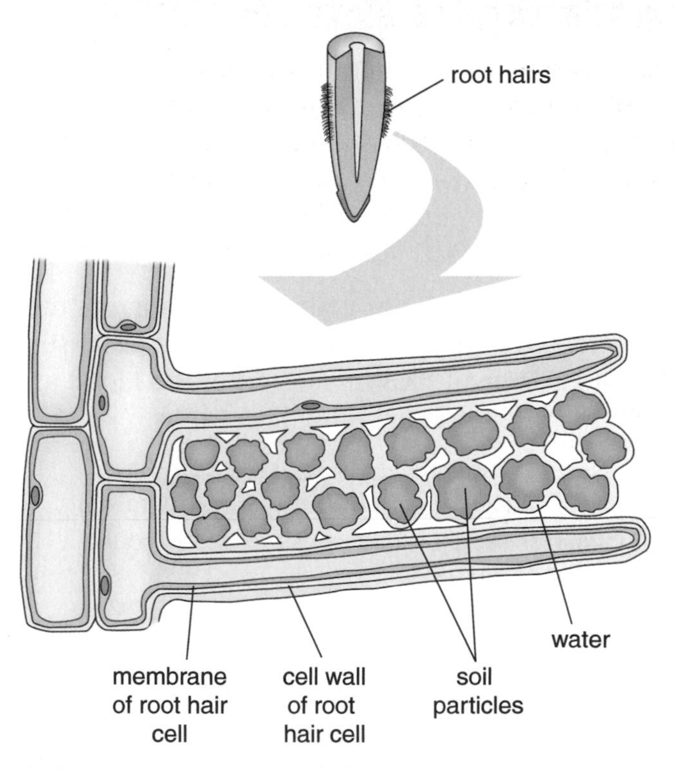

▲ Root hairs are specialised for absorbing water.

Question

a Describe the features of the root hair that increase the efficiency of water absorption.

Absorbing water

The root hairs absorb water by osmosis. The concentration of water in the soil is higher than the concentration of water inside the root hair cell. This causes water to diffuse through the partially permeable membrane down a concentration gradient into the cell. This increases the amount of water in the outer root hair cells. Water moves by osmosis from the root hair cells into other root cells down a concentration gradient.

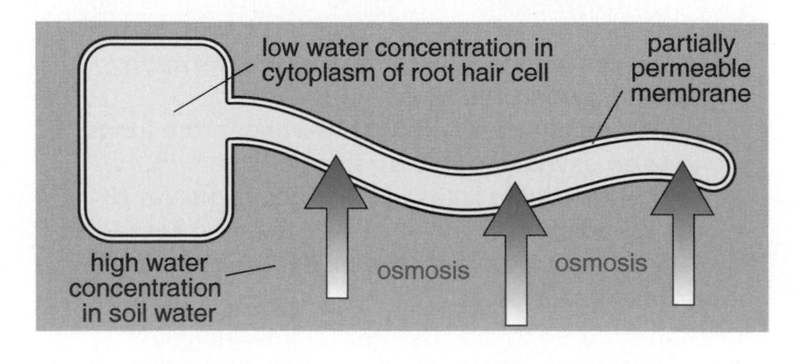

▶ Root hair showing osmosis.

Question

b Explain how water moves from the root hair cell into the adjoining cell.

Absorbing mineral ions

As well as absorbing water by osmosis, root hairs absorb minerals. Ions are absorbed by root hairs by diffusion and by active transport. Some minerals are present in the soil in low concentrations. The concentration of these ions is higher in the root hair cells. Root hair cells can absorb mineral ions against a concentration gradient using active transport. This uses energy supplied by mitochondria in the root hair cells.

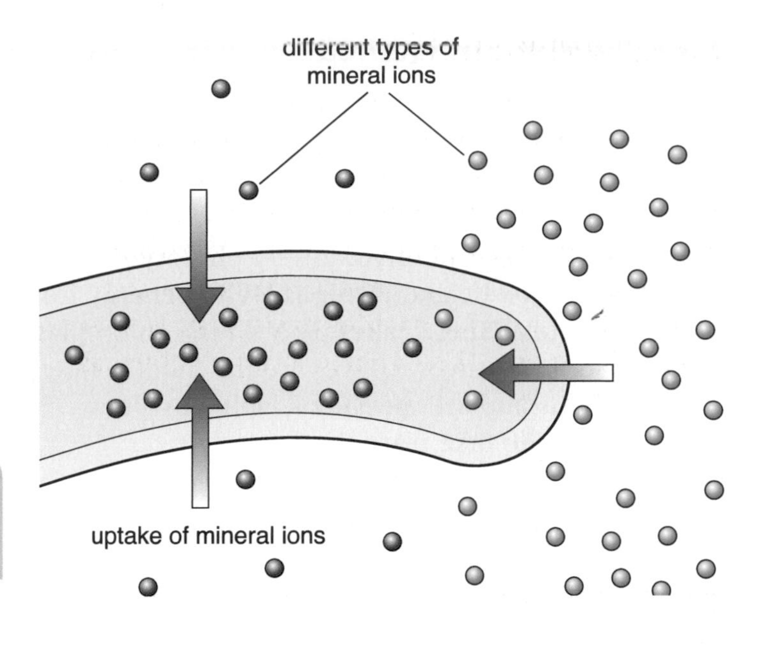

Question

c By what process are the ions shown in red and in green absorbed? Explain your answer in each case.

Investigating the absorption of ions

In an investigation, young plants were grown for several days with their roots in a nutrient solution containing mineral ions. At the end of the investigation, the concentration of different ions was measured in the root hair cells and in the nutrient solution. The results are shown in the bar chart.

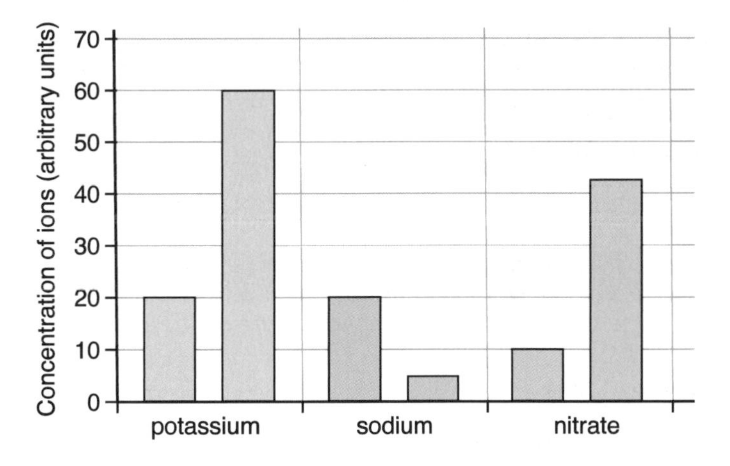

Questions

d By how many times have the nitrate ions become concentrated in the root cell compared with the nutrient solution?

e Name the process which the cell is using to absorb each type of ion shown in the chart. Use the information in the chart to explain your answer.

Key points

- Root hairs increase the surface area of roots to increase the efficiency of absorption.
- Water is absorbed by root hair cells by osmosis.
- Mineral ions can be absorbed by diffusion and by active transport.

Movement in and out of leaves

Using simple molecules

Using simple substances, green plants build up large, complex substances in the process of photosynthesis.

Specialised food factories

Green plants make carbohydrates by photosynthesis using carbon dioxide and water as well as energy from sunlight. Photosynthesis takes place in the chloroplasts inside the cells of leaves. Leaves are the food factories of green plants and are specially adapted to allow photosynthesis to take place.

By having a large surface area, leaves can absorb carbon dioxide very efficiently. The surface area is increased by leaves having a flat surface and many internal air spaces. The diagram shows the path that carbon dioxide takes. Molecules of carbon dioxide diffuse from the air into the leaf through tiny pores called **stomata**. Behind each stoma there is an air space which links with other air spaces in the leaf. Carbon dioxide diffuses through the air spaces to reach photosynthesising cells. The cells around the air spaces provide a large surface area for gas exchange. Carbon dioxide diffuses into the cell, through the cell walls and cell membranes to reach the chloroplasts.

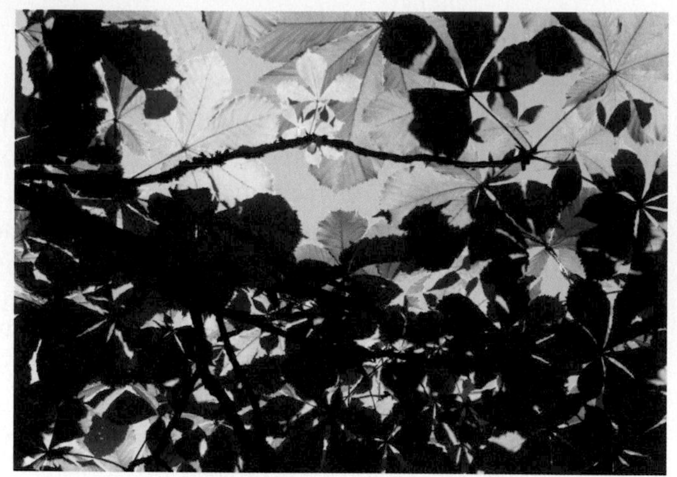

Leaves are the plant's food factories. Leaves can carry out photosynthesis efficiently because they are adapted to absorb large amounts of carbon dioxide and sunlight.

Question

a *In the diagram, identify the parts of the cell labelled A, B and C.*

Losing water

Plants lose water from the surface of their leaves continuously. This is called **transpiration**. The cells lining the air spaces in the leaf are covered with a thin film of water. As this water evaporates the air spaces become saturated with water vapour. This diffuses through the air spaces and then out through the stomata. Most of the water vapour lost by transpiration is through stomata. A small amount of water is also lost from the upper surfaces of leaves.

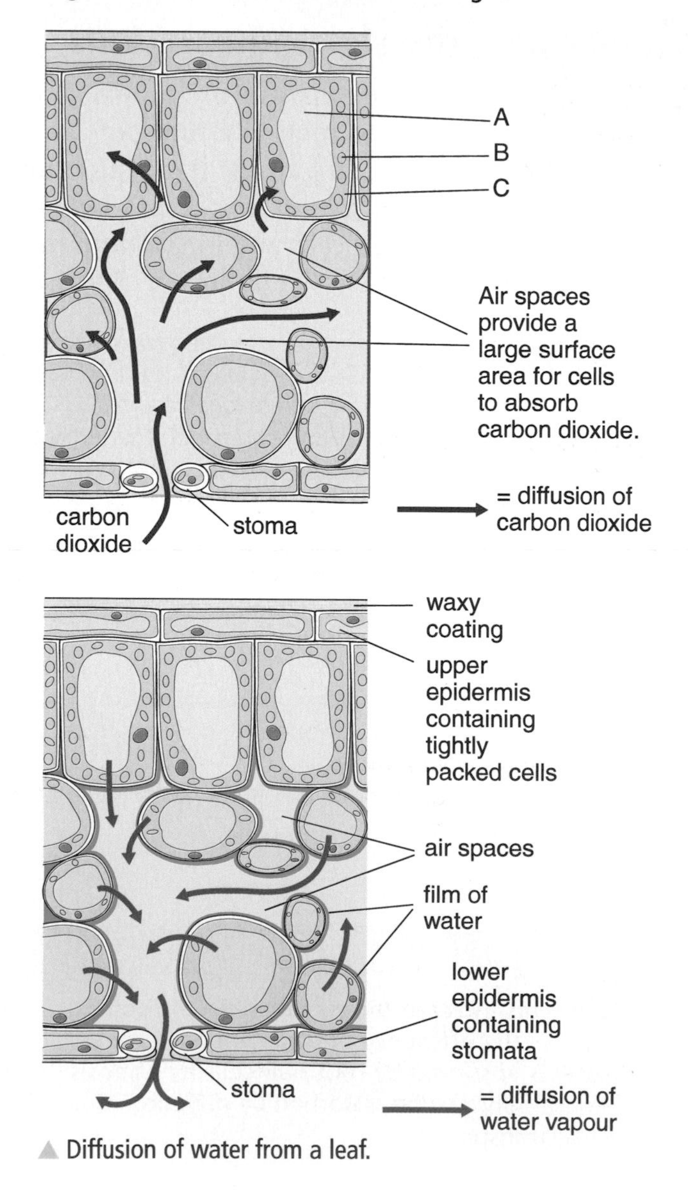

Diffusion of water from a leaf.

Question

b *What features of the upper surfaces of leaves help to reduce the loss of water by transpiration?*

Cutting down water loss

The amount of water lost by transpiration increases in hot, dry and windy conditions. As water is lost by transpiration from the leaves, more water is absorbed into the plant at its roots. When large amounts of water are lost by transpiration, the roots may not be able to absorb enough water from the soil to replace the water lost at the leaf surface. When this happens the plant wilts.

▲ Plants wilt when they become short of water. The leaves droop rather than being held out to absorb sunlight.

To prevent wilting, plants can close their stomata to reduce the rate of transpiration. Each stoma is surrounded by a pair of sausage-shaped cells called **guard cells**. The diagram shows how the guard cells open and close stomata.

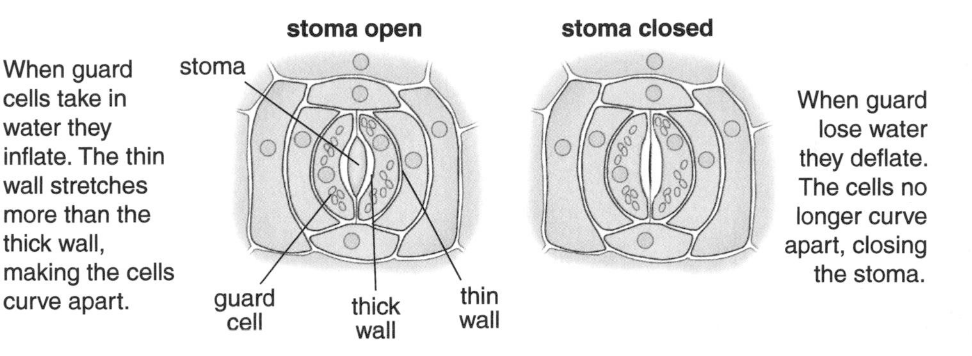
stoma open
stoma closed
When guard cells take in water they inflate. The thin wall stretches more than the thick wall, making the cells curve apart.
stoma
guard cell
thick wall
thin wall
When guard lose water they deflate. The cells no longer curve apart, closing the stoma.

Questions

- **c** *Suggest why stomata are found only on the lower surface of most leaves.*
- **d** *Explain how stomata close when guard cells lose water.*

Living in dry environments

The 'Two-leaf Hakea' is a plant that grows in Australia. It produces two types of leaves – one in the spring when the conditions are cool and wet, and the other in the summer when it is very hot and dry. The table shows different measurements taken from each type of leaf.

	Spring leaves	Summer leaves
Length (mm)	32	50
Width (mm)	10	0.8
Surface area (mm^2)	290	40

Questions

- **e** *Which type of leaf will have the highest transpiration rate? Use the data in the table to explain your answer.*
- **f** *Explain the advantages to the plant of producing the types of leaves formed in the summer months.*

Key points

- Carbon dioxide enters leaves through stomata by diffusion.
- The surface area of leaves is increased by their flattened shape and internal air spaces.
- Plants lose water vapour by transpiration through stomata.
- The size of the opening of stomata is controlled by guard cells.

▲ The 'Two-leaf Hakea' produces two different types of leaf to suit the conditions at different times of the year.

Measuring transpiration

The rate at which a plant takes up water depends on the rate of water loss by transpiration. The faster a plant transpires, the faster it takes up water.

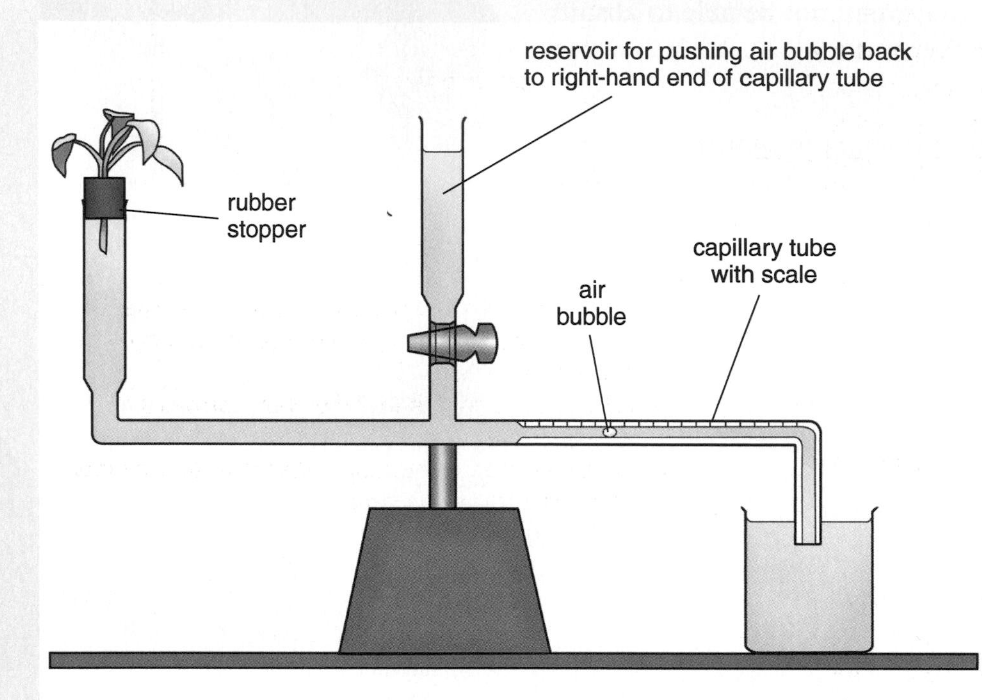

A potometer is used to measure the rate of transpiration in plants.

The diagram shows an apparatus called a **potometer**. This is used to record the rate of transpiration by measuring the rate of water uptake. As the plant loses water by transpiration, more water is taken up, making the air bubble move along the capillary tube. The rates of transpiration in different conditions can be compared by measuring how fast the air bubble moves.

A potometer was used by a group of students to measure the rate of transpiration in different conditions. The same plant was used throughout the investigation. The conditions were altered by using a hair drier to blow hot or cold air over a leafy shoot. The results are shown in the table.

Time (minutes)	Distance moved by air bubble (mm)	
	Still air	Warm air being blown by hair drier
0	0	0
5	1	10
10	3	20
15	4	27
20	5	36
25	8	50
30	10	58

Question

a *The air bubble was positioned at the start of the scale during each recording. Use the diagram to explain how to get the bubble back to the start of the scale.*
(i) Under which conditions is the rate of transpiration the highest?
(ii) Explain why the rate of transpiration was higher under these conditions.

Investigating water loss

An investigation was carried out to measure the rate of transpiration of privet leaves. The rate was measured by weighing the leaves at the start of the investigation and then after three days. Leaves of the same size were taken from the same plant and placed in identical environmental conditions. Some of the leaves were coated with grease to provide a waterproof barrier. The results of the investigation are shown in the table.

Type of treatment	Mass loss in 3 days (g)
No grease applied	12
Grease applied to upper leaf surface	7.5
Grease applied to lower leaf surface	3.5
Grease applied to both leaf surfaces	0.05

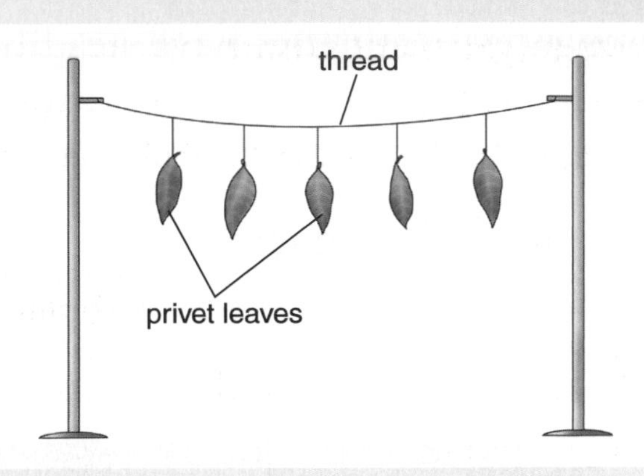

Questions

b *Explain why the leaves used were the same size.*
c *Explain why the leaves were placed in the same environmental conditions.*
d *What is the evidence that the grease provides a waterproof barrier?*
e *Explain how the results show that most transpiration takes place through stomata.*

Living in dry environments

Some plants are adapted to survive in very dry, hot environments. Such plants have special features to reduce water loss. For example, cacti are adapted to live in deserts where it is very hot and dry. By reducing the surface area of their leaves to form spines, cacti conserve water by reducing water loss. The diagram shows the structure of a leaf from a plant found growing in hot, dry conditions.

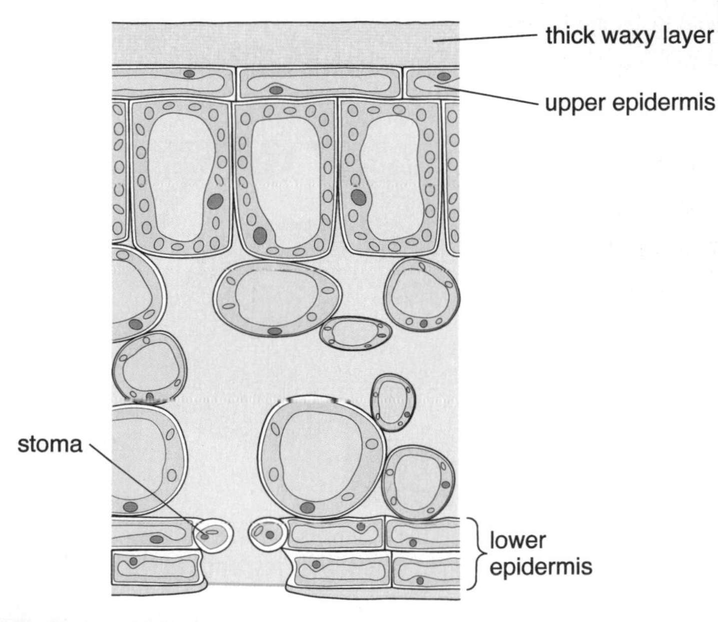

▲ The leaves of plants growing in hot, dry conditions are adapted to reduce water loss.

Questions

f *Explain how having stomata in grooves helps to reduce water loss by transpiration.*
g *Describe two other features that reduce water loss.*

Key points

- Transpiration is the loss of water vapour from leaves.
- Transpiration is more rapid in hot, dry and windy conditions.
- Most transpiration is through stomata.

Record breakers

Performances in athletics have steadily improved, as shown in the record times for the 100 metres. World records continue to tumble as performances keep getting better.

Improving performance

The tables show the top five all-time male and female 100 metre runners, with their fastest times in seconds. It is getting harder and harder for athletes to run faster than these winning times.

Best all-time 100 m runners – men

Year	Athlete	Time
2005	Asafa Powell (Jamaica)	9.77
1999	Maurice Greene (USA)	9.79
1996	Donovan Bailey (Canada)	9.84
1999	Bruny Surin (Canada)	9.84
1994	Leroy Burell (USA)	9.85

Best all-time 100 m runners – women

Year	Athlete	Time
1988	Florence Griffith Joyner (USA)	10.49
1998	Marion Jones (USA)	10.65
1998	Christine Arron (France)	10.73
1996	Merlene Ottley	10.74
1984	Evelyn Ashford	10.76

▲ Understanding how the body works during very strenuous activity has helped to develop training methods to improve athletic performance.

Modern training methods

Scientists have helped in the continuous improvement of performances by developing tests to monitor the training for particular sports.

Professional athletes spend a lot of time getting very fit and improving their techniques and skills. Coaching and training techniques use the understanding of how the body responds to extreme physical exertion to improve an athlete's performance. Modern equipment is used to monitor heart rate and to measure the volume of blood pumped from the heart of athletes as they perform. Measurements can also be made to monitor changes in the content of blood, such as the amounts of oxygen and carbon dioxide. The information obtained is used by coaches to develop ways of improving an athlete's performance.

▲ These sensors provide data about how an athlete's body responds to vigorous exercise. Coaches use this data to devise the best training methods to improve performance.

The effects of training

One example that shows how training affects performance is to compare the amount of oxygen a person normally takes in during breathing. A reasonably fit and active adult can breathe in 12 litres of air in a minute. This provides 3 litres of oxygen to the blood. After the right training, a long-distance athlete can take in 24 litres of air, providing 6 litres of oxygen to the blood, every minute. This provides muscle cells with a lot more energy from respiration.

Record times are running out

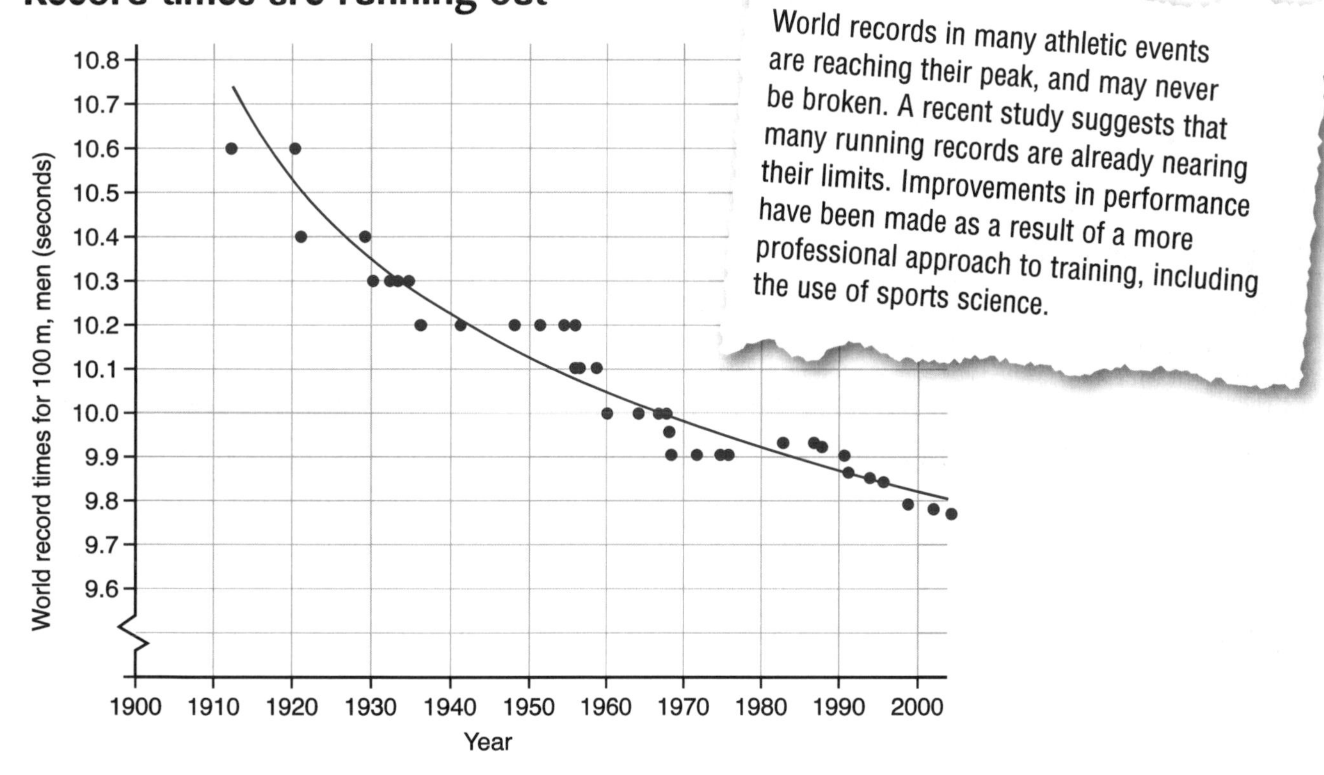
World records in many athletic events are reaching their peak, and may never be broken. A recent study suggests that many running records are already nearing their limits. Improvements in performance have been made as a result of a more professional approach to training, including the use of sports science.

▲ World record progression for the men's 100 metres. As the curve levels out like this, it becomes increasingly difficult to set a new record.

Think about what you will find out in this section

How exercise affects breathing, heart action and circulation.	How oxygen uptake affects aerobic and anaerobic respiration.
How data is used to monitor the effects of exercise on the body.	How lactic acid affects muscles.
How regular exercise improves fitness.	How the body recovers from vigorous activity.

Effects of exercise

Powering muscle cells

When you work hard your muscles need more energy. This energy is released from glucose during respiration, and is used to contract muscles to move your body. The rate of respiration in muscle cells increases during exercise so that more energy can be released. To increase the rate of respiration more oxygen and glucose is needed by muscle cells. The table shows the amount of oxygen that needs to be breathed in for different types of activity.

Activity	Energy used (kJ/min)	Oxygen required (litre/min)
Lying down	4	0.20
Sitting	6	0.30
Walking	30	1.50
Jogging	80	4.00

Questions

a *Explain the relationship between energy used and the amount of oxygen required.*
b *Explain why lying down uses up energy.*
c *Walking fast uses 40 kJ of energy per minute. Predict the amount of oxygen required per minute to carry out this activity. Explain your answer.*

Providing more oxygen

During exercise both the rate and depth of breathing increase so that much more air enters the lungs. Breathing rate is measured by the number of breaths taken each minute. The depth of breathing is measured by the volume of air taken in during each breath. The graph shows how a cyclist's breathing changed at different cycling speeds.

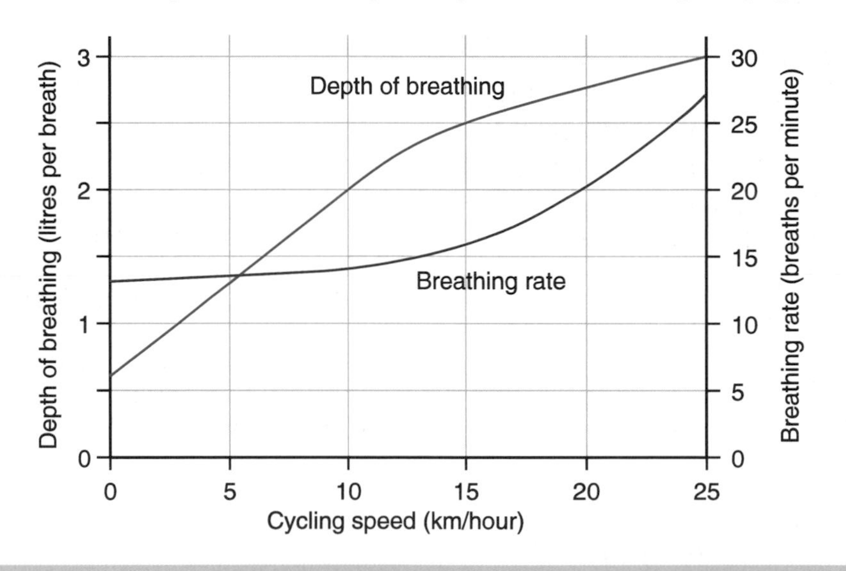

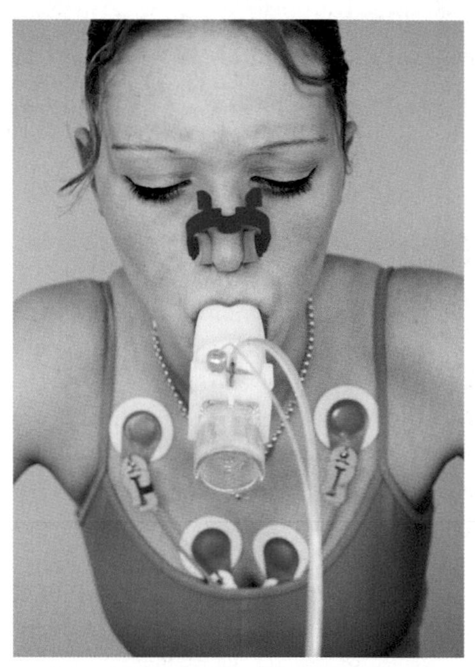

▲ The rate and depth of breathing can be measured accurately as a person runs at different speeds on an exercise treadmill.

Questions

d *How many breaths per minute were taken when the person was cycling at 10 km/hour?*
e *What was the volume of each breath when the person was cycling at 15 km/hour?*
f *Calculate the total volume of air breathed in and out when the person was cycling at 20 km/hour. Show your working.*

Making the heart work harder

As you exercise harder your heart also works harder, pumping more blood to get oxygen and glucose to your muscles. During exercise your heart beats faster and also contracts with more force during each beat. By contracting more strongly a greater volume of blood is pumped out during each beat. The table shows the changes in heart action during different types of activity.

Activity	Rate of heart beat (beats/min)	Volume of blood pumped from heart during each beat (cm^3)
Lying down	72	80
Sitting	82	100
Walking	110	125
Jogging	120	130

The amount of blood pumped from the heart per minute is called the *cardiac output*. This is calculated using the following formula:

$$\text{Cardiac output } (cm^3/min) = \text{rate of heart beat (beats/min)} \times \text{volume of blood pumped during each beat } (cm^3)$$

Question

g *Using the table, how much more blood was pumped out of the heart during each beat when the person was jogging compared to when they were sitting?*

Questions

h *Using the table, calculate the cardiac output when jogging for one minute.*

i *How much more blood was pumped from the heart per minute when jogging compared to sitting?*

Changes to circulation

As well as your heart beating faster during exercise, there are also changes in the amount of blood flowing to different parts of your body. For example, the major arteries supplying muscles dilate, increasing the amount of blood flowing into muscle tissue. This ensures that as much blood as possible flows to muscle, providing the glucose and oxygen that is needed and removing carbon dioxide. The diagram shows the amount of blood flowing to different organs when a person is resting and when they are taking part in exercise.

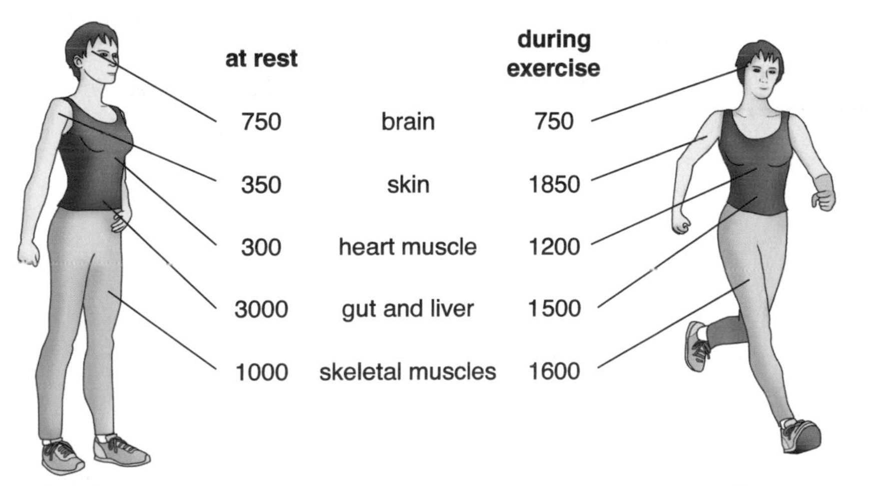

▲ The rate of blood flow to the main regions of the body (shown here in cm^3/min) increases as you work harder.

Questions

j *Explain why the blood flow increases during physical activity in (i) skeletal muscle, and (ii) heart muscle.*

k *Suggest why there is no change in the amount of blood flowing through the brain.*

Key points

- During exercise a number of changes take place:
  - heart rate increases,
  - the rate and depth of breathing increase,
  - arteries supplying muscles dilate.
- These changes increase the supply of glucose and oxygen to muscle.

Releasing energy

Aerobic respiration

Energy is released for use by cells during respiration. When glucose reacts with oxygen, energy is released and carbon dioxide and water are produced as waste products. This process uses oxygen from the air to break down glucose and is called **aerobic respiration**. It is summarised in this equation:

Glucose + oxygen → carbon dioxide + water (+ 2800 kJ energy)

Glucose is the body's main fuel that is used in respiration. During strenuous exercise the glucose in the blood is quickly used up. When this happens, **glycogen** from stores in muscle tissues is used to supply energy. Glycogen is a form of carbohydrate. By eating lots of carbohydrates in the days before an event, athletes build up the glycogen stores in their muscles so that they will have a good energy supply.

Respiration without oxygen

When your muscles are used for long periods of vigorous activity, they become fatigued and stop contracting efficiently. Sometimes even highly trained, fit athletes cannot get enough oxygen to their muscles during very strenuous activity. For example, in a 100 m race athletes are working their muscles so hard that muscle cells become short of oxygen. To provide the energy that is needed, glucose is broken down without using oxygen. This process is called **anaerobic respiration**. **Lactic acid** is produced as a waste product, instead of carbon dioxide and water. Anaerobic respiration is summarised in this equation:

Glucose → lactic acid (+ 120 kJ energy)

During anaerobic respiration glucose is not completely broken down, and some energy is locked in the lactic acid molecules. This means that less energy (120 kJ) is released than in aerobic respiration (2800 kJ).

▲ Some activities require energy to be released so quickly that muscle cells break down glucose without using oxygen. This is anaerobic respiration.

Getting into debt

The lactic acid produced by anaerobic respiration causes muscles to work less efficiently. This is because lactic acid lowers the pH in muscle cells. This increase in acidity has two effects on muscle cells:

- The pH is no longer at the optimum for enzymes to work efficiently. This slows down the rate of respiration in muscle cells.
- The low pH affects muscle contraction, causing cramp and muscle fatigue.

Lactic acid has to be broken down to carbon dioxide and water to enable muscle cells to work efficiently. This is an oxidation reaction and requires oxygen:

Lactic acid + oxygen → carbon dioxide + water

The extra oxygen needed to remove lactic acid is called an **oxygen debt**. This is why you continue to breathe deeply even when an activity is finished. By breathing deeply after an activity, you get the extra oxygen into your body to 'pay back' the oxygen debt.

Marathon runners need to run at a steady pace to prevent lactic acid building up.

Obtaining energy

Athletes get energy from both aerobic and anaerobic respiration. The table shows the percentage of energy that comes from aerobic and anaerobic respiration in different types of running events.

Length of race	Source of energy (%)	
	Aerobic respiration	Anaerobic respiration
100 metres	5	95
1500 metres	55	45
10 000 metres	90	10
Marathon, 42.2 km	98	2

Questions

- **a** *Use the information in the table to explain why athletes do not need to breathe when they run the 100 metres.*
- **b** *Explain why it is important for marathon runners not to run too fast at the start of a race.*
- **c** *Marathon runners eat lots of carbohydrates in the days just before a race. This is called 'carboloading'. Explain the advantage of 'carboloading'.*

Producing and removing lactic acid

The concentration of lactic acid in the blood increases during strenuous exercise, and then decreases as the body recovers. The graph shows the changes during and after exercise.

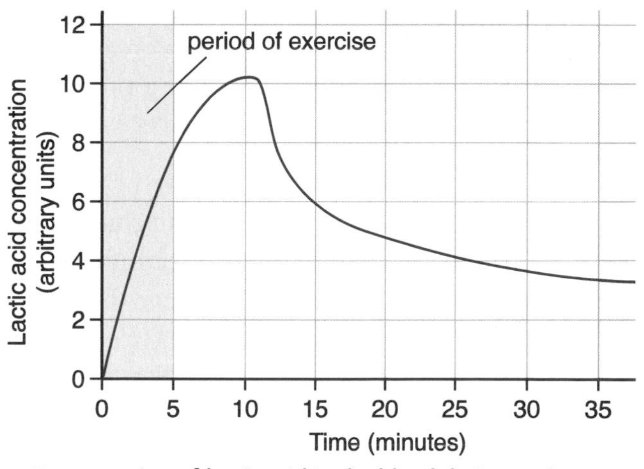

Concentration of lactic acid in the blood during and after exercise.

Questions

- **d** *Explain why the concentration of lactic acid in the blood increases between 1 and 5 minutes after the exercise starts.*
- **e** *Explain why the concentration of lactic acid in the blood continues to rise for a while after the exercise is finished.*
- **f** *Explain why the person will still be breathing heavily 10 minutes after the exercise has finished.*

Key points

- Anaerobic respiration is the incomplete breakdown of glucose and produces lactic acid.
- Anaerobic respiration produces less energy than aerobic respiration.
- An 'oxygen debt' is formed during anaerobic respiration and is 'repaid' as lactic acid is oxidised.

Getting fit

Many people spend a lot of their time going to the local gym or jogging to get and stay fit. Professional athletes need to be very fit to compete against others in their sport. But everyone needs to be fit enough to carry out everyday activities and to stay healthy. Becoming short of breath after climbing a few stairs or lacking the energy to walk to the local shops are signs of being unfit and unhealthy.

Any type of activity which gives you regular exercise is good for you. This is because exercise has several effects on your body. When you exercise regularly:

- your heart muscle becomes stronger so that blood is pumped around your body more efficiently;
- your breathing becomes more efficient so that more air can be taken in with each breath;
- your blood circulation becomes improved so that more glucose and oxygen can be delivered to respiring muscles.

▲ Improving fitness makes it easier to carry out everyday activities as well as improving health.

Measuring pulse rates

Measuring pulse rate during and after exercise and comparing these readings with the resting pulse rate provides a good indication of a person's level of fitness.

The graph shows the pulse rate of a young man who is trying to lose weight and get fitter. To improve his fitness he exercised for one hour every day by cycling to and from work. His pulse rate was measured before he started to cycle every day, and then four weeks later.

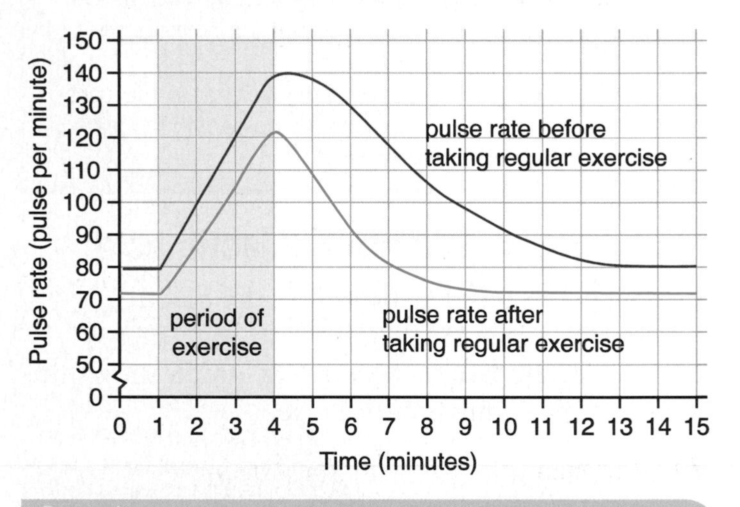

Heart and lung fitness

Regular exercise increases the strength of heart muscles and increases the volume of the chambers in the heart. This allows the heart chambers to fill with more blood and the walls of the chambers to contract more strongly. This means that more blood is pumped out during each heartbeat of a person who is physically fit compared to someone who is less fit.

Questions

a *The time taken for the pulse rate to return to its resting level after exercise is used as a measure of fitness. How long did it take for the man's pulse rate to return to its resting level (i) before and (ii) after taking regular exercise?*

b *In addition to the time taken for the pulse rate to return to normal, describe* ***two*** *other ways in which the pulse rate changed after taking regular exercise.*

The graph shows the effect of training on heart action. The data was obtained by measuring the volume of blood pumped out of the heart during each beat when cycling on an exercise bicycle at different speeds. The measurements were carried out before and after a six-month training period.

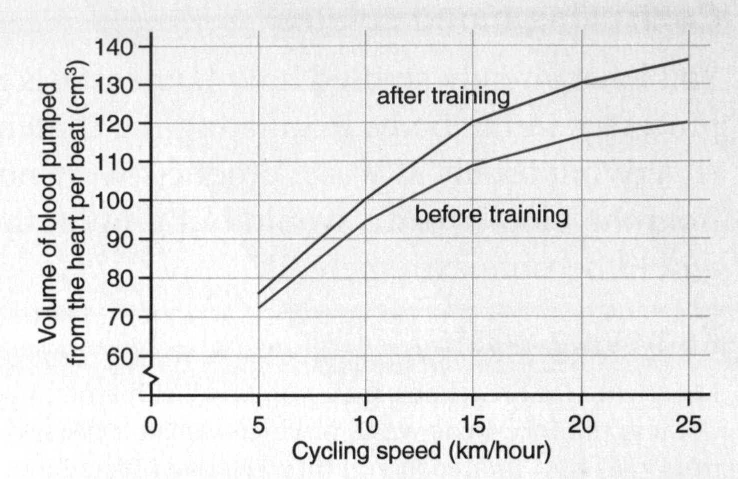

Questions

- ***c*** *What is the increase after training in the volume of blood pumped out during each beat when cycling at 25 km/hour?*
- ***d*** *A person cycling at 25 km/hour needs a cardiac output of 20 000 cm^3 of blood per minute. Calculate the heart rate the person would have (i) before training, and (ii) after training.*

Providing more oxygen

Regular exercise increases the strength of the breathing muscles – the diaphragm and the intercostal muscles – so that more air can be taken in during each breath. This helps to increase the amount of oxygen that can be transported in blood to respiring muscles.

The amount of oxygen that can be taken in increases until it reaches a person's maximum level. The degree of physical activity cannot be increased any more once this maximum oxygen uptake is reached without muscles respiring anaerobically and producing lactic acid. The relationship between oxygen uptake and anaerobic respiration is shown in the graphs.

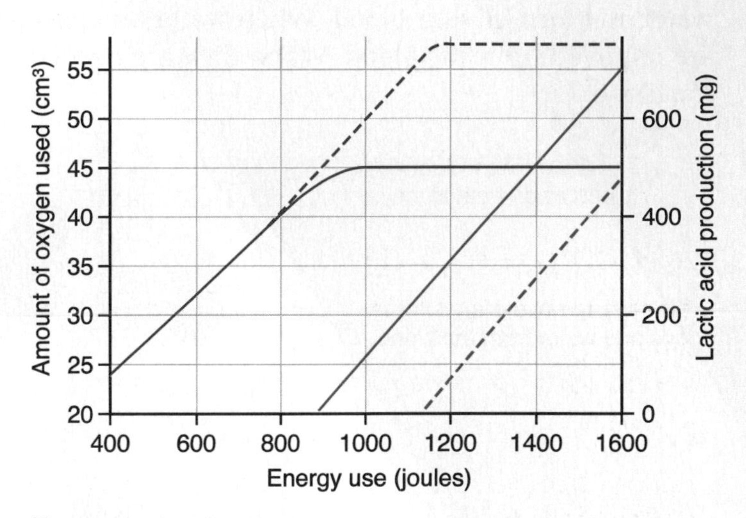

Questions

- ***e*** *Why does oxygen consumption increase with increased use of energy?*
- ***f*** *How much more oxygen is used by a trained athlete when using 1200 joules of energy per minute?*
- ***g*** *Use the information in the graphs to explain why lactic acid production in a trained athlete occurs at a higher rate of energy use.*

Key points

- **If insufficient oxygen is reaching muscles they begin to respire anaerobically.**
- **The effects of regular exercise increase the blood flow to the muscles and so increase the supply of glucose and oxygen.**

Regulating blood content

Controlling blood solutes

You have already studied how homeostasis regulates processes in the body. Regulating the content of the blood is very important. If waste products were not removed from the blood, toxins would build up in the body and we would become extremely ill.

Removing waste

Many chemical reactions are going on all the time in body cells. Some of these produce toxic waste products. Waste substances are transported from cells into the blood and then removed from blood as it flows through the **kidneys**. The job of the kidneys is to remove unwanted substances from the blood and pass them to the bladder.

As well as removing waste, your kidneys control the concentration of water and ions in your blood. When water and ions are in excess, they are removed from the blood. When they are needed, they are retained in the blood.

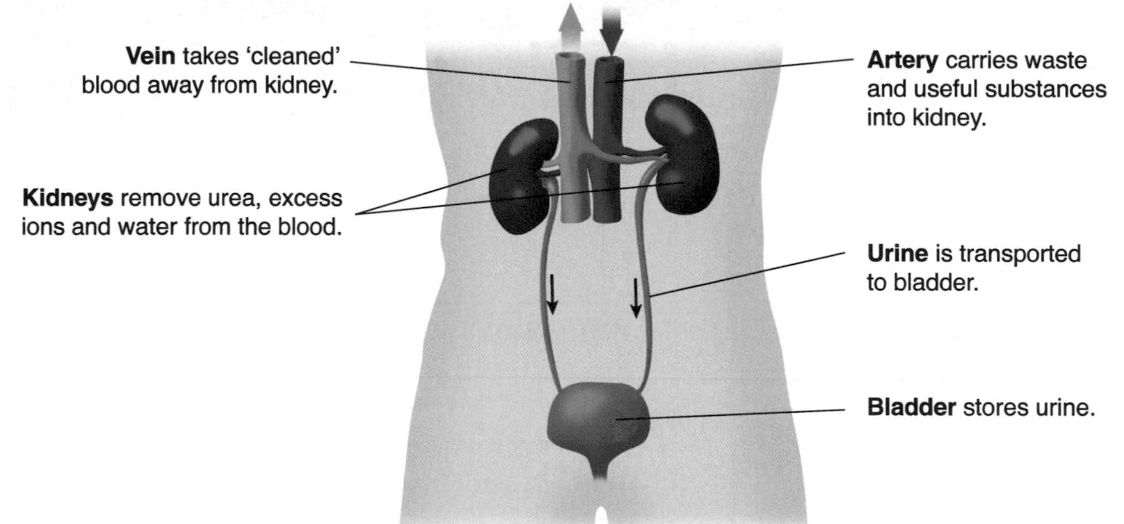
Vein takes 'cleaned' blood away from kidney.
Artery carries waste and useful substances into kidney.
Kidneys remove urea, excess ions and water from the blood.
Urine is transported to bladder.
Bladder stores urine.

Kidney failure

Sometimes the kidneys stop working due to disease or because of injury. A person can live with only one healthy kidney, but if both kidneys fail then the condition is very serious. Toxins will build up in the body, and the blood cannot function normally because it is not being cleaned properly. There are two treatments for kidney failure:

- Kidney **transplant**, in which a functioning kidney from a donor is surgically transplanted into the patient.
- Kidney **dialysis**, which involves circulating the patient's blood outside the body through a series of tubes into a chemical bath to remove waste products.

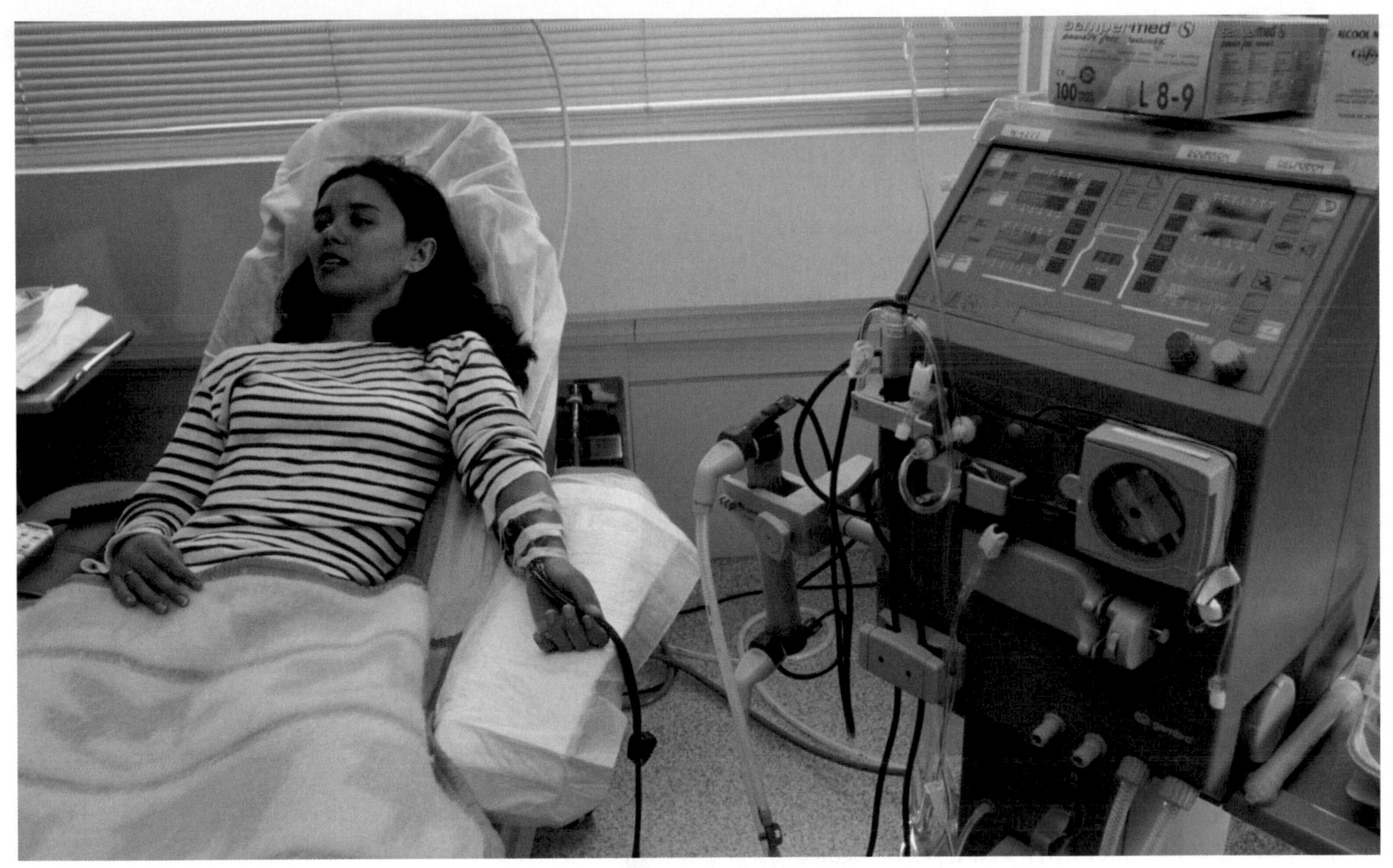

▲ This girl's kidneys no longer work properly and she has to be hooked up to a dialysis machine every 2–3 days.

Lucy's story

Lucy was born with kidneys that did not work properly. At six months old, she started to have dialysis treatment. This meant travelling to a dialysis centre at a hospital three times a week, with each treatment taking about four hours. Lucy is now five years old and has travelled over 200 miles every week to receive life-saving dialysis. Despite her illness, Lucy is full of life and enjoys going to nursery on the days she is not receiving dialysis. Her parents are hoping that Lucy will be able to have a kidney transplant when she is older.

Think about what you will find out in this section

How the kidneys remove waste products from the blood.	How the kidneys control ion and water content of the blood.
How people who suffer from kidney failure can be treated.	How dialysis works to control the concentration of dissolved substances in the blood.
The benefits and risks of having a kidney transplant.	The advantages and disadvantages of treating kidney failure by dialysis or by a transplant.

How your kidneys work

Cleaning the blood

Once waste and toxic substances have entered the blood, it is important that they are removed from the body. This must be done as soon as possible. The kidneys take these substances out of the blood and remove them from the body in urine.

Kidney tubules

Each kidney contains millions of tiny **tubules**. Each tubule is divided into different regions where different processes take place. The diagram shows the processes that remove unwanted substances from the blood and reabsorb substances that are needed by the body.

Filtering the blood

The first stage in removing waste is to filter the blood. Filtration is brought about by blood pressure. As blood flows into a tubule, blood pressure forces small molecules such as water, glucose, urea and ions from the blood into the tubule. Because the walls of the kidney tubules are partially permeable, only small molecules are filtered. The fluid that is formed in the tubule is called the filtrate.

Question

a *Why does blood passing out of the tubule contain only blood cells and molecules such as proteins?*

Reabsorbing useful substances

The filtrate in the tubule contains a mixture of waste and useful substances. As the fluid passes along the tubule small molecules such as glucose, amino acids and ions are reabsorbed back into the blood by active transport. They are then taken to the parts of the body where they are needed.

Urea, excess ions and excess water are not reabsorbed from the tubule. These substances form urine which is stored in the bladder before being removed from the body. Through these processes, the blood is maintained with the proper composition, and excess or unwanted substances are removed from the blood into the urine.

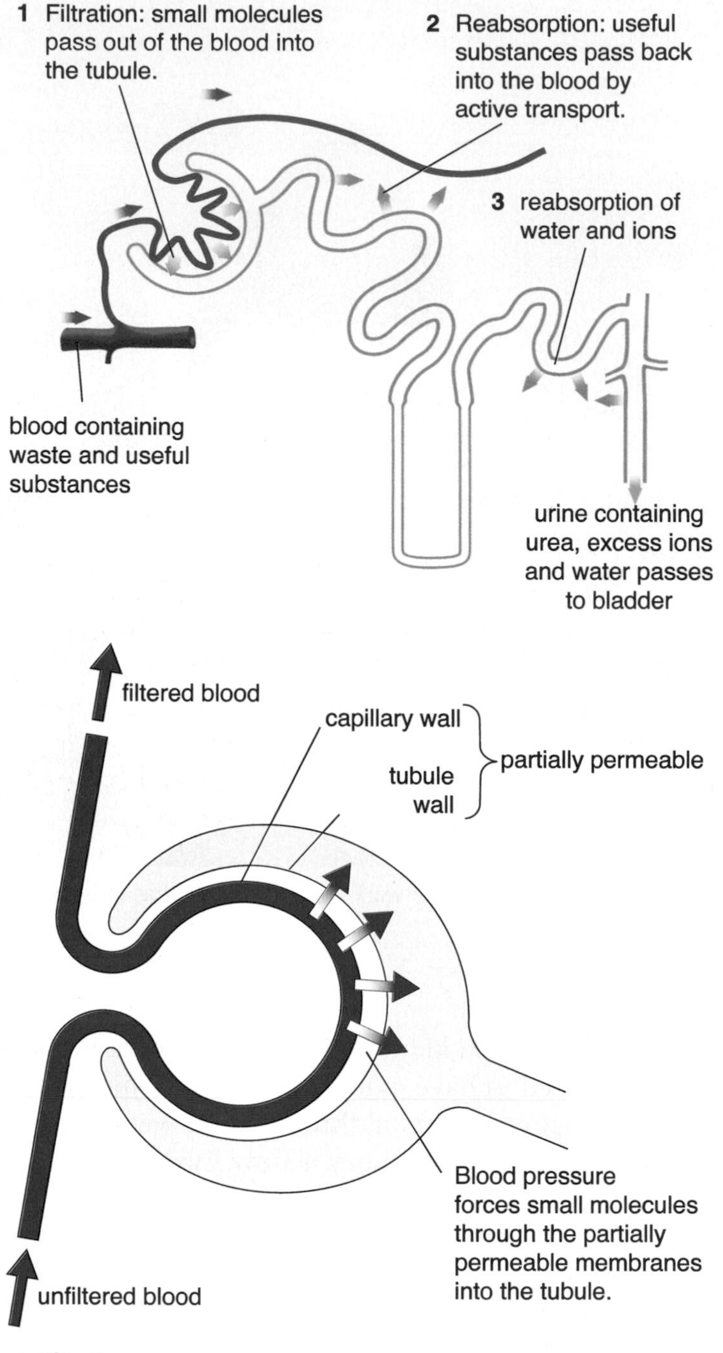
1 Filtration: small molecules pass out of the blood into the tubule.
2 Reabsorption: useful substances pass back into the blood by active transport.
3 reabsorption of water and ions
blood containing waste and useful substances
urine containing urea, excess ions and water passes to bladder
filtered blood
capillary wall
tubule wall
partially permeable
Blood pressure forces small molecules through the partially permeable membranes into the tubule.
unfiltered blood

Filtration.

The wall of the tubule is adapted to reabsorb useful substances. The features of the wall which allow molecules to be reabsorbed efficiently are as follows:

- The tubule wall is only one cell thick.
- The cells lining the wall contain large numbers of mitochondria.
- The cells lining the wall contain **microvilli**.

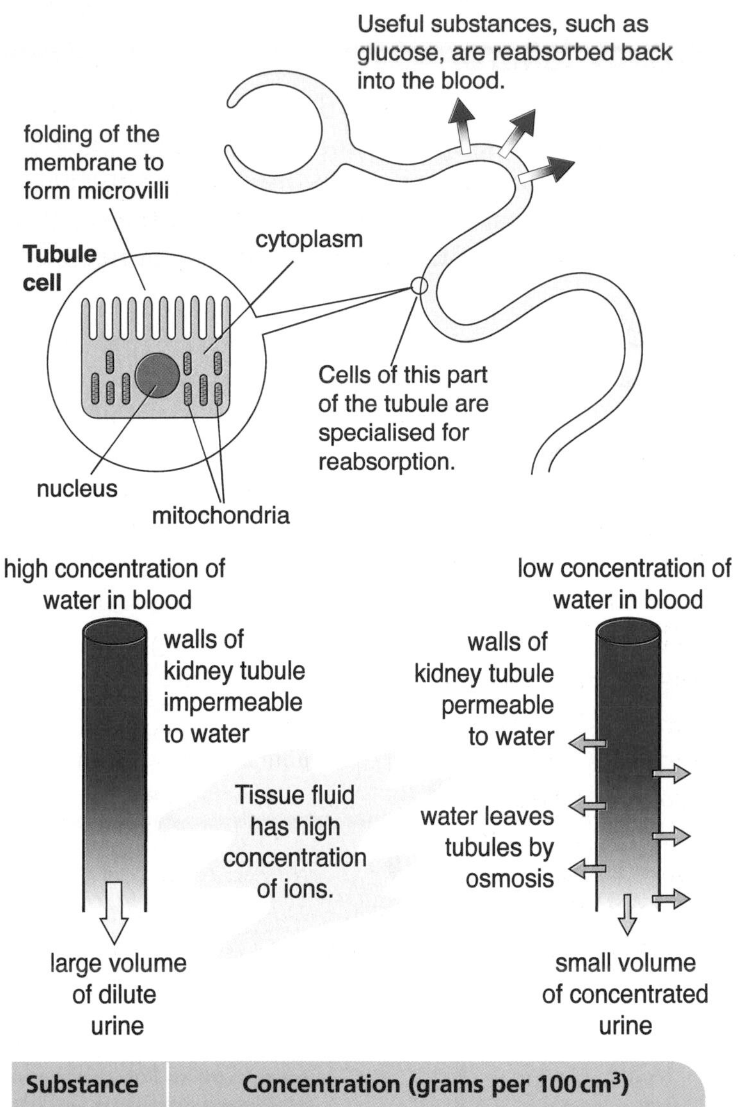

Question

b Describe the features of the tubule wall that increase the efficiency of reabsorption of useful substances.

Water balance

Your body has to balance the amount of water you take in with the amount it gets rid of. As the filtrate passes along the last part of the tubule, any water needed by the body is reabsorbed back into the blood, by osmosis, and any excess water is removed in urine. When the body is short of water it is reabsorbed from the tubule, and small amounts of concentrated urine are produced. This is why you urinate less when you do not drink much fluid.

When you drink a lot of fluid, your blood becomes diluted. When blood contains excess water, very little water is reabsorbed from the tubule. As a result your kidneys remove more water, producing large quantities of dilute urine.

The table shows some of the concentrations of substances found in different fluids.

Substance	Concentration (grams per 100 cm^3)		
	Blood plasma	Kidney filtrate	Urine
Glucose	0.10	0.10	0
Mineral ions	0.43	0.43	1.18
Protein	6.5	0	0
Urea	0.02	0.02	2.0
Water	91.0	91.0	96.0

Questions

c Which substances pass from the blood into the kidney tubule?
d Which substance is reabsorbed from the tubule into the blood?
e Explain why no protein appears in the kidney filtrate.
f Suggest why people with low blood pressure have a lower concentration of dissolved substances in the kidney filtrate.

Key points

- The kidneys produce urine, containing urea, excess ions and excess water.
- Urine is produced by first filtering the blood, then reabsorbing substances needed by the body, including glucose and some water and mineral ions.
- Glucose and dissolved ions may be actively absorbed against a concentration gradient.

Treating kidney failure

Kidney failure

Kidney failure occurs when the kidneys stop working properly. Your kidneys are working continuously, keeping you alive and well. Many causes of kidney failure can be treated, and the kidney function will return to normal with time. When the kidneys fail completely, dialysis or kidney transplantation will then become necessary.

Dialysis

Dialysis is carried out by pumping a patient's blood through an 'artificial kidney' machine, containing partially permeable membranes. As blood flows between the membranes, waste products such as urea are removed from the blood, but useful substances such as glucose and mineral ions are not. The diagram below shows how treatment by dialysis restores the concentrations of dissolved substances in the blood to normal levels. Dialysis is usually carried out in a specialised dialysis unit attached to a hospital (see the photograph on page 187). It takes a few hours and needs to be repeated every 2–3 days.

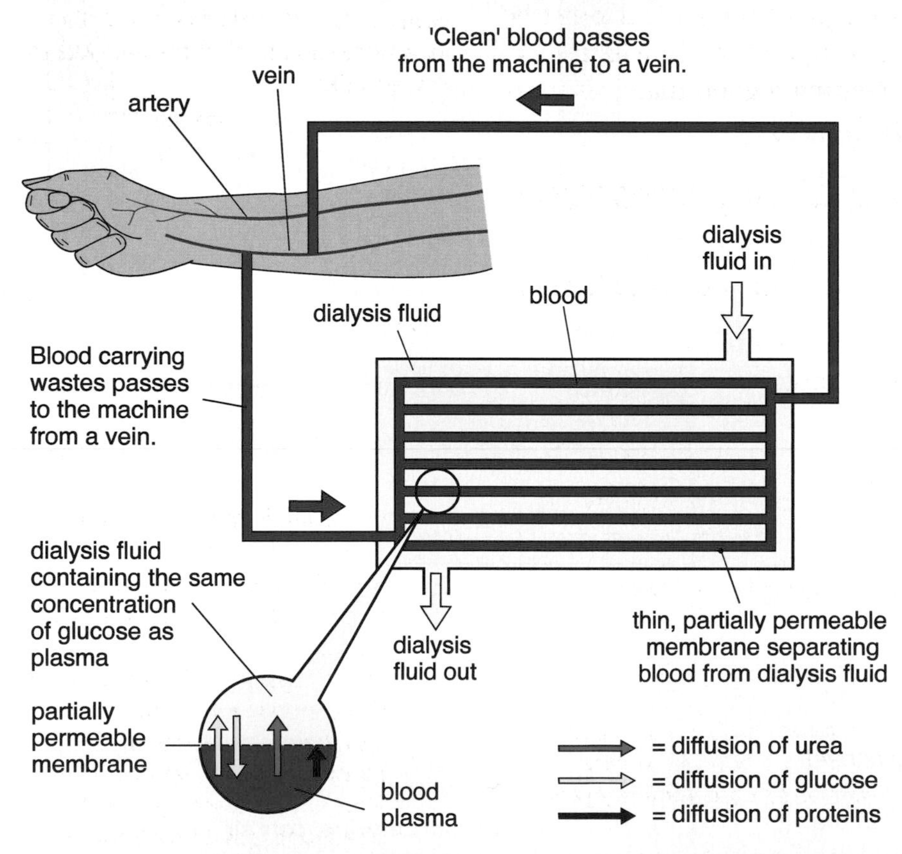

Question

a *(i) By what process does urea pass from the patient's blood into the dialysis fluid?*

(ii) Why do proteins not pass into the dialysis fluid?

(iii) Explain why glucose diffuses in both directions.

Kidney transplants

In a kidney transplant, a healthy kidney from another person is surgically removed and put into the body of someone with kidney failure. The new kidney does the work of the failed kidneys, so the patient no longer needs dialysis. The kidneys come from a donor who is either a living relative or someone who has recently died. A living donor will have only one kidney after the operation but can still lead a full and active life.

A transplant gives most people with kidney failure a healthy life again. They will no longer have to rely on repeated hospital visits for dialysis. However, not all transplants are successful, and even following successful transplants, patients need to take certain drugs for the rest of their lives to prevent the transplanted kidneys being 'rejected'. An organ is rejected when the body's immune system recognises the donor organ as foreign, and attacks it. This is why the patient is treated with drugs to control the immune system.

For a transplant to be successful, the tissue type of the patient must be similar to that of the donor, otherwise the recipient's body will 'reject' the kidney. This is called a 'good match'. Getting a good match is more likely when the organ is donated by a family relative.

Donor Card
I would like to help someone to live after my death.

I request that after my death
*A. my *kidneys, *corneas, *heart, *lungs, *liver, *pancreas be used for transplantation, or
*B. any part of my body be used for the treatment of others
*(DELETE AS APPROPRIATE)
Signature ________ Date ________
Full name ________
(BLOCK CAPITALS)
In the event of my death, if possible contact:
Name ________ Tel. ________

▲ There are currently over 7000 people on the national organ waiting list, and nearly 6500 of these people are waiting for a kidney transplant. (Data from Kidney Research UK)

Advantages and disadvantages of kidney transplants

Both dialysis and kidney transplantation are very expensive treatments. There are currently 37 000 patients with kidney failure in the UK. Nearly 20 000 are on dialysis.

Advantages of transplantation	Disadvantages of transplantation
• No need for frequent dialysis treatment • Better quality of life • Better health • Reduced medical costs • No diet restrictions	• Regular medical checks to monitor 'new' kidney • Pain and discomfort of major surgery • Risk of transplant rejection • Prone to infections • Need to take 'anti-rejection' drugs for life

Questions

b *The cost-benefit of a kidney transplant compared to dialysis is £14 200 per patient each year. In 2004, 1783 patients benefited from having a kidney transplant. Calculate how much the NHS saves in dialysis costs for every year these patients survive.*

c *Imagine you were the parent of a young child with kidney failure who is receiving dialysis treatment. How would you explain, giving reasons, why a successful kidney transplant would make their life better than by treatment on a dialysis machine?*

Key points

- In a dialysis machine, urea passes out of the blood into dialysis fluid, and glucose and useful salts remain in the blood.
- Treatment by dialysis restores the concentrations of dissolved substances in the blood to normal levels.
- There are advantages and disadvantages of treating kidney failure by dialysis or kidney transplant.
- Kidney transplants allow people to live normal lives, but there are not enough donors.

The gift of life

Transplants save lives

Today more than 7000 people in the UK need an organ transplant that could save or dramatically improve their life. Most are waiting for a kidney transplant, others for a heart, lung or liver transplant.

Transplants save lives and depend entirely on the generosity of donors and their families who are willing to make this life-saving gift to others. There is a desperate need for more donors. In 2004, more than 400 people died while waiting for a transplant. The Department of Health says that improved road safety and medical advances mean fewer potential donors are dying. There are between three and four patients waiting for every organ that becomes available. The NHS has a confidential database, called the Organ Donation Register, which now has nearly 13 000 people willing to donate their organs.

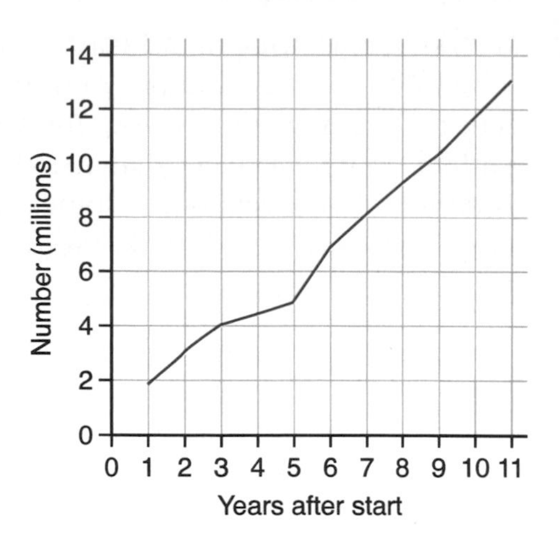

Number of people on the NHS Organ Donation Register (started in 1994).

In the UK in 2004–05, a total of 1783 people received a kidney transplant, of whom 475 were given their kidney by a friend or relative – the highest number of living kidney transplants ever recorded in the UK. Transplants from living donors are significantly more successful than those using kidneys from people who have recently died. Living donors who are close relatives can make an excellent tissue match, which is an added bonus for the person needing a transplant.

Making the decision

The decision to become a living donor is not easy – it is one of the most difficult decisions a person is likely to make. The following case study gives a first-hand account of what it is like to be involved in a living donor kidney transplant.

Question

a *Explain why an organ donated from a close relative is less likely to be rejected following a transplant.*

Mike's story

"My doctors told me that I would probably need dialysis treatment within the next couple of years because my kidneys were not working properly because of disease. The doctor also told me about transplants and suggested that I discussed the issue with my family. I really found it hard to raise the subject and it took several weeks before I discussed it with them. I felt guilty about putting someone I love through a major operation to make me fit and well again. When I told my family they were amazingly positive – all my three brothers offered to donate one of their kidneys.

My elder brother Jon was the most persuasive. He said that my quality of life would be so much better with not having to go for regular dialysis treatment, and my two young sons and my wife would see me leading a normal life again. I knew a transplant made sense but it was Jon's eagerness to help that finally convinced me to go ahead."

Jon gave his brother, Mike, the 'gift of life' by becoming a live kidney donor.

Match making

To check whether a transplant is possible, a blood sample is taken from the donor and from the person receiving the kidney to check that their blood groups are compatible. The table shows the blood groups that 'match'. The O blood group is called the 'universal donor' blood group because it can be used with any of the four blood groups.

Blood group of person receiving transplant	Required blood group of donor
A	O or A
B	O or B
AB	O or A or B or AB
O	O

Following this first test, Mike and Jon had a meeting with a transplant coordinator. Jon remembers this first meeting. "One of my impressions of this time was that much of the information we were given seemed to be about things that could go wrong, despite the superb success rates of transplants. I found out later that this was to make sure that, as the donor, I was as clear as possible about what was involved, and that I wanted to donate my kidney without any pressure being put on me. We then went through five months of tests to make sure that there was a good tissue match and everything was OK for the operation to go ahead."

Jon and Mike were operated on in a specially designed twin theatre. After the kidney was removed from Jon, it was transferred to the adjoining theatre and transplanted to replace Mike's diseased kidneys.

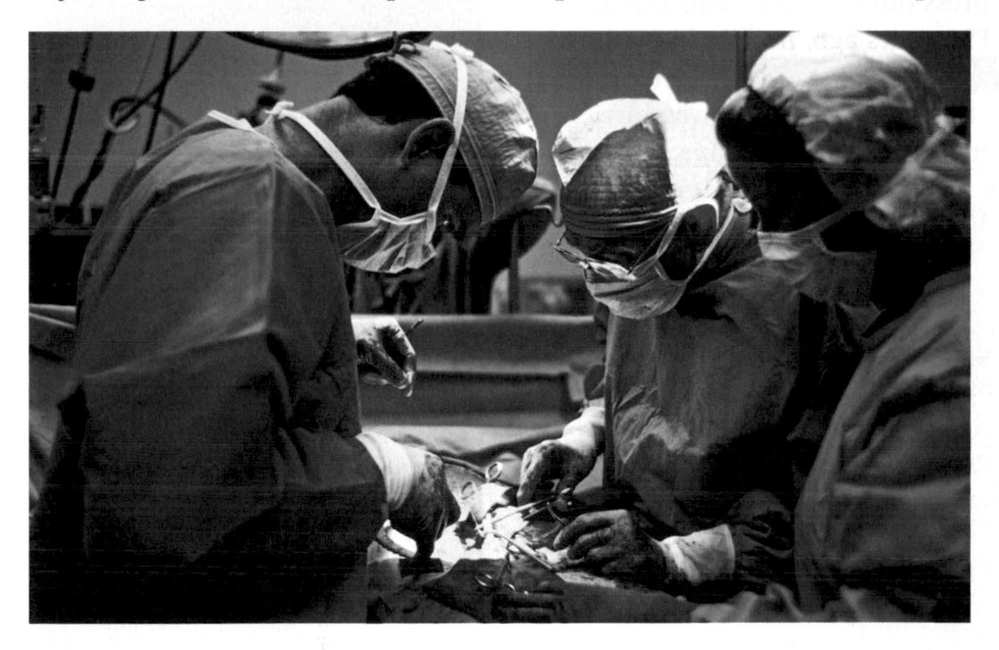

Question

b *Suggest reasons why some people with kidney failure may not be suitable for a transplant.*

No more dialysis

Following his operation, Mike found that his life was very different. "I have found I have more energy to do things, and I can eat and drink whatever I want. I know I need to look after my new kidney by taking anti-rejection drugs regularly. I can now go on holiday with my family. When I was on dialysis, I was unhappy about visiting a different hospital for treatment. I feel like a new person and I can't thank Jon enough for what he has given me."

Key points

- Transplants save lives, but transplantation from a living donor is a very difficult decision for families to make.
- There are not nearly enough donors.
- Transplants allow patients to live almost normal lives.
- You should encourage your family to sign up to the donor register online.

Fungi

When fruit goes mouldy, the moulds are seen as very fine threads called **hyphae** that cause the mouldy area to rot. Like mushrooms, yeasts and moulds are types of fungi.

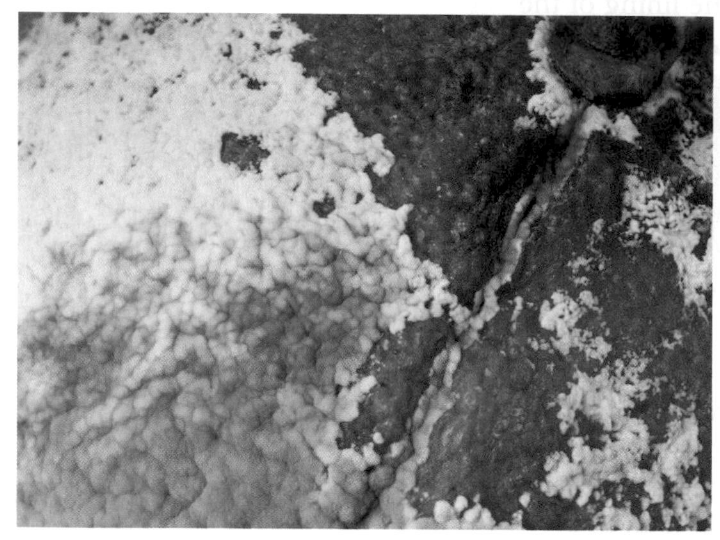

▲ Moulds are a type of fungus.

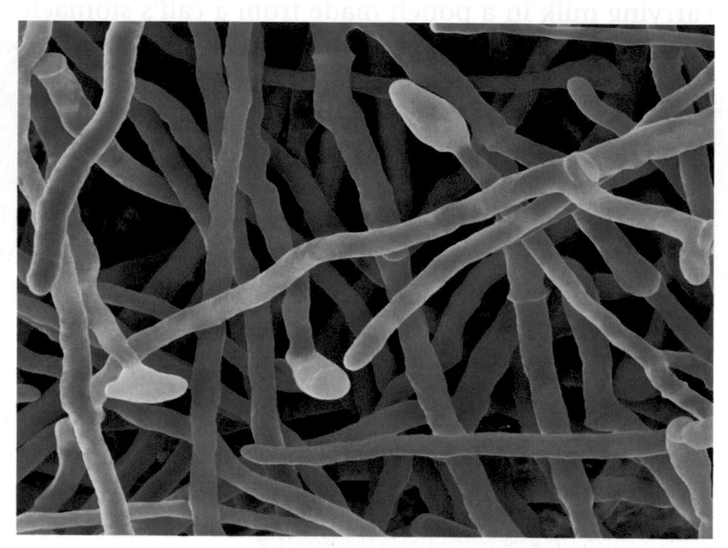

▲ Moulds are made up of hyphae.

Yeasts

Yeasts are found in dust, soil, water and milk, and even on some of the inside surfaces of our bodies. Under a microscope they are oval-shaped cells. Mature yeast cells usually have smaller yeast cells attached to them that have been 'budded-off' by asexual reproduction. Under the **electron microscope** we can see the internal structure of a yeast cell.

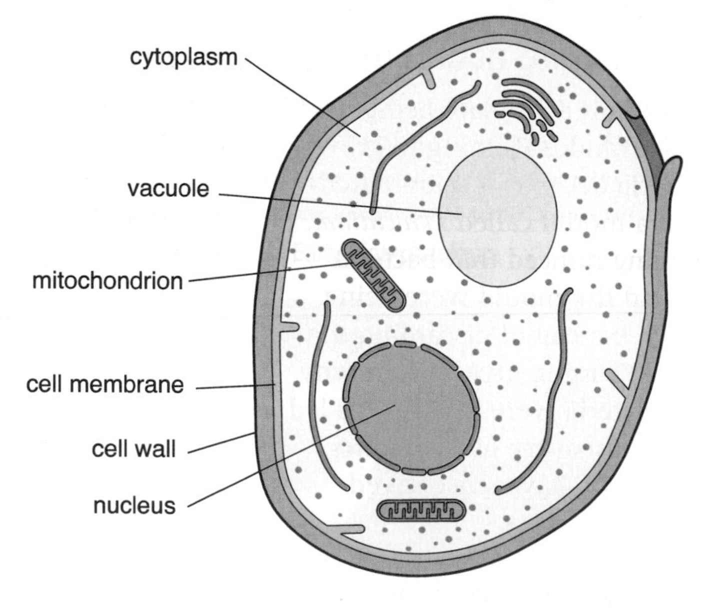

▲ Structure of a yeast cell.

Question

a *In what ways is a yeast cell*
- *(i) similar to a plant leaf cell?*
- *(ii) different from a plant leaf cell?*

Feeding microorganisms

If you wanted to grow a microorganism, it would need food. Most microorganisms are similar to animals in that they do not possess chlorophyll, so they cannot produce or **synthesise** (by photosynthesis) their own food. But bacteria and yeasts cannot be fed on solid foods like animals. They have to be provided with food that they can absorb through their cell membranes.

All the microorganisms used in the home or in industry need carbohydrates such as sugars. Like all other living organisms, microorganisms use carbohydrates as an energy source. They also use them to produce new cells.

Like plants, microorganisms, such as yeast, grow if provided only with sugar and mineral ions. Like plants, yeasts need ions such as nitrate, potassium and phosphate to produce proteins for growth. They also produce their own supply of vitamins. Moulds like those used to produce antibiotics will not grow if given only sugar and ions. These moulds cannot produce their own proteins and vitamins, so these must be supplemented or provided for the moulds.

Culturing microorganisms

Microorganisms are usually cultured (grown) in a **culture medium** which is either broth or nutrient **agar**. In industry, they are usually grown in broth. This is like soup – it contains all the nutrients they need.

Nutrient agar is often used in laboratories for growing microorganisms. Agar is a jelly-like substance that melts when heated and sets when cooled, but does not contain any nutrients. The nutrients needed by the microorganisms are added and dissolved into molten agar. The nutrient agar sets when poured into a Petri dish and allowed to cool.

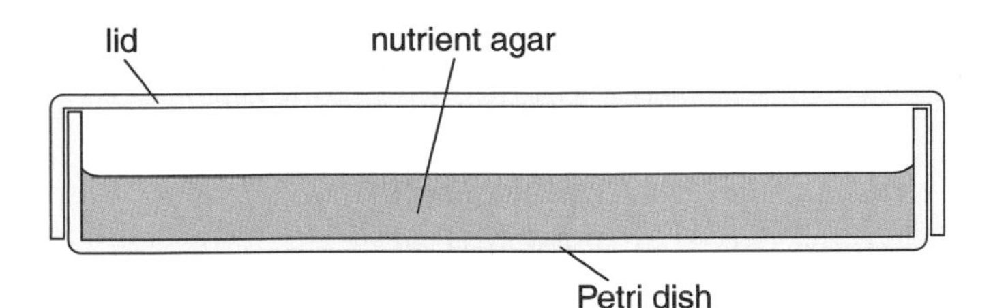

▲ A Petri dish for growing microorganisms.

Finally, the microorganisms are **inoculated** onto the surface of the nutrient agar where they grow and divide to produce a **colony** which can be seen with the naked eye.

Question

c *Suggest one advantage of growing bacteria on a jelly such as agar rather than in a broth.*

Question

b *(i) Name the process that green plants use to make food.*
(ii) Some moulds can grow if they are given starch, but they grow much faster if given sugar rather than starch. Explain why.
(iii) Name the process in all living things that releases energy from sugars.
(iv) What substance can plants, and yeast, produce from nitrates and sugars?

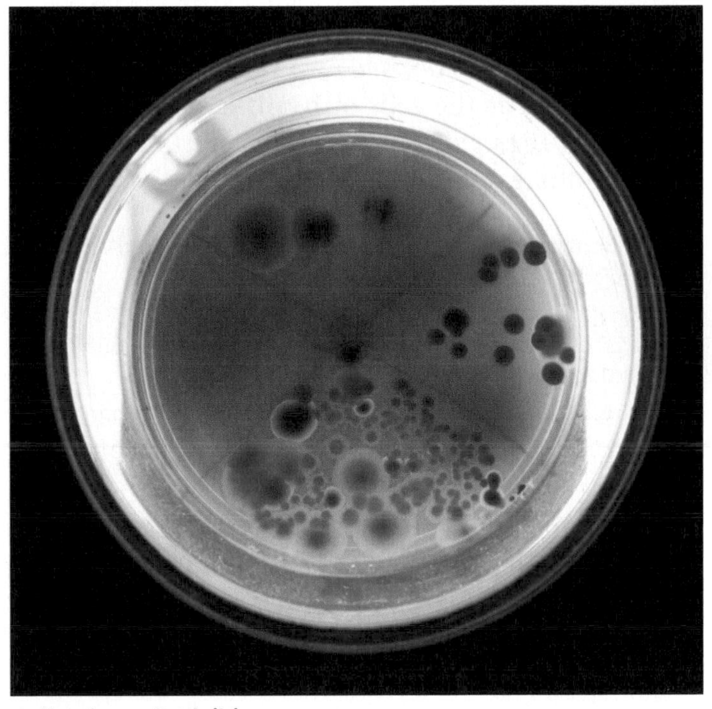

▲ Fungi on a Petri dish.

Key points

- Yeast cells have a nucleus, cytoplasm and a membrane surrounded by a cell wall.
- Microorganisms can be grown in culture media containing:
  - carbohydrate as an energy source
  - mineral ions

  and in some cases:
  - proteins
  - vitamins.
- These nutrients are often contained in agar jelly in a Petri dish.

Keeping things sterile

The importance of sterile techniques

Many different types of microorganisms will grow in nutrient broth or agar. However, most of these do not make useful products, and some produce harmful substances. It is only safe to use microorganisms if we have a pure culture, so it is important to prevent cultures of microorganisms from becoming contaminated with unwanted microorganisms. There are two main ways of doing this:

- Using **sterile** equipment and nutrient materials
- Preventing contamination from the air.

The type of autoclave used in many schools.

Sterilisation

In adverse conditions, many bacteria and fungi produce tiny reproductive structures called spores. These are extremely resistant to changes in conditions, so we must use techniques that will kill spores as well as the cells or **hyphae** of microorganisms. Sterilisation equipment uses either dry heat or moist heat (steam). The table shows the conditions needed to be sure of killing microorganisms.

Form of microorganism	Time required to kill bacteria	
	Moist heat	Dry heat
Bacterial cells, yeast cells, fungal hyphae	10 minutes at 65 °C	60 minutes at 100 °C
Fungal and yeast spores	25 minutes at 90 °C	60 minutes at 115 °C
Bacterial spores	10 minutes at 120 °C	60 minutes at 150 °C

These conditions can be produced in an autoclave. This works on the same principle as a pressure cooker. Heating water under pressure produces steam at a temperature of 120 °C.

Question

a *Which is the quickest method of ensuring that culture media and equipment are completely sterile?*

Transferring microorganisms

A common piece of equipment used for transferring microorganisms is the inoculating loop. This is a loop of nichrome wire fixed in a glass or metal rod. The loop is held in a Bunsen flame until it glows. At this temperature, all microorganisms are killed. The loop is then allowed to cool before use.

Question

b *Why must the loop be cooled before use?*

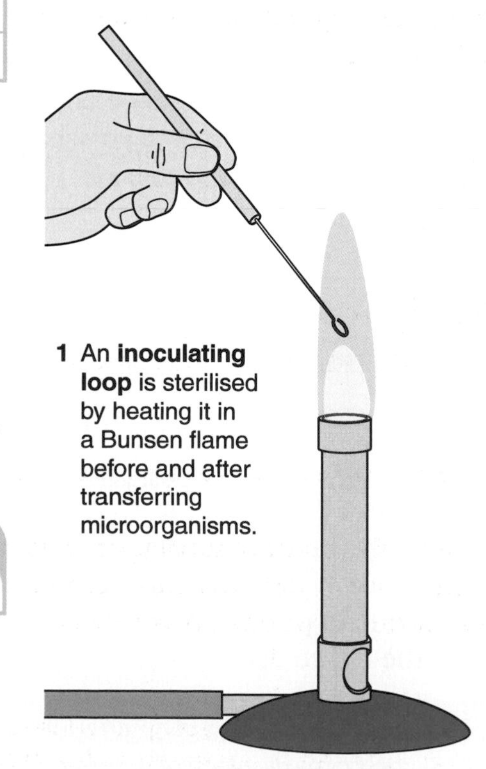

1 An **inoculating loop** is sterilised by heating it in a Bunsen flame before and after transferring microorganisms.

Sterilising an inoculating loop.

Collecting microorganisms

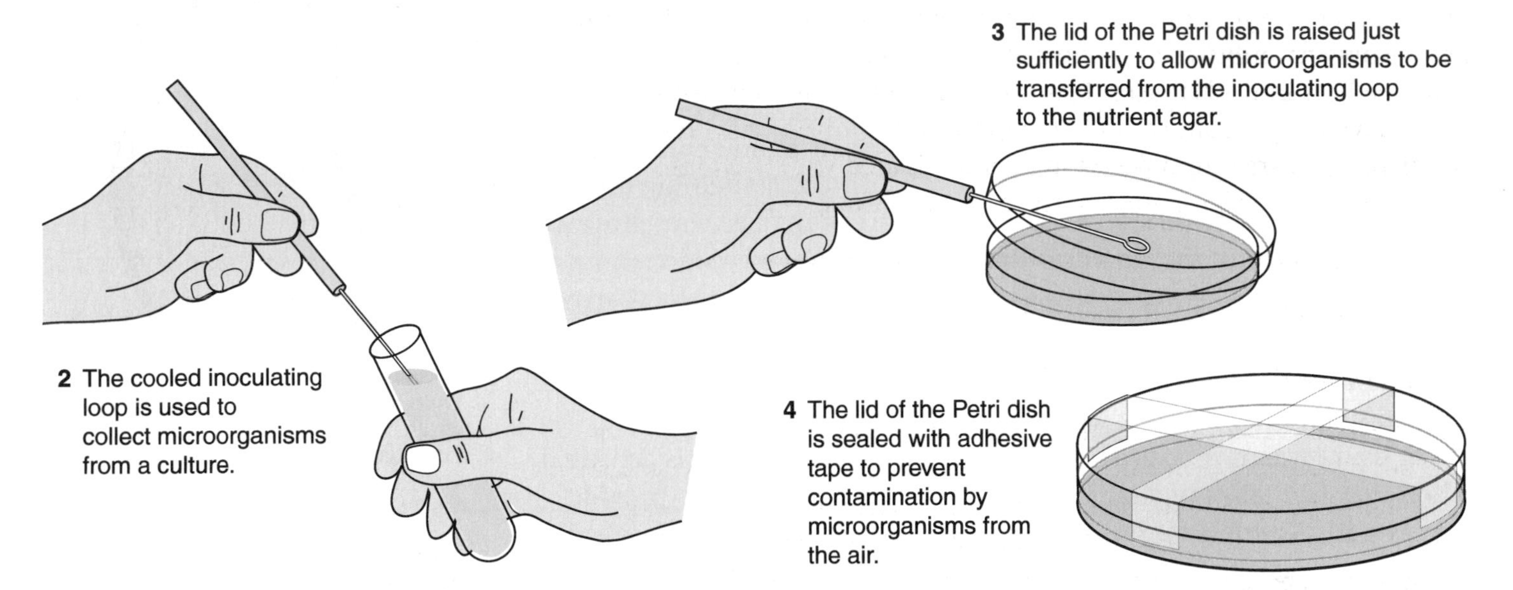
3 The lid of the Petri dish is raised just sufficiently to allow microorganisms to be transferred from the inoculating loop to the nutrient agar.
2 The cooled inoculating loop is used to collect microorganisms from a culture.
4 The lid of the Petri dish is sealed with adhesive tape to prevent contamination by microorganisms from the air.

▲ Collecting microorganisms, inoculating an agar plate and sealing the Petri dish.

After sterilisation in a Bunsen flame and cooling, the inoculating loop is used to collect microorganisms from a culture. The lid of the Petri dish is raised just sufficiently to allow microorganisms to be transferred from the loop onto the nutrient agar. The lid is then sealed with adhesive tape.

Questions

c *Explain why it is important not to touch the nichrome wire with your fingers.*
d *Explain how the last two techniques will help to ensure a pure culture on the nutrient agar.*

Safety first

Danger! Bacteria that cause disease in humans grow best at about 37 °C.

Question

e *Explain why disease-causing bacteria have evolved to grow best at 37 °C.*

In school laboratories, bacteria must be grown at a maximum temperature of 25 °C. At this temperature useful bacteria grow quickly but harmful bacteria grow slowly.

Question

f *In the sections that follow you will see that temperatures much higher than 25 °C are used in some industrial processes. Explain why higher temperatures are used in industry.*

Key points

- Equipment and culture material must be sterilised to prevent the growth of unwanted microorganisms.
- Transfer techniques must ensure that cultures are not contaminated with unwanted microorganisms.
- In schools, bacteria must be cultured at 25 °C or less to reduce the rate of growth of disease-causing bacteria.

B3 4.3 Making beer and wine

Barley – the beer cereal

Beer is made by using yeast. The source of carbohydrate for the yeast in this process is barley.

The diagram shows a section through a barley grain. The **embryo**, which can grow to form stems, roots and leaves, is at one end of the grain. Most of the remainder consists of a food store containing mainly starch. Yeast cannot use starch, so the first part of the brewing process is the breakdown of the starch to simpler sugars.

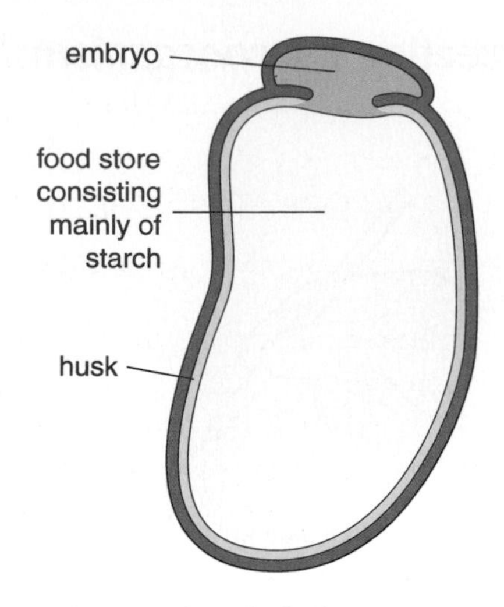

▲ Cross-section of a barley grain.

Malting

Before a barley grain can **germinate**, it must convert starch from the food store into sugars. This reaction is called **hydrolysis** – breaking down using water. It is done by the enzyme amylase which is activated when the seed takes in water. The embryo uses the sugars to provide energy and materials for growth. In malting, barley grains are soaked in water to begin this process. After about three days the grain is heated to stop growth, but the temperature used is not high enough to **denature** the enzyme amylase. The barley is now called malt.

▲ Malting barley.

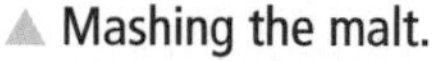

▲ Mashing the malt.

Mashing

The malt is crushed, then placed in a mash tun and covered with water at about 65 °C. The amylase that was produced during malting now breaks the starch down into sugars. This process takes about an hour. The liquid is now called wort.

Question

a *In what way does this amylase differ from most enzymes that you have studied?*

After about an hour the amylase has broken the starch down into sugars that can be used by yeast. The wort is then filtered to remove the barley husks and finally boiled for about another hour.

While the wort is boiling, hops are added to it. Substances from the hops give additional flavour to the beer.

Question

b *Suggest why the wort is boiled.*

Where the alcohol comes from

The wort is cooled below 20 °C, then run into deep fermentation tanks. Now the yeast is added. At first the yeast respires aerobically, in the same way as most other living organisms:

sugar + oxygen → carbon dioxide + water (+ energy)

This type of respiration is called aerobic respiration, because oxygen is used. The energy released is used in growth and reproduction of yeast cells. The surface of the liquid in the fermentation tank soon becomes covered with a thick foam containing carbon dioxide. This cuts off the oxygen supply to the yeast, which responds by respiring anaerobically.

Anaerobic respiration is respiration without oxygen. You have already met it in muscle earlier in this book. In yeast, it is similar in one respect – not as much energy is released as in aerobic respiration. But the end products are completely different. In muscle, anaerobic respiration produces lactic acid, but in yeast it produces ethanol (alcohol) and carbon dioxide. The word equation for anaerobic respiration in yeast is

glucose → ethanol + carbon dioxide

After a few days the yeasts respire all the sugar, and fermentation stops. The yeast is then separated from the beer. Some of the yeast is used to ferment the next batch of wort. The rest is sold, to be used mainly in foodstuffs. The beer is left for a few days to become clear, then it is bottled, canned or stored in barrels.

▲ Where sugar is turned into alcohol and carbon dioxide.

Making wine

Wine is produced when yeast ferments the natural sugars present in grapes. The yeast respires anaerobically just as in beer production. Modern wine making consists of the following steps:

- Grapes are crushed mechanically.
- Grape juice is separated from the skin and stems.
- Sulfur dioxide is added to kill other microorganisms.
- Wine yeast is added and fermentation begins.
- The wine is separated from the yeast, then bottled.

▲ Vats for making wine.

Question

c *Explain why malting and mashing are not required in wine making.*

Key points

- Yeast can respire aerobically (using oxygen): sugar + oxygen → carbon dioxide + water (+ energy)
- The energy released during aerobic respiration is used mainly for growth and reproduction.
- Yeast also respires anaerobically (without oxygen): glucose → ethanol + carbon dioxide
- Anaerobic respiration releases less energy than aerobic respiration
- Beer and wine making use yeast to ferment sugars in barley (for beer) and grapes (for wine) to produce ethanol.

Making yoghurt

The 'birth' of yoghurt

Why the Arab merchant thought he could carry milk in a pouch made from a calf's stomach across a desert for days, we'll never know. That night when he ate the curds, so the story goes, yoghurt was born.

The Ancient Greeks and Romans realised that they did not have to wait for milk to go sour to make yoghurt. They speeded up the process by adding sour milk to fresh milk. Yoghurt is the Turkish word for sour milk.

Adding bacteria

We now know how this works. Milk goes sour when lactic acid bacteria called *Lactobacillus* ferment the milk sugar, **lactose**, to form lactic acid. These bacteria are present on the skin of cows, which is how they easily get into milk.

lactose + water → lactic acid

The lactic acid produced by the bacteria lowers the pH of the milk sufficiently to make the milk curdle. Yoghurt used to be made by adding yoghurt to fresh milk, but yoghurt factories now use pure cultures of bacteria.

The milk for the process is usually sterilised, with added milk proteins.

From milk to yoghurt

After sterilising and cooling to 45 °C, a starter culture of bacteria is added. The most common are *Streptococcus thermophilus* and *Lactobacillus vulgaris*. Both are needed and both produce lactic acid. *Streptococcus* works first, lowering the pH of the milk to 5.0. Then *Lactobacillus* lowers it further, to about 4.0. The change in pH makes the milk proteins go solid and the whole of the milk goes solid. This process takes about 5 hours.

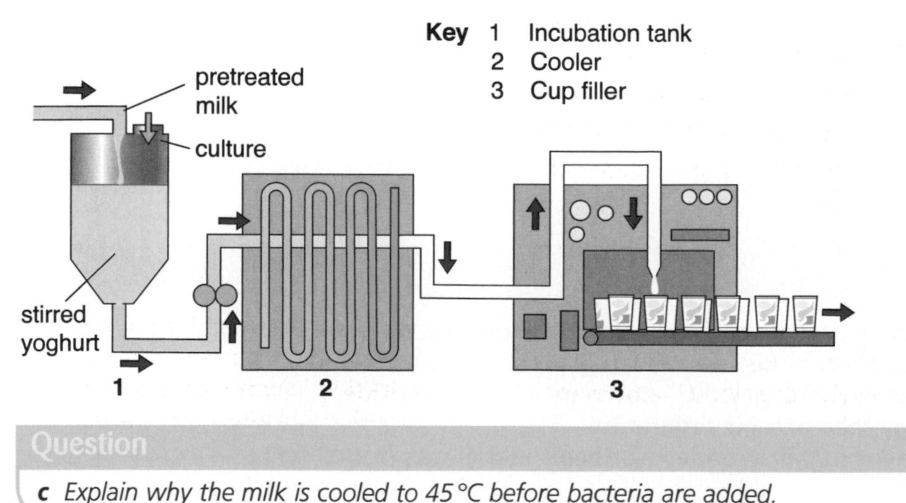

Question

c Explain why the milk is cooled to 45 °C before bacteria are added.

Traditional yoghurt making.

Questions

a Explain why pure cultures of bacteria are used rather than adding old yoghurt.

b Suggest why:

(i) the milk is sterilised

(ii) extra protein is added to the milk.

When the pH reaches 4.0, the yoghurt is cooled quickly. It is then put into pots and sealed.

Yoghurt has a shelf life of about 2–3 weeks if refrigerated. After that it begins to taste sour.

Question

d *Explain why the yoghurt is rapidly cooled when the pH reaches 4.0.*

▲ Modern yoghurt making.

Do yoghurt bacteria make you more healthy?

In countries where yoghurt has been eaten for centuries, many people live to be 90 or even 100 years old. This has led many people to believe that they will live longer if they eat lots of yoghurt. A whole new industry has developed, selling bacterial cultures for people to drink. These drinks are called '**probiotics**'.

Question

e *From the information on the left, what evidence is there that eating yoghurt makes you live longer?*

An investigation was sponsored by a Dairy Marketing group to find out whether yoghurt bacteria improved the health of elderly people. Thirteen elderly subjects were given 180 cm^3 (180 ml) of whole milk supplemented with lactic acid bacteria twice daily for six weeks. As a control group, 12 other elderly subjects were given 180 cm^3 (180 ml) of low-fat milk twice daily for six weeks. At the end of the six-week trial, the researchers reported that the group who had been given the bacterial supplement had white blood cells which were far more active than those in the control group.

Question

f *(i) Does this investigation give valid, reliable results? Explain the reasons for your answer.*
(ii) How credible are the results of this investigation? Explain the reasons for your answer.

Key points

- Yoghurt is made by the action of bacteria on warm milk.
- Bacteria ferment the sugar in milk to produce lactic acid.
- Lactic acid causes the milk to solidify and form yoghurt.

Using microorganisms to make drugs

Microorganisms are used to make drugs. These have extremely complicated molecules and are produced only in tiny amounts by microorganisms.

To make large quantities of drugs, microorganisms have to be grown in very tightly controlled conditions. They are grown in large tanks called **fermenters** like the one shown in the diagram.

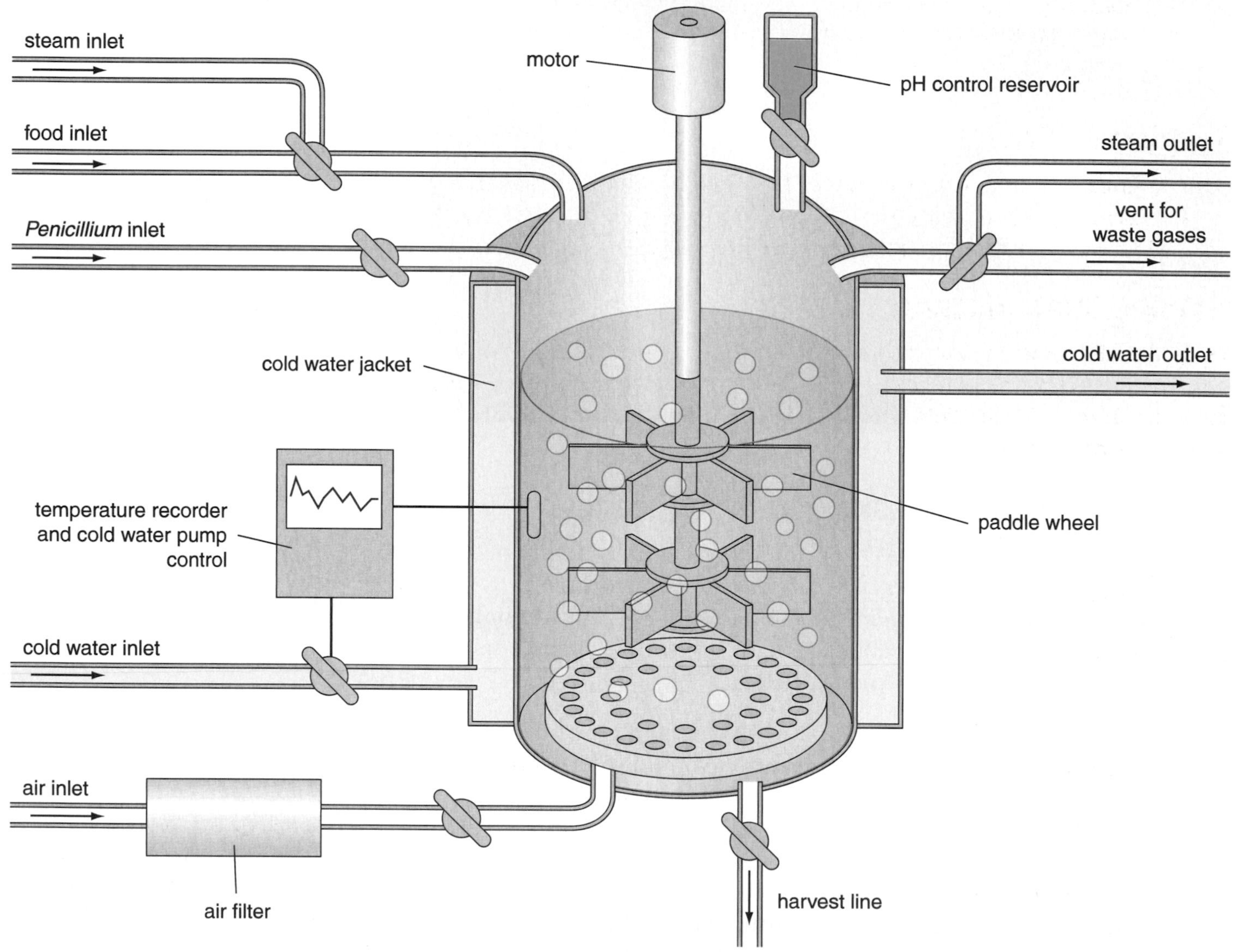

▲ Fermenter.

Microorganisms are grown in fermenters, which may hold up to 2 million litres. The cells are supplied with nutrients and oxygen. The conditions inside the fermenter are carefully controlled to keep the microorganisms at the desired stage in their growth. Nutrients are fed in at a controlled rate, again to keep the microorganisms at the correct stage in their growth.

Oxygen supply

Most organisms used in fermenters are aerobic – they require oxygen. In these cases the fermenter is well aerated by first filtering the air, then feeding in the air at the base of the fermenter. The air then bubbles up through the liquid.

Question

a *(i) Which process in microorganisms requires oxygen?*
(ii) Explain why air should be sterilised before entering the fermenter.
(iii) Suggest two advantages of bubbling the air in from the bottom of the fermenter.

Keeping it stirred

Most fermenters have stirrers composed of paddles. These ensure that all the microorganisms in the tank receive oxygen and nutrients. They also help to maintain the same temperature in all parts of the tank.

Keeping the temperature constant

When the fermenter is set up, steam is blown through the contents to raise the temperature to the optimum. The steam supply is then turned off. After a while, cold water has to be pumped around the cooling jacket to prevent the temperature rising higher than 24 °C.

Question

b *The temperature in the fermenter would continue to rise after the steam supply is shut off were it not for the cold water jacket.*
(i) Explain why the temperature would continue to rise.
(ii) Suggest what might happen to the microorganisms if the temperature was allowed to continue to rise.

Monitoring conditions

The contents of the fermenter are monitored by probes. These record:

- temperature
- pH
- oxygen concentration
- carbon dioxide concentration.

Question

c *From your work in Additional Science biology, explain why it is important to keep a constant, optimum pH inside the fermenter.*

Key points

- Microorganisms can be grown in large vessels called fermenters.
- They are supplied with nutrients, and oxygen for respiration. The contents of the fermenter are kept well mixed by stirrers. A water-cooled jacket prevents the fermenter from overheating.
- Temperature and pH in the fermenter are monitored by probes.

Making penicillin

Antibiotics

Moulds like *Penicillium* produce antibiotics in the wild to kill competitors. In the wild they live on dead matter. Spores land on the dead matter and produce digestive hyphae. These secrete antibiotics to kill other microorganisms that might be feeding on the material. We grow these moulds to produce antibiotics that we use to kill pathogens.

The species of *Penicillium* which Fleming worked on, *Penicillium notatum*, does not make much penicillin. It makes less than 1 μg (0.000001 g) of penicillin for each gram of sugar supplied. Since then, several species of *Penicillium* with higher yields of penicillin have been discovered.

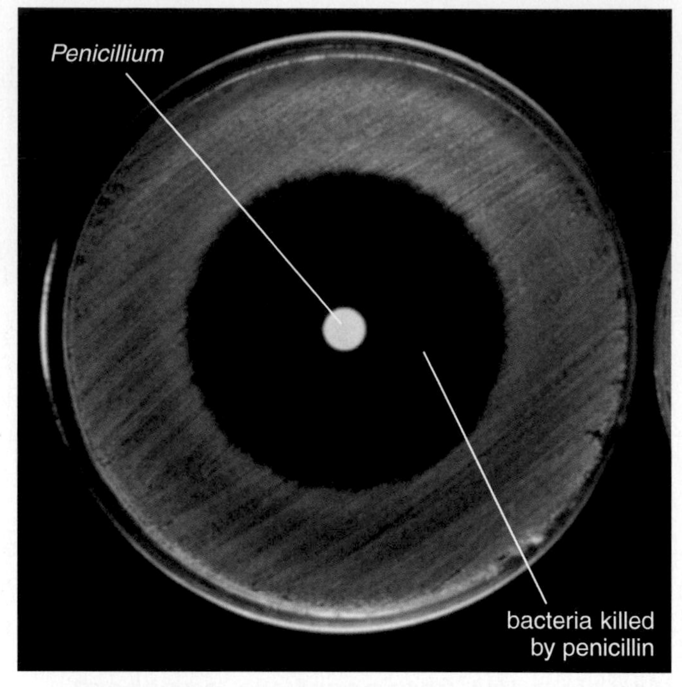

▲ Penicillin is lethal to most bacteria.

Question

a *(i) Scientists are continually screening new species of moulds for antibiotic production. Suggest two reasons why.*
(ii) Outline an investigation to compare penicillin production by different species of Penicillium. *In your outline, refer to:*
– the independent variable
– the dependent variable
– any controls
– the method of measuring the dependent variable.
(iii) Would your investigation lead to valid, reliable results? Explain the reasons for your answer.

Conditions inside the fermenter

The *Penicillium* is provided with the conditions needed for optimum growth in the fermenter. This can have a capacity as large as 150 000 litres. The food material is usually a substance called corn-steep liquor. This is a thick, sticky waste product of corn processing that contains sugar and other nutrients needed by the *Penicillium*. These nutrients include nitrogen compounds. The liquor also contains a **precursor** of penicillin – which speeds up penicillin production.

▲ *Penicillium* fermenters.

Question

b *Penicillium uses nitrates in a similar way to green plants.*
(i) Which compounds can green plants synthesise from nitrate?
(ii) What is the function of these compounds?

Conditions inside the fermenter are kept optimum for the growth of the fungus:

- high oxygen concentration
- slightly alkaline pH
- constant temperature of 24 °C.

Why does *Penicillium* make penicillin?

No further nutrients are added once growth has begun. Fermentation takes about a week altogether. Penicillin is made by the fungus from about 20 hours onwards – when **exponential growth** (2, 4, 8, 16, 32, 64, etc.) of the fungus has ceased. This is when the fungus has used up most of the nutrients.

Question

c Suggest why Penicillium *produces penicillin only when nutrient supplies are low.*

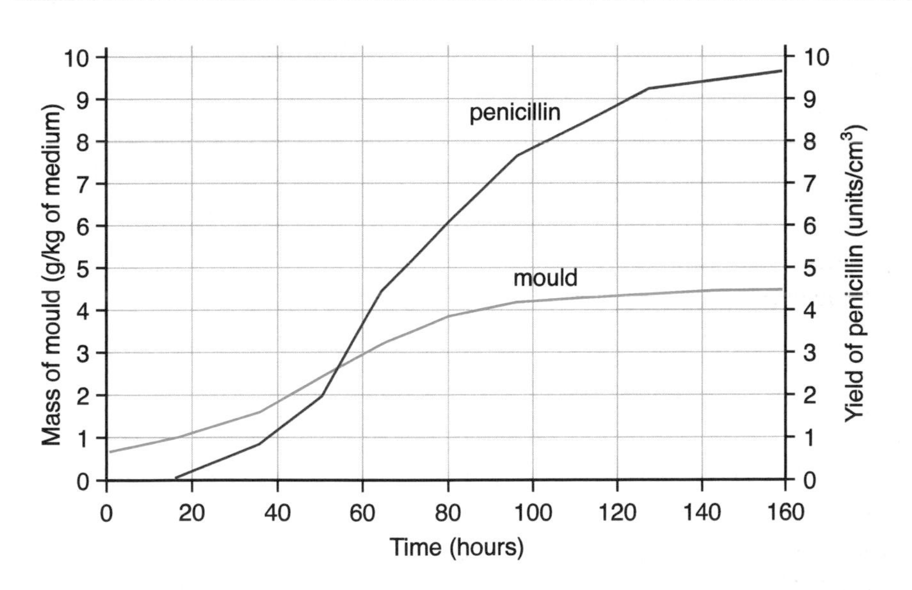

The graph shows how the yield of penicillin production is related to the mass of the mould.

Question

d (i) Describe how the mass of the mould varies with time.
(ii) Describe how the yield of penicillin production varies with time.
(iii) Suggest an explanation for the relationship between the yield of penicillin and the mass of the fungus.

Once fermentation is complete, the contents are removed from the fermenter and filtered to separate off the fungus. Penicillin is then extracted from the liquid by allowing crystals to form from the solution.

Key points

- Penicillin is made by *Penicillium* mould in fermenters.
- The nutrient source is corn-steep liquor which contains sugars and nitrogen compounds.
- The *Penicillium* grows rapidly at first, then more slowly as nutrients run out.
- Penicillin is not made by the fungus until most of the nutrients have been used up.

Making mycoprotein

Food for vegetarians

In 1968, scientists discovered a fungus called *Fusarium venetum* in a wheat field in Buckinghamshire. Their research led to the marketing of protein made by the fungus – **mycoprotein**. Mycoprotein is ideal for vegetarians because it has no connection whatsoever with animals. Recently the company that manufactures mycoprotein was sold for £172 million. Who says that science doesn't pay?

▲ The £172 million fungus.

The world population is increasing rapidly. This has resulted in a rising demand for protein for both people and animals. Microorganisms as a source of protein are one solution to this problem. The single cell of a microorganism is a perfect protein factory. In a fermenter, the culture of single cells can transform simple substances into protein. A 500 kg bullock can yield 1 kg of new protein per day; 500 kg of soya beans can yield 40 kg of new protein per day; but 500 kg of microorganisms could yield 50 000 kg per day.

Protein factories

▲ Mycoprotein fermenters.

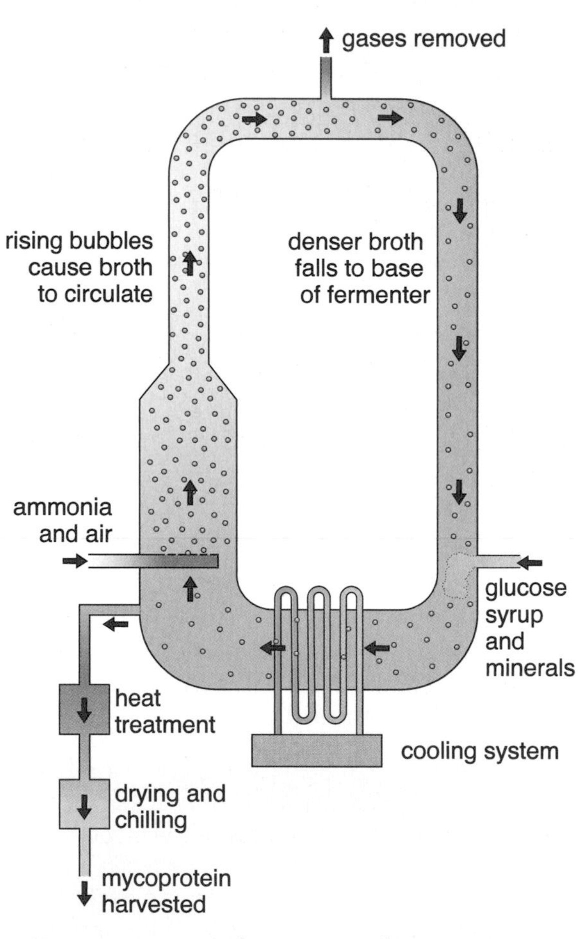

▲ How a mycoprotein fermenter works.

Mycoprotein is manufactured in fermenters. The fermenters currently being used to manufacture mycoprotein are twice as high as Nelson's column. The two fermenters run continuously for six weeks. They are then sterilised and prepared for the next run. During a run, nutrients are fed into the fermenter continuously and mycoprotein removed continuously.

Compare the mycoprotein fermenter with the penicillin fermenter on page 206.

Question

a *Apart from shape, what is the major structural difference between the two fermenters?*

In the mycoprotein fermenter, circulation of the broth is achieved by pumping air in near the base. This air at the base reduces the density of the broth, which therefore rises. Gases, including carbon dioxide, leave the fermenter at the top. This increases the density of the broth, which descends in the other limb of the fermenter.

Question

b *Air is used to mix the broth. Give one other reason why air is pumped into the fermenter.*

Nutrients

Apart from oxygen, carbohydrate is required. This is provided as a syrup, containing approximately 95% glucose. The glucose is obtained mainly from maize starch which has been broken down using amylase enzymes. Nitrogen is also needed; this is supplied as ammonia.

Questions

c *Why is glucose supplied to the fungus rather than starch?*
d *Why is nitrogen needed by the fungus?*

Other mineral ions, including potassium, magnesium and phosphate, are also supplied, as well as trace elements.

Making mycoprotein into food

After emerging from the fermenter, the mycoprotein is heated to 65 °C. This treatment breaks down substances in the mycoprotein which can cause kidney problems if left in the food.

The material is then dried in huge **centrifuges**. It now looks rather like pastry and has a slight mushroom-like smell. Flavourings and other ingredients are then added to it to produce a palatable product which is then purified. The fibrous texture of the mycoprotein makes it slightly chewy, similar to meat and fish.

Question

e *The table shows the composition of chicken pieces, beefburgers, and some Quorn products (food made from mycoprotein). In which ways:*
(i) are Quorn burgers healthier than beefburgers?
(ii) are chicken pieces healthier than Quorn pieces?

Per 100 g	Quorn pieces	Chicken pieces	Quorn burgers	Beef-burgers
Energy (kJ)	355.0	621.0	490.0	1192.0
Protein (g)	12.3	24.8	12.8	15.0
Carbohydrate (g)	1.8	0.0	5.8	3.5
Oil/fat (g)	3.2	5.4	4.6	23.8
of which saturates (g)	0.6	1.6	2.3	10.0
Fibre (g)	4.8	0.0	4.1	0.4
Sodium (g)	0.2	0.1	0.5	0.5

Key points

- The fungus *Fusarium* is used to make food called mycoprotein, which is particularly suitable for vegetarians.
- The fungus is grown on waste carbohydrates in aerobic conditions in fermenters.
- Ammonia is supplied as a source of nitrogen to help the microbes build proteins.

Where do microbes come from?

Biogenesis or spontaneous generation?

Although humans have used microorganisms for at least 3000 years, it is only in the last 200 years that scientists have investigated their structure and functioning. Investigation of microbes began during a debate among scientists about two competing theories.

Spontaneous generation was the theory according to which fully formed living organisms sometimes arise from non-living matter. The Ancient Greek philosopher Aristotle laid it down as an observed fact, citing the following 'evidence'.

- Plant lice arise from the dew which falls on plants.
- Fleas are developed from decaying matter.
- Mice come from dirty hay.

For nearly 2000 years, no one questioned this theory!

A theory based on science

In 1668, Francesco Redi, an Italian physician, did an experiment with flies and wide-mouthed jars containing meat. This was a true scientific experiment – many people say this was the first real experiment. It was generally believed at that time that rotten meat turned into maggots, then flies. Redi's hypothesis was that only flies can produce more flies.

Redi's experiment and others led to the alternative theory, of **biogenesis** – that living organisms can be produced only by other living organisms.

A close look at microbes

Microorganisms were first discovered about 250 years ago. Antonie van Leeuwenhoek built one of the earliest microscopes. He used his microscope to look at lots of different things, including drops of water. When he looked at water drops magnified many times, he saw microorganisms.

Throughout the 1700s there was a great debate among scientists about the origin of the microorganisms found in decaying meat and fermenting fruit juices. The question was whether the microorganisms *cause* decay and fermentation, or whether they are *produced* by decay and fermentation.

1 One jar was left uncovered. The second jar was covered.

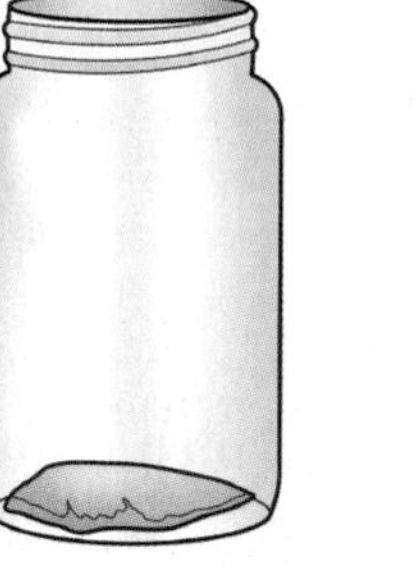

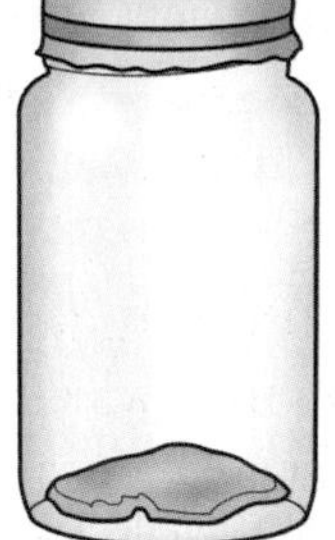

2 Maggots appeared on the meat in the open jar.

3 No maggots appeared in the covered jar.

▲ Redi's experiment.

Question

a *Write a conclusion to Redi's experiment.*

Evidence to support the theory

In the late 1700s Lazzaro Spallanzani did an experiment to try to settle the debate. He put meat broth into two flasks and sterilised them both by boiling the broth. One of the flasks was left open to the air. The other flask was sealed to keep out any organisms that might be floating in the air. Microorganisms developed only in the uncovered flask.

Question

b *Where did the microorganisms in the open flask come from?*

Unfortunately, many scientists were not convinced by Spallanzani's experiment. They believed that air needed to enter the flask in order for life to be created from non-living materials. It was another 100 years before Spallanzani was proved correct, when another scientist called Louis Pasteur performed a similar experiment.

Question

c *How did Pasteur's results convince the scientists who doubted Spallanzani's conclusion?*

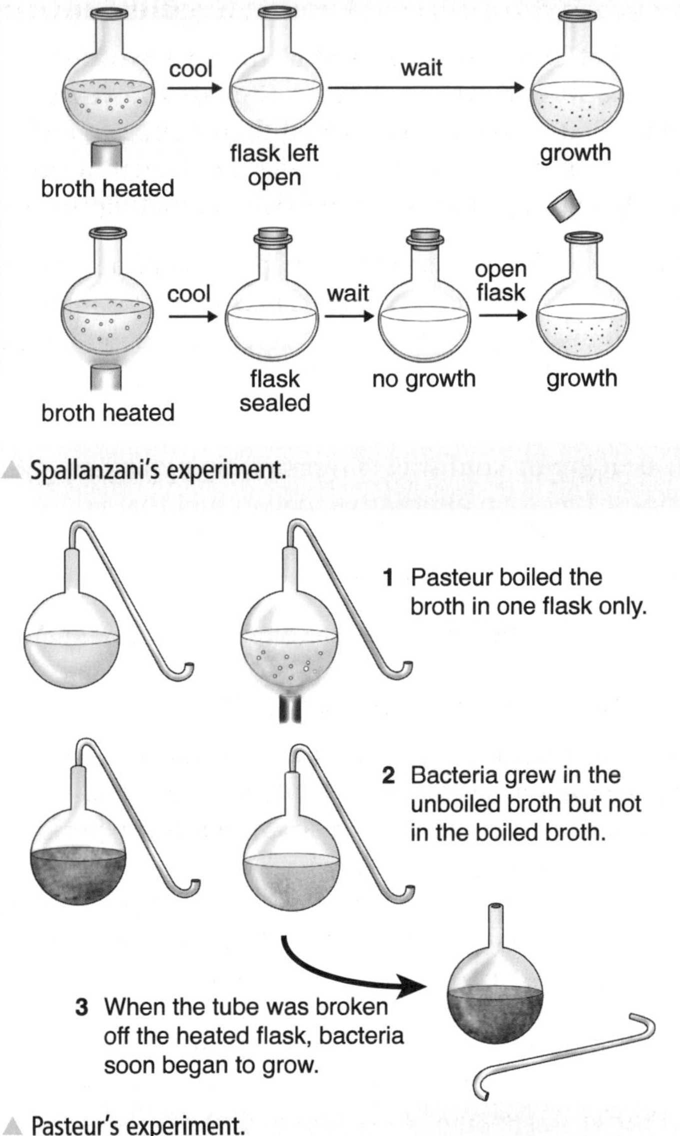

Spallanzani's experiment.

1 Pasteur boiled the broth in one flask only.
2 Bacteria grew in the unboiled broth but not in the boiled broth.
3 When the tube was broken off the heated flask, bacteria soon began to grow.

Pasteur's experiment.

Schwann establishes the theory

Theodore Schwann modified Spallanzani's experiment. He passed the air entering the unsealed flask through a tube heated to 357 °C. This time, no microorganisms appeared in the unsealed flask.

Question

d *Explain Schwann's results.*

Schwann's results were the final nail in the coffin for the theory of spontaneous generation. Biogenesis became an established theory. Schwann then went on to do a series of experiments on yeast and fermentation. He showed that yeast is the microorganism that is responsible for fermentation in the manufacture of both beer and wine.

Key points

- Spontaneous generation was a theory based on non-scientific ideas rather than scientific evidence. Evidence from a controlled scientific experiment led Redi to propose the hypothesis of biogenesis – that living organisms were produced only by other living organisms.
- Spallanzani's experiment, which prevented air from entering broth, supported this hypothesis. Pasteur provided further evidence to support the hypothesis from an experiment which allowed air, but not microbes, to enter broth.
- Schwann heated air entering broth to kill any microbes. This experiment was sufficient to make biogenesis a theory rather than a hypothesis.

B3 5 Sustainable fuels

Most developing countries are short of fuel for cooking. Often, no fossil fuels are available. Wood is a renewable fuel. But too often, trees are chopped down for fuel without replanting. So wood is often scarce – and because of this its price is high.

▲ Wood is often very scarce.

Sustainable fuel for developing countries

In developing countries, there is little money to import fuels. An alternative, cheap fuel that is readily available is the dung produced every day by farm animals. For some people this provides their only income.

Dung sticks are a **sustainable** fuel source – they are made from animal waste products. However, burning them does harm the atmosphere. The carbon dioxide released when they are burned replaces the carbon dioxide taken in during photosynthesis.

Fuel made of cow dung has become the sole source of energy for many poor families in the Sunamganj district of Bangladesh. Members of about 400 families in different parts of the district make and sell cow dung sticks. About 60% of the people in the district use the sticks as fuel, 20% use firewood, 18% use jute, straw and dry leaves, and only 2% use natural gas for cooking.

◀ Selling dung sticks.

The demand for cow dung sticks is increasing each day due to shortage of natural gas and firewood. Until a few years ago, there were plenty of trees. These were used for firewood and house building. But due to indiscriminate felling of trees the supply of firewood has now been used up; prices also shot up. As a result, many families have switched to using these sticks as fuel.

Sumita Rani, a widow with four children, lost everything she had due to flash floods last year. She had nothing to feed her children. So she began collecting cow dung and made fuel sticks for cooking. She sold these in the market and earned some money. Now, selling dung sticks is her livelihood. Making fuel sticks from dung is very easy. Dry grasses are mixed with dung, and sticks are coated with this mixture. The sticks are then dried in the sun to make fuel.

Biogas

People have been using biogas for over 200 years. In the days before electricity, biogas from the underground sewer pipes in London was burned in street lamps. Biogas is a mixture of gases, usually containing carbon dioxide and methane. It is produced by microorganisms in anaerobic conditions.

Animals that eat a lot of plant material, such as cows, produce large amounts of biogas. The biogas is produced not by the cows themselves, but by billions of microorganisms living in their digestive systems. Biogas also develops in bogs and at the bottom of lakes, where decaying organic matter builds up under wet and anaerobic conditions. Now, biogas generators are used by individual families in the developing world. Large-scale generators are used by some cities as a method of obtaining fuel from refuse.

▲ Two common sources of biogas.

▲ This bus runs on fuel produced from soybeans.

Nature's fuel

Developing countries can now 'grow' their own fuel for cars. Importing oil and petrol is very expensive for countries with only agricultural produce to export, so many have decided to produce fuel from crops. They grow sugar cane or corn, and then obtain ethanol by fermentation followed by distillation.

Think about what you will find out in this section

How do microorganisms help us to produce fuels?	What are the advantages and disadvantages of different types of biogas generator?	How do we evaluate economic and environmental data relating to the production of ethanol-based fuels?

B3 5.1 Biogas for families

How microorganisms produce fuel

Faeces from cows contain methane-producing bacteria. In anaerobic conditions, these bacteria work at their optimum rate, fermenting the complex carbohydrate cellulose, which produces methane.

The residual material is known as **effluent**. It is very rich in nutrients thanks to the action of the bacteria.

Question

a *Which part of a plant cell is made from cellulose?*

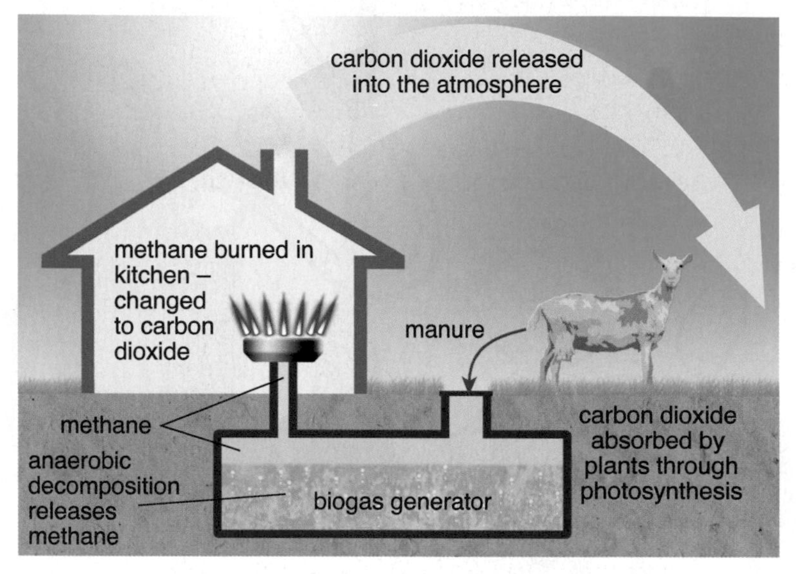

▲ Biogas recycling.

Gas works

Instead of burning faeces on fires, they can be used to produce biogas. Flammable biogas can be collected using a simple tank, as shown above. Animal manure is stored in a closed tank where the gas accumulates. It makes an excellent fuel for cooking stoves and furnaces, and can be used in place of regular natural gas, which is a fossil fuel.

The diagram on the right shows a biogas generator suitable for a small farm in the developing world. A trench about 10 m long and 1 m deep is dug in the ground, and a heavy-duty plastic bag is then placed in the trench. The bag is fitted with inlet and outlet pipes and a valve to draw off the gas. To start up the digester, the bag is two-thirds filled with water, and then topped off with the exhaust gases from a car. Then a mixture of water and faeces is fed through the inlet pipe. It takes about 18 kg of faeces per day to provide enough gas for the farmer's needs. The methane produced is released through the valve at the top. Effluent to fertilise the crops can be run off by lowering the outlet pipe into the ditch.

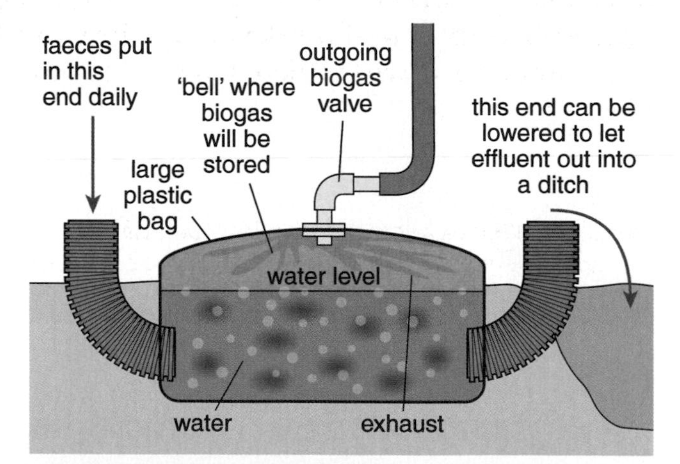

▲ Simple biogas production in the developing world.

Question

b *(i) Name the main gas in car exhaust fumes.*
(ii) Why is the plastic bag topped up with this gas?
(iii) List the factors that influence the rate at which biogas is produced in this setup. Explain how each factor will have its effect.

Faeces-powered computers

The generator on the opposite page provides enough biogas for a family to cook its food. By building bigger biogas generators, enough biogas can be produced to run an electricity generator.

A biogas generator at a school at Myeka in a remote part of South Africa uses human and animal faeces to produce gas for cooking, science experiments, and driving an electricity generator. The installation consists of two 20 000-litre digesters. These receive the faeces from 16 school toilets. The toilets are arranged in two concentric circles around the digesters. The installation also incorporates two cow dung inlet ports to increase the output from the digesters and to cover weekends and holidays when the students are not in school.

Another dimension to being excused for the loo?

Question

c *How is the construction of this type of biogas generator different from the one shown on the previous page?*

The installation at Myeka High School delivers an estimated 13 cubic metres of gas daily. This is enough to produce 18 kWh of electricity per day for computers and lighting. To operate efficiently, biogas digesters like the one at Myeka need to be located in a warm climate where there is a good supply of water.

Constructing the digesters at Myeka.

Question

d *Explain why there needs to be (i) a warm climate, and (ii) a good supply of water, for this type of biogas generator.*

Key points

- Methane can be produced by the fermentation of cellulose in faeces.
- Biogas is a natural product made from animal manure.
- These substances are being used to bring energy to developing countries for technological developments.

Large-scale biogas generation

Farmers can get together to produce biogas. The town of Tillamook Bay in Canada has constructed a digester to process the manure from 10 000 dairy cows. The digesters are about 40 m × 10 m × 4 m in size. Each digester is equipped with heating and insulation.

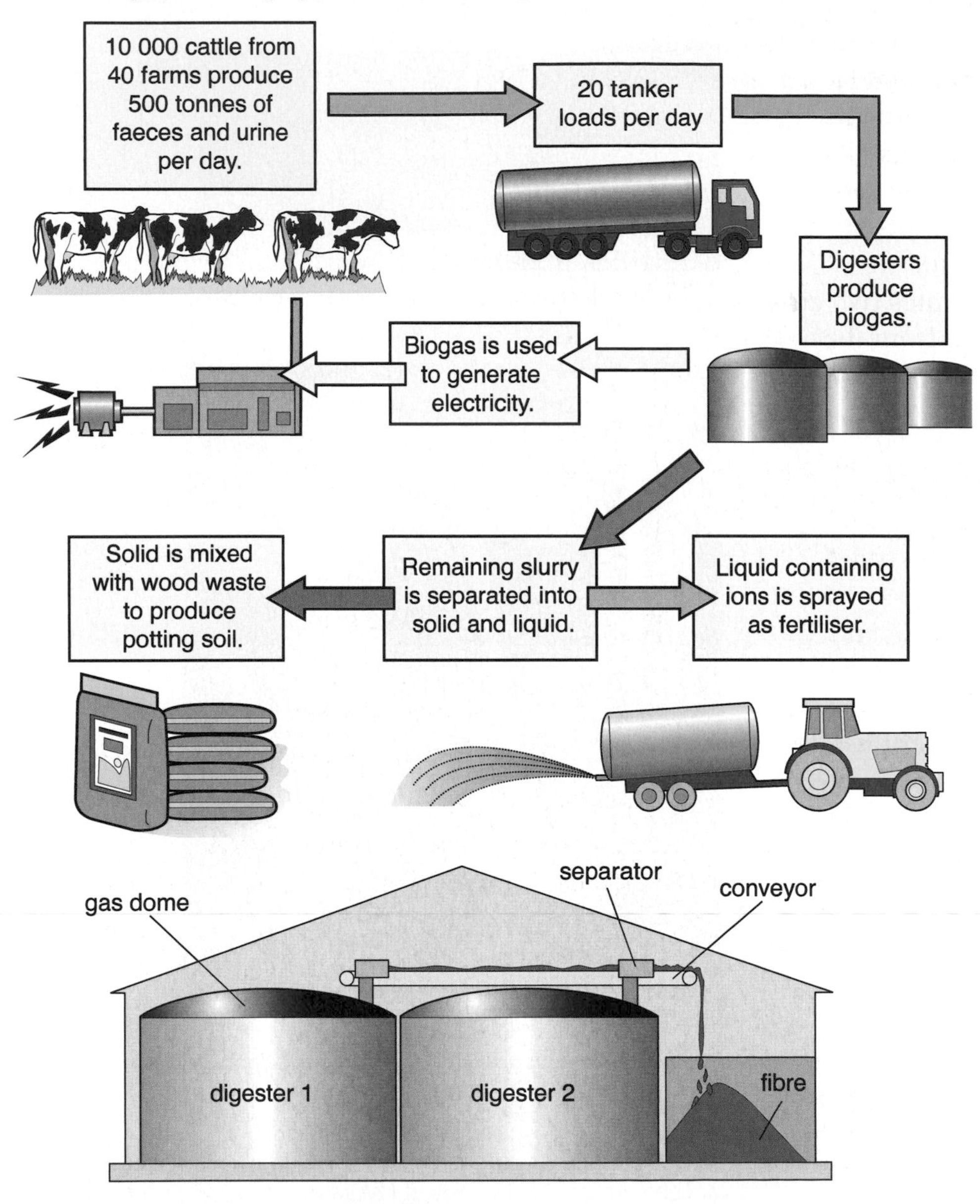

▲ Cow manure digester at Tillamook Bay.

This installation cost about £1 000 000 to build. The benefits are as follows:

- It provides enough electricity for 150 average homes.
- It produces 23 000 tonnes per year of high-quality potting soil.
- It returns 64 million litres of liquid fertiliser per year to the farmers' fields.

Question

a *Suggest why this digester needs both insulation and heating.*

This type of generator has many advantages. The biomass used is abundant and renewable. All of the waste products from the generator are either saleable or useful. The main disadvantage is the high capital cost and the time taken to recover the cost before a profit is made.

Question

b *Explain why this type of generator is unlikely to be built in a developing country.*

Making better use of domestic waste

We can generate biogas on a large scale from manure and organic waste. The diagram shows a scheme that operates in Kristianstad in Sweden. The biogas is produced in the digester. Most of the gas is piped to a district heating supply in the town.

The table shows this plant's annual inputs, outputs and energy statistics.

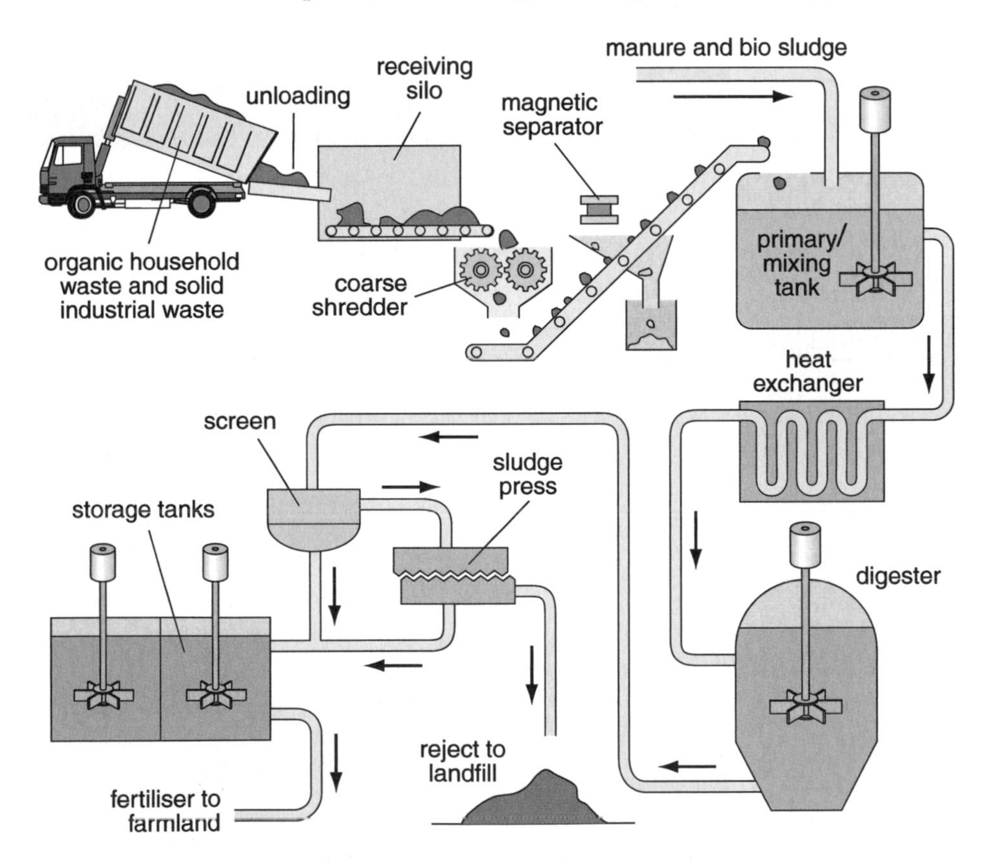

▲ Producing gas from city refuse. The gas is drawn from the primary/mixing tank and the digester.

Input (tonnes)	
Manure	41 200
Household waste	3100
Abattoir waste	24 600
Distillery waste	900
Vegetable waste	1400
Output (tonnes)	
Liquid biofertiliser	67 150
Biogas	4000
Waste	50
Energy statistics (MWh)	
Gross biogas production	20 000
Biogas used to heat plant	2100
Biogas sales to district heating plan	17 900
Electricity purchased to operate biogas plant	540

Question

c *For the Swedish biogas digester:*

(i) Calculate the percentage mass of the input material that is converted into biogas.

(ii) Calculate the proportion of the biogas output that is needed to heat and operate the plant.

(iii) List the advantages and disadvantages of producing fuel in this way rather than using North Sea gas.

Key points

- **Many different sorts of waste can be used to generate biogas.**
- **Initial costs of installing digesters may be high.**
- **The biomass used as a raw material is plentiful and renewable.**

Landfill and lagoons

Getting biogas from garbage

For many years, rubbish was simply dumped into open areas of ground. When the site was full, earth was bulldozed over the top and the area left to settle. A lot of rubbish is still disposed of in this way. The organic waste dumped in such a **landfill site** will decompose with time. If there is an impervious surface below the waste, the waste will become waterlogged and anaerobic fermentation will occur. This will lead to production of mostly methane gas from the waste. This methane will slowly work its way up through the waste and be vented into the atmosphere. If a landfill is covered after use, this gas will slowly seep through the earth covering and be released into the atmosphere.

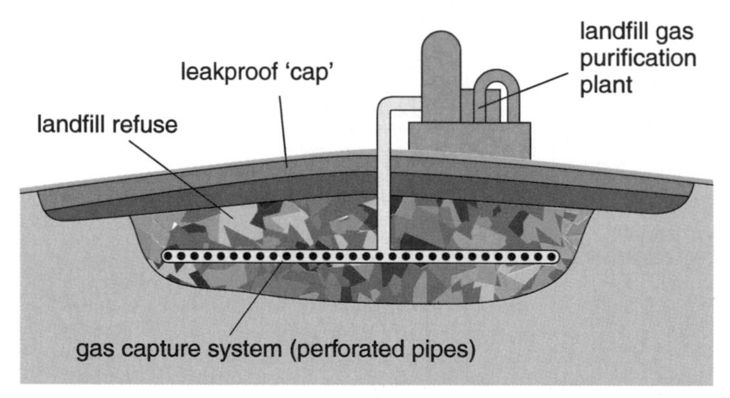

▲ Fuel from rubbish.

▲ Landfill dumping.

Questions

a *Give one effect of methane gas escaping into the atmosphere.*

b *What are the advantages and disadvantages of generating biogas from a landfill site?*

Instead of letting the methane escape, it can be collected and used as a fuel. Perforated pipes (pipes with holes in) are used to collect the gas. A cover prevents gas escaping into the atmosphere. The production of gas from many landfill sites can continue for around 20–30 years, though there is a gradual reduction after about 10 years.

Biogas from lagoons

Farms with large numbers of animals have problems with disposing of all the faeces and urine produced. These are often run into large lakes or **lagoons** and left to decay naturally. But the stench can be terrible! However, these lagoons can be fairly easily adapted to collect the biogas.

▶ Natural disposal of animal waste.

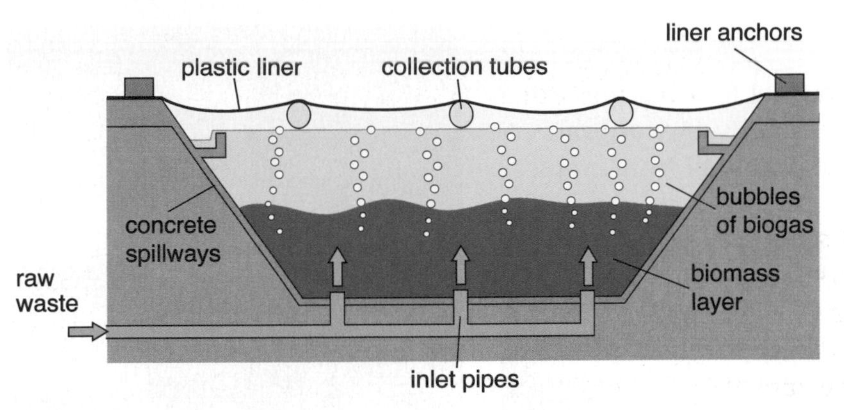

▲ Biogas from anaerobic lagoon.

Basically, all that is done is to cover the lagoon with heavy-duty plastic sheet to prevent the biogas escaping, and to put in tubes to collect the gas. The gas is then used as a fuel on the farm.

Question

c (i) Apart from collecting biogas, suggest one other advantage of covering lagoons on farms.
(ii) These lagoons are being built extensively in countries south of the UK, but there are comparatively few in the UK. Suggest an explanation for this.

Biogas statistics

The table shows how much biogas and energy we can get from various types of animals and their manure.

	Hen	Pig	Horse	Sheep	Cow	Unit of measurement
Mass of each animal	2	70	400	60	500	kg
Manure						
Total	0.19	5.88	20.40	2.40	43.00	kg per day
Total manure/kg animal	0.085	0.084	0.051	0.040	0.086	kg/kg animal per day
Output						
Total biogas	0.0167	0.3338	1.62	0.1975	1.24	m^3 per day
Total power	0.0045	0.0904	0.44	0.0535	0.34	kW
Biogas/kg manure	89.43	56.77	79.44	82.28	28.91	litres/kg manure per day
Power/kg manure	24.22	15.38	21.51	22.28	7.83	W/kg manure per day
Biogas/kg animal	7.60	4.77	4.05	3.29	2.49	litres/kg animal per day
Power/kg animal	2.06	1.29	1.10	0.89	0.67	W/kg animal per day

Question

d (i) What is the relationship between the mass of an animal and the mass of manure produced?
(ii) What volume of biogas is produced from 1 kg of pig manure per day?
(iii) What is the relationship between total power and total biogas?
(iv) Which animal manure produces the largest amount of energy per kg of animal?
(v) What is the general relationship between biogas/kg animal and the mass of the animal? Are there any anomalies in this relationship?

Key points

- **Organic waste, including sewage, is decomposed by microbes, often by fermentation which produces methane to use as biogas.**
- **Landfill sites also produce methane.**
- **Lagoons of sewage allow animal waste to decay naturally to produce methane.**

Alcoholic cars

Developing countries can now 'grow' their own fuel for cars. Importing oil and petrol is very expensive for countries with only agricultural produce to export, so many have decided to produce fuel from crops.

▲ Sugar cane being harvested.

▲ Crushing sugar cane.

▲ Gasohol on sale in Brazil.

Brazil has done this most successfully by growing sugar cane to produce ethanol. This is mixed with petrol to make a fuel called gasohol. Car engines have been adapted to run on gasohol or on pure ethanol – though a little petrol is added to the ethanol to stop people from drinking it! Sugar cane grows quickly in tropical climates, producing two crops per year.

Question

a *Explain why sugar cane is not grown as a crop in the UK.*

The cane is taken to mills and crushed by enormous rollers to extract the juices. These are then refined to produce the 'cane sugar' that we put in tea and coffee. There are two waste products – bagasse and molasses. Bagasse is the fibrous remains of the plants after crushing. Molasses is a syrupy liquid left after cane sugar has been extracted from the juices. These waste products can be used to make ethanol.

Making ethanol from sugar cane

Molasses contains lots of sugars. Adding yeast to molasses has the same effect as adding it to wort in brewing. The yeast respires anaerobically, producing ethanol and carbon dioxide. The fully fermented liquid contains about 10% ethanol. Water is then removed from it by **fractional distillation**. When the liquid is heated, alcohol evaporates first because it has a lower boiling point than water. The vapour is cooled to condense the ethanol.

The thermal energy for distillation is provided by burning the bagasse. The purified ethanol is either used as a fuel by itself or mixed with petrol. In Brazil, gasohol contains 22% ethanol, but in most other countries no more than 10% ethanol is added.

▲ Industrial ethanol distillation columns.

Ethanol from bagasse

Sixty per cent of the sugar cane plant is bagasse. Until recently the only use of bagasse was as a fuel. This is because the carbohydrate fibres in bagasse are built up from sugars containing a different number of carbon atoms from glucose. The yeasts used to ferment cane sugar anaerobically cannot ferment these particular sugars. But some soil bacteria can produce carbohydrase enzymes which will break down these sugars. Scientists have used these bacteria and genetic engineering techniques to produce yeasts that can ferment these carbon sugars.

There are many industries which produce organic waste materials. Examples include paper mills and food-processing factories. Scientists are now researching methods of fermenting the waste products from these.

Questions

b *Outline the method scientists could use to produce yeasts that can ferment these carbon sugars.*

c *Explain how the work of these scientists is contributing to sustainable development.*

Ethanol from corn

In the USA, ethanol is produced from maize rather than sugar. Maize is a cereal crop, like barley, so it stores carbohydrates as starch rather than sugar. The diagram shows how ethanol is produced from starch.

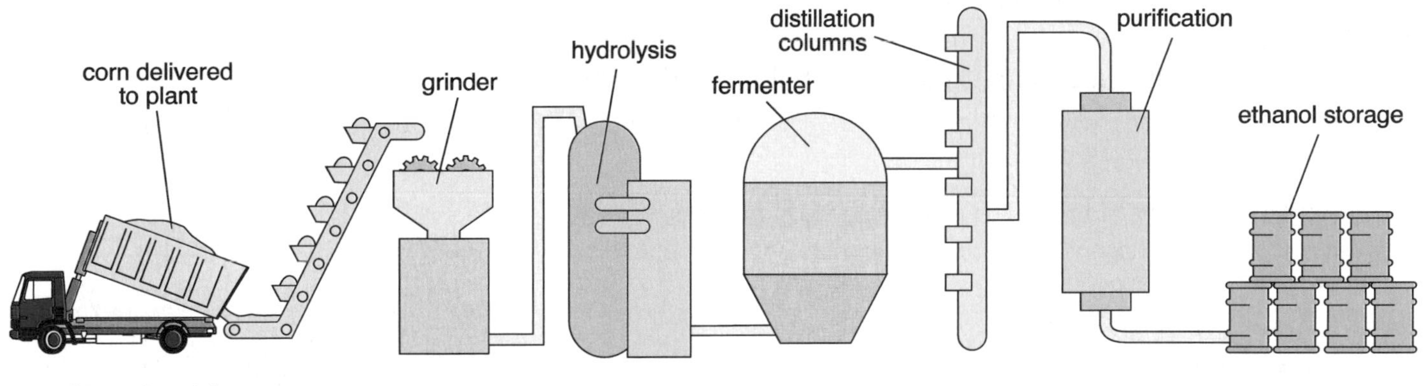

Making ethanol from corn.

Question

d *What additional process is required to produce ethanol from maize rather than sugar cane?*

After grinding, the starch has to be converted into sugars that the yeast can ferment. The ground corn is hydrolysed to sugar by special amylase enzymes that have an optimum temperature of 90 °C. The sugars are then fermented using yeast. After purification the yeast is ready to be mixed with petrol to form gasohol.

Question

e *What is the advantage of using amylases that have an optimum temperature of 90 °C?*

Key points

- In developing countries, it is cheaper to produce fuels from ethanol than to import petrol for fuels.
- The alcohol for these fuels is produced from the waste products of sugar cane cultivation.
- Ethanol can also be produced from corn and other crops by using carbohydrase enzymes to break down starch into sugars.
- The ethanol is produced by fermentation of sugars, followed by distillation.

Ethanol – the green fuel?

Producing ethanol for use as a fuel is beneficial because, unlike oil, the source is renewable. There are several other advantages. Burning fossil fuels releases carbon dioxide into the atmosphere. So does burning ethanol – but this is only replacing the carbon dioxide that the sugar plants took in during photosynthesis.

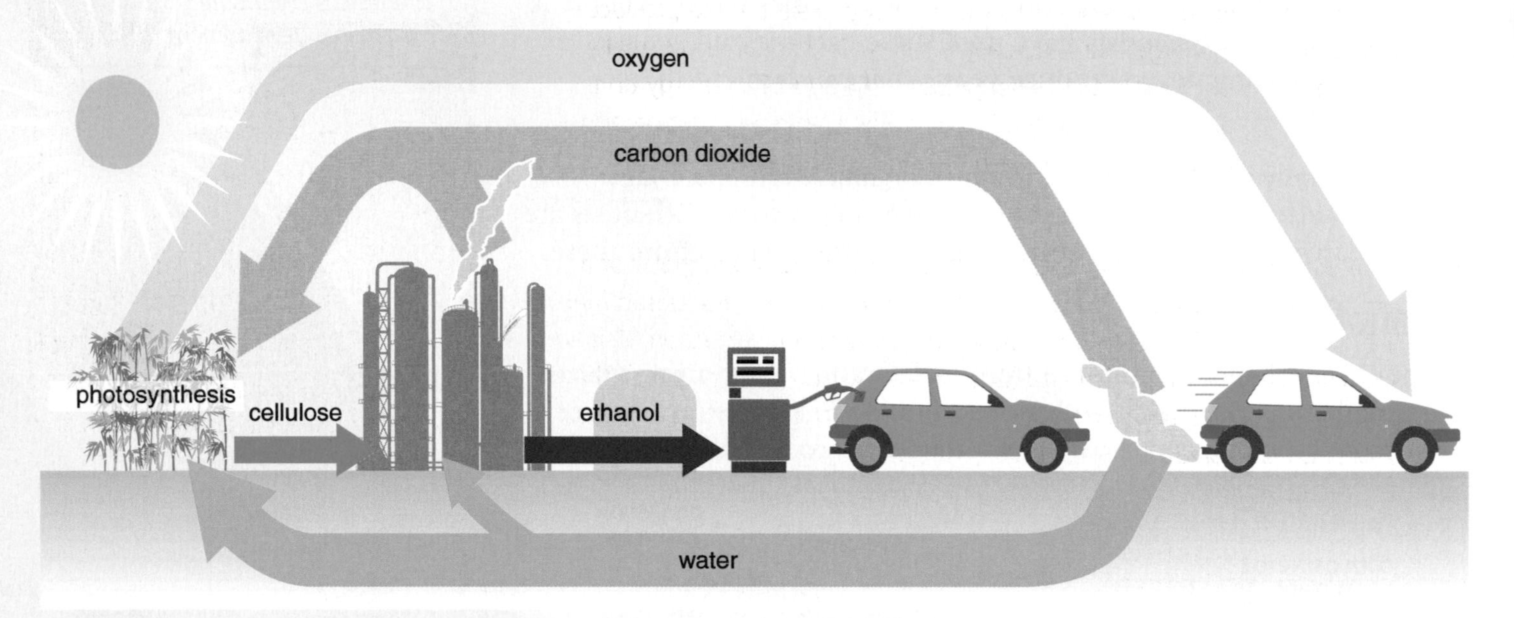

▲ Ethanol – the green fuel.

Question

a *What is expected to happen if the carbon dioxide content of the atmosphere continues to rise?*

Burning ethanol also produces less air pollution than burning oil. The table shows the changes in pollutant levels in Brazil since the introduction of gasohol. The data show average emissions per car of CO (carbon monoxide), HC (hydrocarbons) and NO (nitrous oxides) in grams per kilometre travelled.

Year	Fuel	Pollutant (g/km)		
		CO	HC	NO
<1980	Petrol	54.0	4.7	1.2
1986	Gasohol	22.0	2.0	1.9
	Ethanol	16.0	1.6	1.8
1990	Gasohol	13.3	1.4	1.4
	Ethanol	10.0	1.3	1.2
1995	Gasohol	4.7	0.6	0.6
	Ethanol	3.2	0.4	0.3

Question

b *(i) Describe the trends in the data.*
(ii) Explain the effects these changes will have on the environment in Brazil.

Deforestation debate

There are some problems in replacing oil with ethanol. Many people are concerned that rainforests are being cleared to provide sugar cane plantations, and that small farmers are being displaced. Others argue that the ethanol would be used more effectively in the chemical industry to make plastics.

What price fuel?

The US government is encouraging the use of ethanol-based fuels.

An acre of US maize yields about 328 gallons of ethanol. But planting an acre of maize requires about 140 gallons of fossil fuel. So even before maize is converted to ethanol, the **feedstock** (raw materials) costs \$1.05 per gallon of ethanol. About 70% more energy is required to produce ethanol than the energy that is actually in ethanol. Every time you make a gallon of ethanol, there is a net energy loss of 54 000 BTU (energy units). Ethanol from maize costs about \$1.74 per gallon to produce compared to about 95 cents to produce a gallon of petrol.

The average US car travelling 10 000 miles a year on pure ethanol (not a petrol–ethanol mix) would need about 852 gallons of the maize-based fuel. This would take 11 acres to grow, based on net ethanol production. This is the same amount of cropland required to feed seven Americans. If all the cars in the United States were fuelled with 100% ethanol, a total of about 97% of US land area would be needed to grow the maize feedstock. Maize would cover nearly the total land area of the United States.

Other researchers dispute these figures. They say that the analysis relies on outdated energy data for ethanol processing and on exaggerated irrigation and fertiliser requirements for farming. Also, say the researchers, future ethanol plants will burn unfermentable parts of crops (stems and leaves) to fuel the plant's power needs, further reducing fossil fuel use. They also point to a 2004 study by the US Department of Agriculture which reported that ethanol production returns 67% more energy than it consumes.

▲ Is this the future for gasohol in the USA?

Question

c *(i) Suggest pressure groups that might be opposed to the use of more ethanol-based fuel. Explain why they would be opposed to this.*
(ii) Use the information in the passages to evaluate the case for using ethanol-based fuels in the USA.

Key points

- Burning ethanol produces less pollution than burning fossil fuels.
- Ethanol is a renewable fuel.
- The economics and environmental gains from producing ethanol fuels are not as clear-cut in developed countries as they are in developing countries.

End of unit questions

1 The diagram shows some alveoli and a blood capillary in the lung.

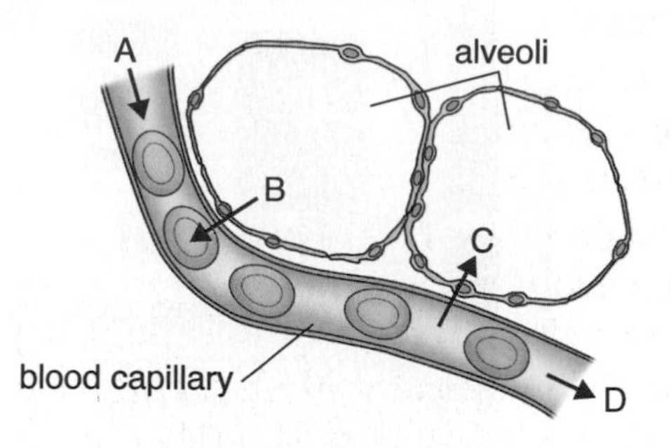

a Which of the letters A–D shows
- **i** oxygenated blood?
- **ii** the diffusion of oxygen?
- **iii** the diffusion of carbon dioxide? *(3 marks)*

b Describe **two** features of the alveoli that help gas exchange. *(2 marks)*

c Explain how oxygen that diffuses into the blood is transported around the body. *(3 marks)*

2 The diagram shows the structure of a root hair cell.

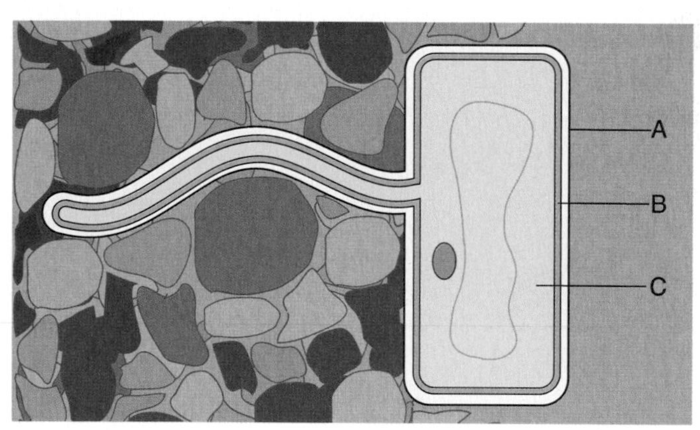

a Which of the structures A–C is partially permeable? *(1 mark)*

b Explain what is meant by the term 'partially permeable'. *(1 mark)*

c Use the diagram to describe **two** features of the root hair cell which make water uptake more efficient. *(2 marks)*

d Explain how water is absorbed from the soil into the root hair. *(3 marks)*

3 Four leaves were removed from the same plant. Petroleum jelly (a waterproofing agent) was spread onto some of the leaves, as follows:

Leaf A – on both surfaces
Leaf B – on the lower surface only
Leaf C – on the upper surface only
Leaf D – none applied.

The leaves were placed near a window and weighed at intervals. The results are shown on the graph.

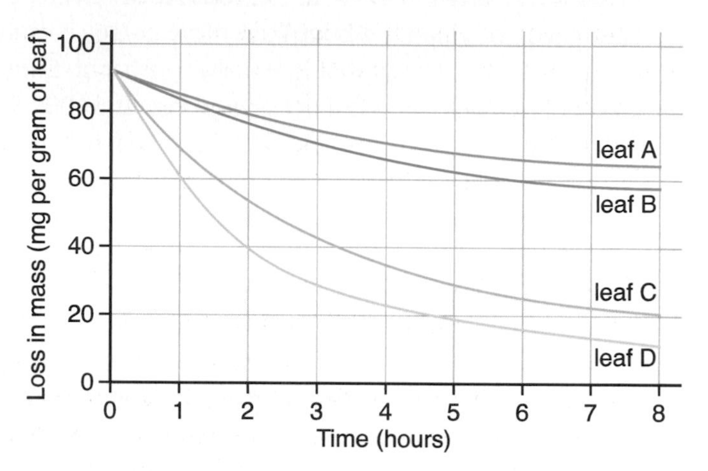

Each surface of the leaf was observed using a microscope. The diagrams below show the appearance of the upper and lower surface of the leaves.

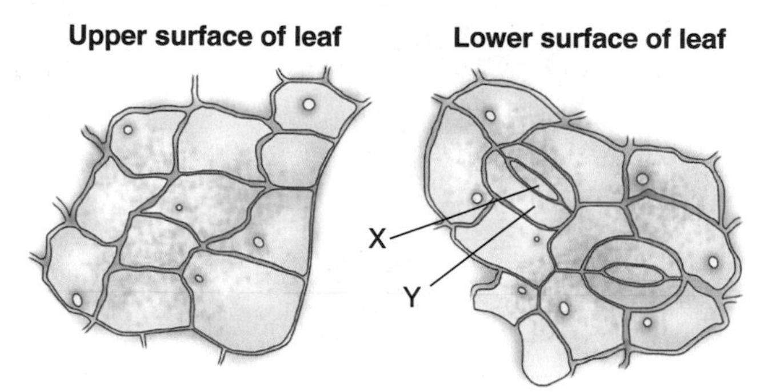

a What process causes the loss in mass? *(1 mark)*

b From which surface is the most mass lost? Use the results of the investigation to explain the evidence supporting your answer. *(3 marks)*

c Suggest why the units to show the loss in mass were in mg per gram of leaf. *(2 marks)*

d Name space X. *(1 mark)*

e Name cell Y. *(1 mark)*

f Use the diagram to explain the difference in the results obtained for leaves B and C. *(2 marks)*

AQA 2004

4 The graph shows the concentrations of lactic acid in the blood of a person after a period of exercise, and also when the person was resting.

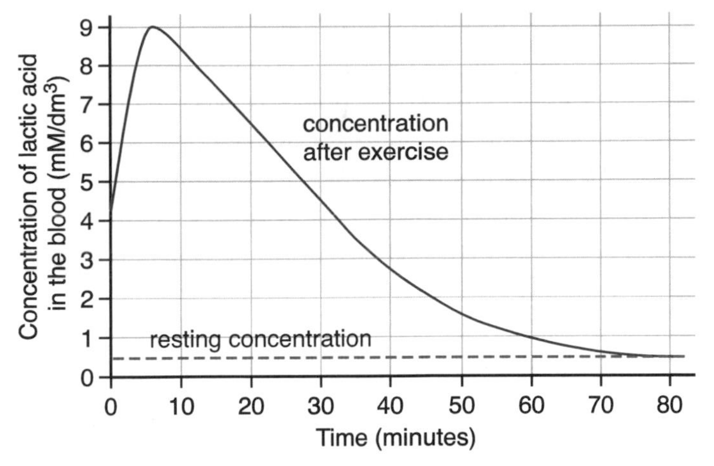

a How high above the resting level was the maximum concentration of lactic acid in the blood? *(1 mark)*

b How long did it take for the lactic acid to return to its resting level after exercise? *(1 mark)*

c What process produces lactic acid? *(1 mark)*

d Explain why the concentration of lactic acid in the blood decreases as the person recovers from the period of exercise. *(3 marks)*

e Explain why strenuous exercise causes an increase in lactic acid concentration in the blood. *(3 marks)*

5 The following table shows the percentage of blood flowing to various parts of the body during two types of activity.

Part of the body	Percentage of blood flowing to parts of the body	
	Walking	Running
Muscles attached to the skeleton	45	85
Brain	10	3
Heart muscle	5	5
Liver and gut	24	3

The next table shows the total volume of blood flowing from the heart as the level of exercise increases.

	Walking	Running
Total volume of blood flowing from the heart (cm^3/min)	9000	30 000

a Calculate the volume of blood flowing to the brain when the person was **i** walking, **ii** running. *(2 marks)*

b Calculate the volume of blood flowing to the muscles attached to the skeleton when the person was **i** walking, **ii** running. *(2 marks)*

c Explain why blood flow to the muscles attached to the skeleton increases as the level of exercise increases. *(3 marks)*

d Use the information from the tables to suggest why a person should not exercise just after eating a meal. *(2 marks)*

6 An investigation was carried out to find out how breathing rate changed during exercise. The breathing rate of a student was measured every minute during the investigation. The student rested, then exercised, and then rested again. The results are shown on the graph.

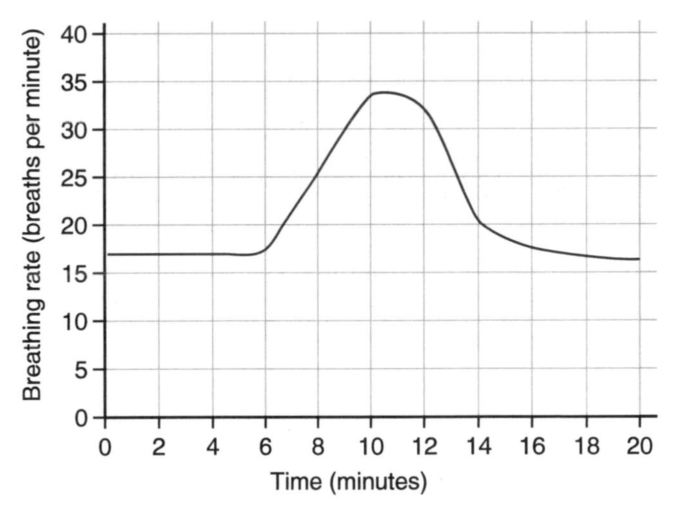

a What was the student's breathing rate at rest? *(2 marks)*

b What was the maximum increase in breathing rate during exercise? *(1 mark)*

c How long did it take for the student's breathing rate to return to its resting level after exercise? *(1 mark)*

d The student's heart rate also increased during exercise. Explain how increasing breathing rate and heart rate enables muscle cells to produce more energy for contraction. *(3 marks)*

In a second investigation, a student exercised at different rates on an exercise cycle. The table shows the effect of exercise on breathing rate and on the volume of each breath.

Exercise rate (arbitrary units)	Breathing rate (breaths/min)	Volume of each breath (dm^3)
10	14.0	0.75
20	15.5	1.5
30	16.0	2.5
40	19.5	2.9
50	20.0	3.2

e What changes occur in the student's breathing as the exercise rate increases? *(1 mark)*

f Calculate the total volume of air taken into the lungs in one minute at an exercise rate of 30 units. Show your working. *(2 marks)*

7 The table shows the results of an investigation carried out to find out how breathing changes during exercise.

Activity	Mean volume of each breath (cm^3)	Breathing rate (breaths per minute)
20 step-ups per minute	500	18
30 step-ups per minute	750	25
40 step-ups per minute	1100	32

a **i** How many breaths did the person take per minute when exercising at 20 step-ups per minute?

ii How many more breaths were taken when the person was doing 30 step-ups per minute? *(2 marks)*

b Make a copy of the table and add a third column with the heading 'total amount of air breathed in'. Add appropriate units to the column, and then calculate the values to complete the table. *(4 marks)*

c The measurements in this investigation were obtained using a student who does very little exercise. Explain why someone who exercises regularly will take fewer breaths when doing the same activities. *(2 marks)*

8 The table shows the amounts of substances filtered from the blood by the kidneys, and the amounts found in urine, over a 24-hour period.

Substance	Amount in kidney filtrate	Amount in urine
Water	180 000 cm^3	1600 cm^3
Mineral ions	1500 g	12 g
Glucose	200 g	0 g
Urea	55 g	30 g
Protein	0 g	0 g

a **i** Which substance is completely reabsorbed from the filtrate by the kidney? *(1 mark)*

ii Calculate the percentage of ions that were reabsorbed. *(1 mark)*

b Explain why there is no protein present in the kidney filtrate. *(2 marks)*

c Explain how substances present in the blood are filtered into the kidney tubule. *(3 marks)*

9 A kidney contains a large number of tiny tubes called kidney tubules. Some of the cells which line the tubules are able to reabsorb glucose. The diagram shows how these cells reabsorb glucose from the tubule and secrete it into the blood.

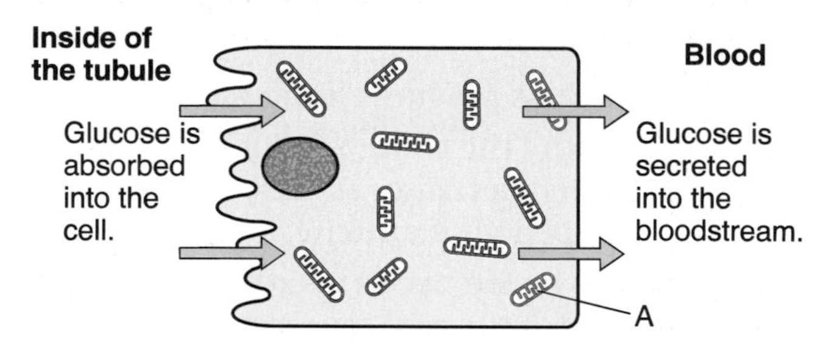

a **i** Name structure A.

ii Give the full name of the process which takes place in structure A. *(2 marks)*

b Use the diagram to describe **three** ways in which the cell is adapted for absorbing glucose from the tubule. *(3 marks)*

c There is a higher concentration of glucose in the blood than in the cell. Explain how glucose is passed from the cell into the blood. *(3 marks)*

10 The diagram shows how a dialysis machine works. The blood of a patient is separated from the dialysis fluid by a partially permeable membrane. The dialysis fluid contains the same concentration of glucose as a healthy person's blood.

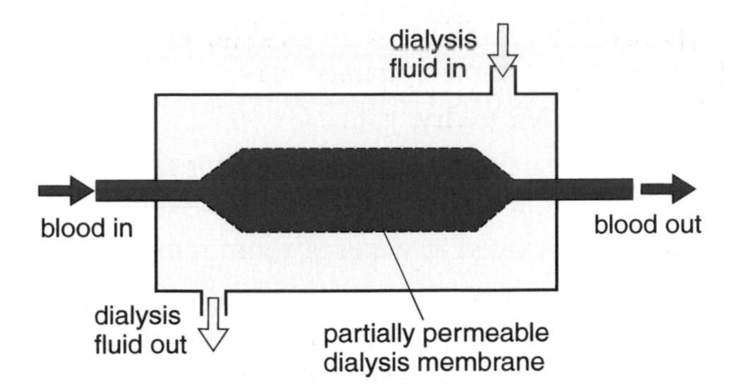

a By what process do substances pass from the blood into the dialysis fluid? *(1 mark)*

b Explain why urea passes out of the blood into the dialysis fluid. *(3 marks)*

c Explain why no glucose is lost from the blood into the dialysis fluid. *(2 marks)*

d Doctors sometimes give patients dialysis treatment rather than a kidney transplant. Suggest **four** reasons for this. *(4 marks)*

11 Anna is 15 years old. Three years ago Anna had a very serious illness and had to spend several weeks in hospital. Even though she recovered from the illness, Anna's kidneys stopped working. Since then Anna has been receiving dialysis treatment.

Her parents have recently received a phone call to let them know that a donor kidney is available with an excellent tissue match with Anna.

Anna says she does not want to go back into hospital for an operation, and she is quite happy receiving dialysis instead.

a Explain why a good tissue match is needed for a successful transplant. *(2 marks)*

b Explain, giving three reasons, why a transplant would make Anna's life better. *(3 marks)*

12 a Use words from this list to complete the sentences.

bacteria moulds viruses yeast

Beer is made using ___________ .

Antibiotics are made using ___________ .

Yoghurt is made using ___________ . *(3 marks)*

b The drawing shows a yeast cell as seen through an electron microscope. Name parts A to D. *(4 marks)*

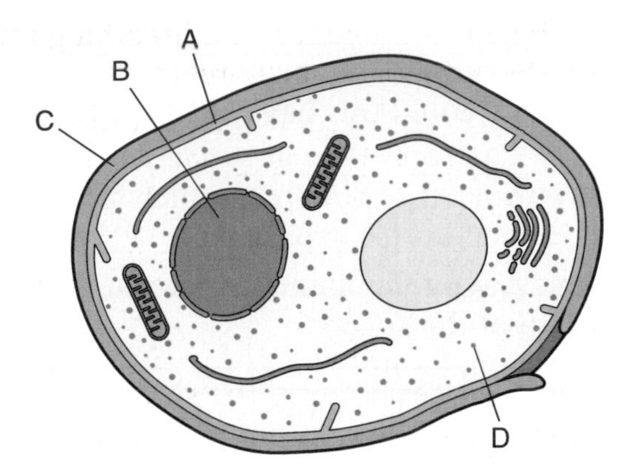

c Yeast respires to produce energy.

i Give **two** ways in which yeast uses this energy. *(2 marks)*

ii Give the **two** products of anaerobic respiration in yeast. *(2 marks)*

13 Yeast ferments different sugars at different rates. The diagram shows apparatus used to investigate these rates.

The rate of fermentation is measured by recording the position of the meniscus every 15 minutes.

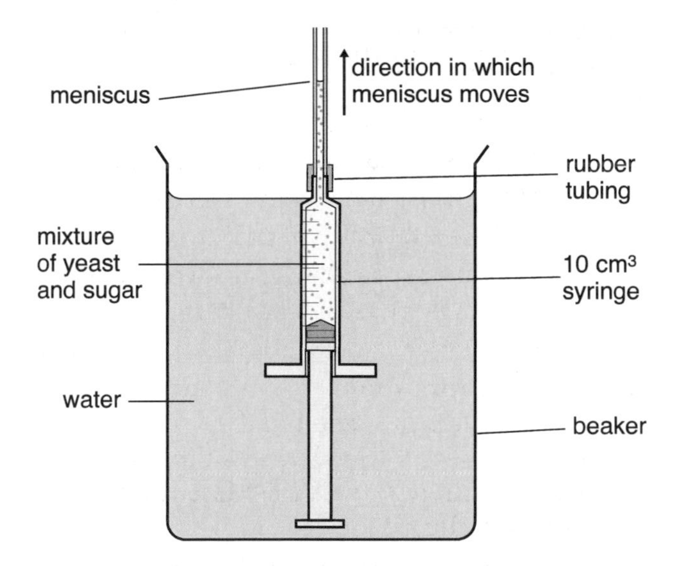

The table shows results for three sugars: glucose, lactose and maltose.

Sugar	Distance (mm) travelled by meniscus after					
	15 min	30 min	45 min	60 min	75 min	90 min
Glucose	0	1	10	15	45	92
Lactose	0	0	0	0	0	0
Maltose	0	0	1	3	6	13

a In this experiment, which was

i the independent variable?

ii the dependent variable? *(2 marks)*

b **i** Which variable was controlled by placing the syringe in a beaker of water? *(1 mark)*

ii How could this have been checked? *(1 mark)*

c Give **four** other variables which should be controlled in this experiment. *(4 marks)*

d Explain why the meniscus moved in the case of two of the sugars. *(3 marks)*

e Describe how the rate of fermentation changed over 90 minutes in the experiment using glucose. *(3 marks)*

f **i** How did the rate of fermentation of maltose by yeast differ from that of glucose? *(1 mark)*

ii Explain why this rate was different. *(1 mark)*

g Suggest why lactose was not fermented in this experiment. *(2 marks)*

h Suggest how the results of this experiment could be used by a manufacturer of alcohol. *(2 marks)*

14 The drawing shows a mould being transferred from one Petri dish (B) to another (C).

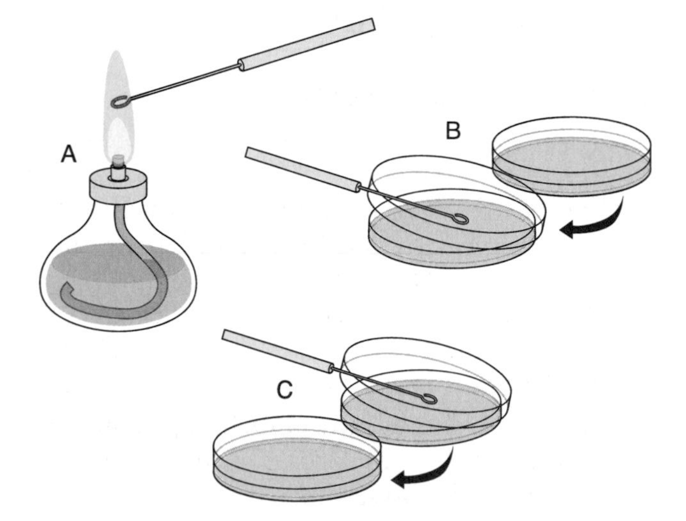

a Explain why the inoculating loop is heated at A. *(1 mark)*

b Explain why the lids of the Petri dishes are not completely removed at B and C. *(2 marks)*

c **i** What is the maximum temperature at which microorganisms should be incubated in a school laboratory? *(1 mark)*

ii Explain the reason for this maximum. *(2 marks)*

15 Three barley grains were given different treatments:

Grain A was left to dry.

Grain B was soaked in boiling water for an hour, then in water at room temperature for 48 hours.

Grain C was soaked in water at room temperature for 48 hours.

All three grains were then cut in half and placed, cut surface down, on agar containing starch in a dish as shown in Figure 1.

After a further 48 hours the grains were removed and the dish was flooded with iodine solution. The results are shown in Figure 2.

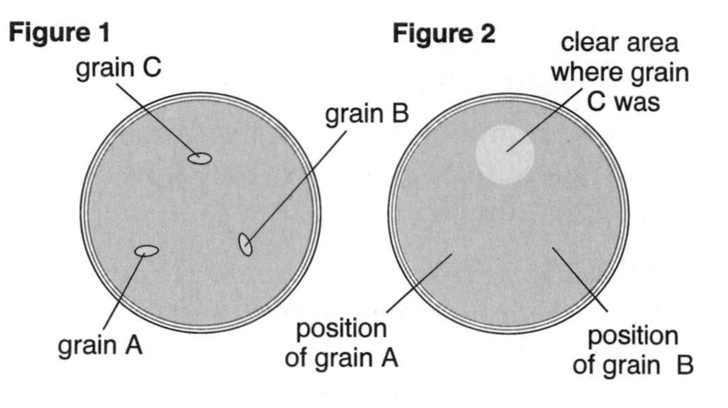

Explain the results for:

a grain A. *(2 marks)*

b grain B. *(2 marks)*

c grain C. *(2 marks)*

16 The flowchart shows how beer is produced.

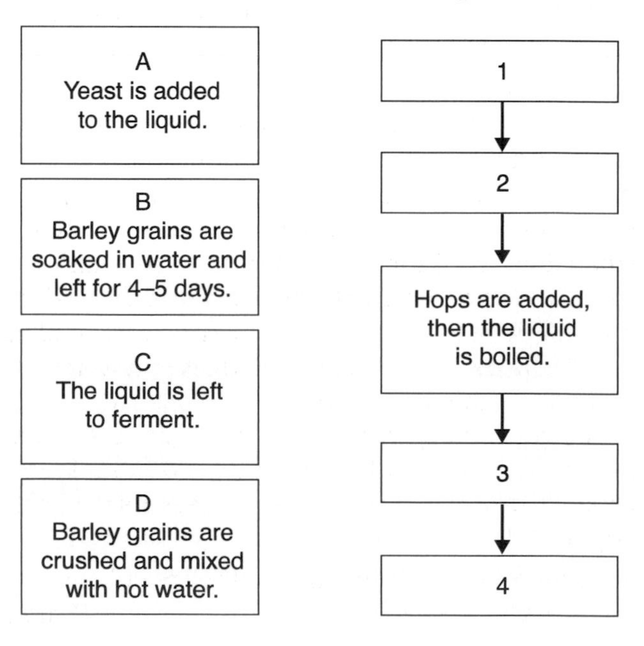

Match the statements A, B, C and D with the correct boxes 1, 2, 3 and 4. *(3 marks)*

17 Wine is produced from grapes.

a Name the organisms which ferment sugars in grape juice. *(1 mark)*

b Write a word equation for fermentation. *(2 marks)*

c Give **three** differences between wine making and the brewing of beer. *(3 marks)*

18 Yoghurt is made from milk.

a Name the sugar found in milk. *(1 mark)*

b Name the bacteria which are used to produce yoghurt. *(1 mark)*

c Explain why milk goes solid during yoghurt production. *(3 marks)*

19 *Penicillium* produces the antibiotic penicillin.

a Explain why some microorganisms produce antibiotics. *(2 marks)*

b **i** Name **three** substances which are supplied to *Penicillium* by corn-steep liquor in a fermenter. *(3 marks)*

ii Give the function of each of the substances. *(3 marks)*

c Describe how penicillin is separated from the rest of the contents of the fermenter. *(2 marks)*

20 Kefir is a fermented, fizzy milk product produced from a complex mixture of bacteria and yeast. Read the passage about the production of kefir.

The inoculum is 'kefir grains', which are small clusters of milk protein, carbohydrate and microorganisms. Like yogurt, kefir contains much less lactose than milk. It has an alcohol content of about 0.01–0.10%. The milk is heat-treated at 95 °C for 10–15 minutes. The milk is then cooled to 18–22 °C, and 2–5% kefir grains are added. This mixture is incubated at 18–22 °C for 24 hours, after which the kefir grains are sieved out, and the product is chilled and packaged. Kefir grains can be rinsed and reused, or dried and stored for later use. The final kefir product can be flavoured in a manner similar to yogurt.

a Suggest what is meant by 'the inoculum'. *(1 mark)*

b Explain why kefir contains less lactose than milk. *(2 marks)*

c Explain why the milk is heated at 95 °C for 10–15 minutes. *(2 marks)*

d Explain why the milk is kept at 35 °C for 24 hours. *(1 mark)*

e Kefir contains small amounts of alcohol, while yoghurt does not. Suggest an explanation for this. *(2 marks)*

f Suggest why kefir is fizzy. *(2 marks)*

21 The diagram shows a fermenter used to produce penicillin.

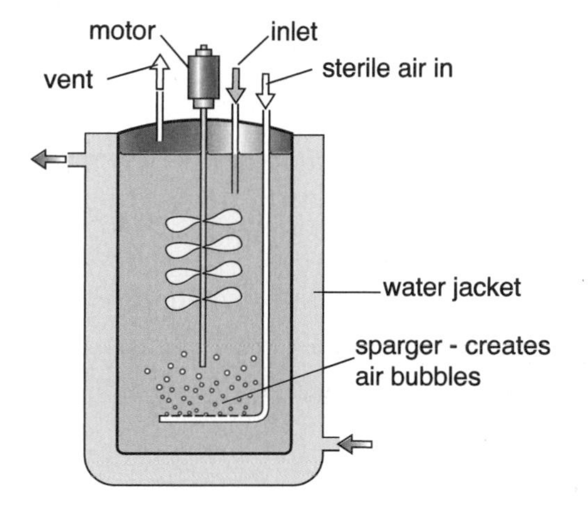

a Air is bubbled in near the base of the fermenter. Give **two** functions of the air bubbles. *(2 marks)*

b When the process is started up, warm water is circulated around the water jacket. Later, this is replaced by cold water. Explain:

i Why warm water is used at the start of the process. *(2 marks)*

ii Why cold water is used later. *(2 marks)*

c The graph shows how penicillin production and the rate of respiration of the mould vary during production.

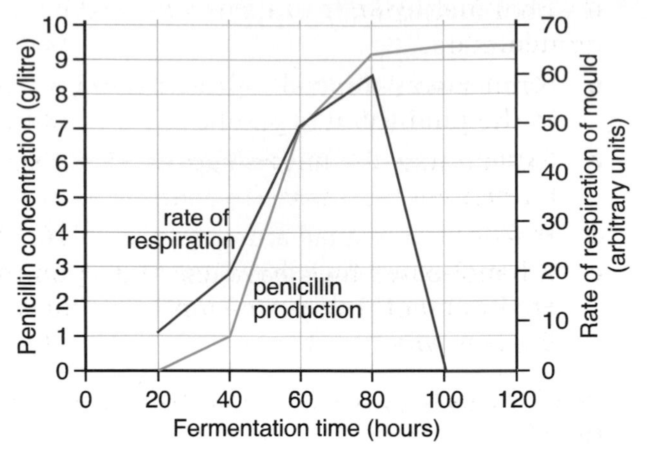

i Describe how the rate of respiration changes during the process. *(3 marks)*

ii Explain why the rate of respiration changes in this way. *(2 marks)*

iii Explain the relationship between the rate of penicillin production and the rate of respiration of the mould. *(3 marks)*

22 a Explain what is meant by biogenesis. *(1 mark)*

b The diagram shows a repeat of an investigation into biogenesis using modern apparatus.

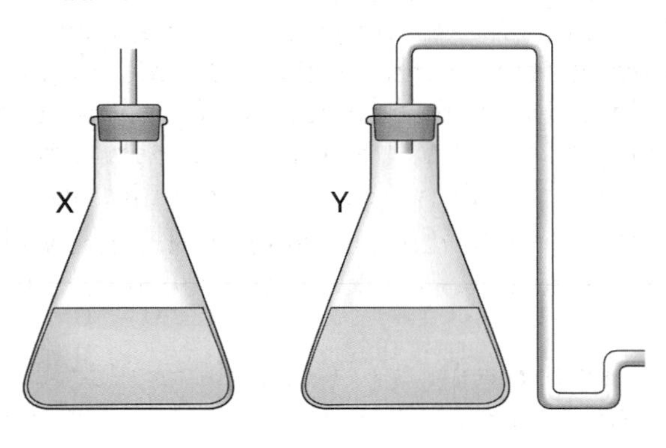

Both flasks contain boiled meat broth.

Flask X is open to the air via a short, straight tube. Flask B is open to the air via a tube with a U-bend.

After 4 days at room temperature the contents of flask X go cloudy, but the contents of flask Y remain unchanged.

i Which scientist performed this experiment originally? *(1 mark)*

ii Why was the meat broth boiled in both flasks at the start of the experiment? *(1 mark)*

iii Explain why the results of the experiment support the theory of biogenesis. *(3 marks)*

23 The flowchart shows how mycoprotein is produced.

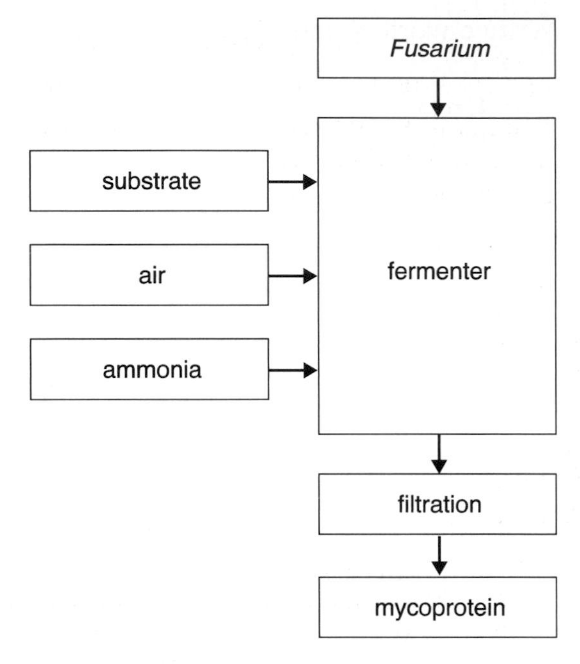

a What type of organism is *Fusarium*? *(1 mark)*

b Name one type of nutrient which is used in this process. *(1 mark)*

c Explain why ammonia is necessary for this process. *(2 marks)*

d Sales of foods made from mycoprotein are increasing rapidly. Suggest an explanation for this. *(2 marks)*

24 The diagram shows a biogas generator on a farm in a developing country.

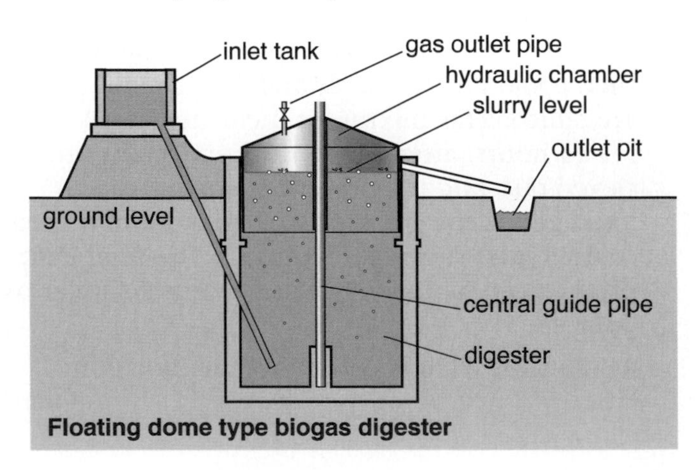

a Suggest what is fed into the inlet tank. *(1 mark)*

b Which gas is the main constituent of biogas? *(1 mark)*

c Suggest the function of the hydraulic chamber. *(2 marks)*

d Suggest why the material in the outlet pit is useful to the farmer. *(2 marks)*

e The graph shows how temperature affects the volume of biogas produced by a biogas generator.

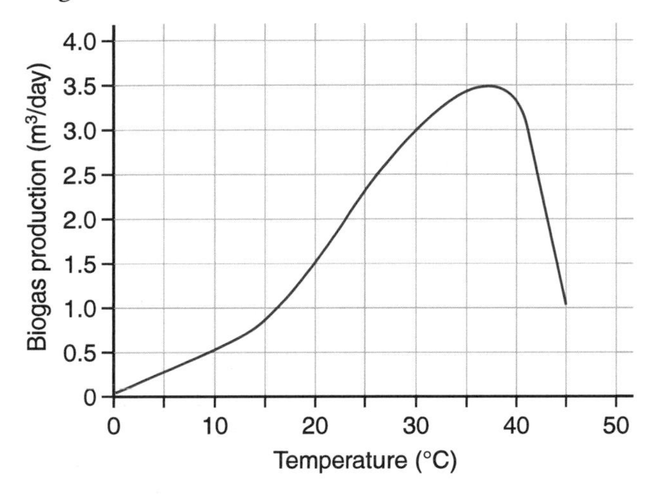

i Describe the effect of temperature on biogas production. *(3 marks)*

ii Explain the effect of increasing the temperature above 35 °C on biogas production. *(2 marks)*

iii The mean winter temperature in this developing country is 15 °C. The farmer uses the biogas mainly for cooking. Explain one disadvantage of this type of biogas generator for the farmer. *(2 marks)*

25 Gasohol fuels contain ethanol produced by fermentation.

a Name **two** different crops which can be used in the production of gasohol. *(2 marks)*

b Explain how the alcohol produced during fermentation is separated from water. *(2 marks)*

c Ethanol-based fuels have less impact on the environment than petroleum-based fuels. Explain why. *(3 marks)*

26 Read the passage about producing ethanol fuel in the USA.

David Pimentel and Tad Patzek co-wrote a recent report which estimates that making ethanol from corn requires 26% more fossil energy than the ethanol fuel itself actually contains.

The two scientists calculated all the fuel inputs for ethanol production – from the diesel fuel for the tractor planting the corn, to the fertiliser put in the field, to the energy needed at the processing plant – and found that ethanol is a net energy-loser. According to their calculations, ethanol contains about 19 million joules per litre, but producing that ethanol from corn takes about 24 million joules. For comparison, a litre of petrol contains about 24 million joules per litre. But making that litre of petrol – from drilling the well, to transportation, through refining – requires around 5 million joules.

In other words, more ethanol production will increase America's total energy consumption, not decrease it.

a Give **three** ways in which the production of ethanol requires energy. *(3 marks)*

b Compare the amount of energy used in producing ethanol with that used in producing petrol. *(2 marks)*

c The US government wants to increase ethanol production despite the data produced by the two scientists. Suggest **two** reasons for the government's decision. *(2 marks)*

Glossary

active transport Movement of molecules against a concentration gradient using energy from respiration.
adaptations A feature or features that make a structure or organism more suitable for its function.
addicted Dependent on a drug.
aerobic respiration Chemical reactions in cells which use glucose and oxygen to release energy. Carbon dioxide and water are produced as waste products.
agar A jelly obtained from seaweeds.
alleles Different forms of the same gene.
alveoli Tiny air sacs in the lungs where oxygen diffuses into the blood and carbon dioxide diffuses out.
amino acids The basic units from which proteins are made.
amylase A type of digestive enzyme that breaks down starch into glucose.
anaerobic respiration Incomplete breakdown of glucose to release energy in the absence of oxygen.
antibiotics Chemicals produced by microbes, used to destroy bacteria in the body.
antibodies Substances produced when white blood cells detect the presence of a particular antigen.
antitoxins Substances produced by white blood cells to neutralise poisons produced by microbes.
arteries Blood vessels which carry blood, under high pressure, away from the heart.
arthritis A painful inflammation of the joints.
asexual reproduction Reproduction that does not involve the formation of gametes.
bacteria Very small single-celled organisms. Each bacterial cell has cytoplasm surrounded by a membrane, but has no distinct nucleus. Bacteria can reproduce very quickly and some cause disease.
balanced diet A diet made up of a variety of different foods to provide the energy and all the nutrients needed to stay healthy.
biogenesis The theory that living organisms are produced only from other living organisms.
biomass The mass of living organisms in an area.
capillaries Blood vessels with walls one cell thick to allow substances to move in and out of blood.
carbon monoxide Poisonous gas produced by incomplete combustion of carbon compounds.
carcinogens Substances that cause cancer.
cartilage A smooth tissue that covers the ends of bones at joints and so allows easy movement.
catalysts Substances, such as enzymes, that speed up a reaction without themselves being used up.
causal An occurrence or variable that results in a change to another variable.
cell membrane The part of a cell that controls the entry and exit of materials.
cell sap The liquid that fills the vacuole in a plant cell.
cell wall The outer part of a plant cell, giving it shape and strength.
cellulose A carbohydrate which is found in the cell walls of green plants.
central nervous system The brain and spinal cord.
centrifuges Instruments that spin materials at high speeds to separate them into components.
characteristics Distinguishing features.
chlorophyll A green pigment in plant cells that captures light energy for photosynthesis.
chloroplasts Structures in a plant cell that contain chlorophyll.
cholesterol A fatty substance made in the liver.
chromosomes Long threads containing many genes, found in the nucleus of a cell.
clinical trial Set of tests on people that provide data on the effectiveness of a drug.
clones Genetically identical organisms.
colony A large group of microbes.
combustion Rapid oxidation causing heat to be released so rapidly that a substance burns.
compost Decaying remains of plants.
concentration gradient When molecules move from a higher to a lower concentration, they are moving down a concentration gradient. When molecules move from a lower to a higher concentration, they are moving up a concentration gradient. Cells use energy from respiration to move molecules up a concentration gradient.
conductor A substance that is good at conducting thermal energy.
contraceptive pill A pill containing hormones that prevent an egg being released.
controls Parts of an experiment that show that effects are due to changes in the independent variable.
correlation A mutual relationship of interdependence between two or more things.

culture medium A solution or jelly containing nutrients for the growth of microorganisms.

cystic fibrosis An inherited diseases caused by a faulty recessive allele which causes excessive production of sticky mucus.

cytoplasm The substance in cells where many of the chemical reactions occur. These reactions are catalysed by enzymes.

decay (organic) The breakdown of dead and waste material.

decomposers Organisms that break down waste material and dead material.

deficiency The lack of a substance in the diet.

dehydration Removal of water.

denature To alter the shape of an enzyme molecule in such a way that it is prevented from working.

deoxygenated Blood which contains a low concentration of oxygen and a high concentration of carbon dioxide.

dependent variable The variable that is measured for each value of the independent variable.

detritus The dead remains of organisms.

diabetes A disorder in which the pancreas fails to produce enough insulin to control blood glucose concentration.

dialysis A process which separates small molecules from large molecules. Only small molecules can pass through a dialysis membrane by diffusion.

differentiated cells Cells which are specialised to carry out a specific job in the body.

diffusion The movement of molecules and ions from a region of higher concentration to a region of lower concentration.

digestion Breaking down large food molecules into smaller ones so that they can be absorbed.

DNA Deoxyribonucleic acid. A gene is a section of DNA.

dominant An allele which shows up in the phenotype when an organism is heterozygous or homozygous for it (see also recessive).

double circulatory system Blood in humans is pumped from the heart to the lungs, and then from the heart to the rest of the body. This keeps oxygenated and deoxygenated blood separate.

effector An organ or cell that brings about a response to a stimulus.

effluent A liquid that flows out after a process.

egg The female sex cell (gamete).

electron microscope A microscope that uses a beam of electrons rather than light rays.

embryo An undeveloped organism.

embryo transplants Transferring embryos from one organism and implanting them into another uterus (womb).

enzymes Biological catalysts that speed up the rate of reactions taking place in and around cells.

epidemic A disease that spreads rapidly across a country.

exponential growth Growth that gets more and more rapid.

faeces Waste matter remaining after food has been digested, which is discharged from the body.

fatty acids Substances formed when fats and oils are digested.

feedstock Plant material used in a chemical process.

fermentation A process carried out by microbes in the absence of oxygen.

fermenter A large vessel for growing microbes.

fluid Any substance which can flow: liquids and gases are fluids.

fossil record Evidence for evolution obtained from the fossil remains of plants and animals.

fossils Preserved remains of plants and animals that have not decayed.

fractional distillation A method of separation involving boiling a liquid, then condensing the vapour at different temperatures.

fungi Group of microorganisms including moulds, mushrooms and yeasts.

gametes Specialised sex cells involved in sexual reproduction in plants and animals.

gene therapy Treating inherited diseases by replacing a faulty gene.

genes Parts of a chromosome that control an inherited characteristic.

genetic cross The mating of organisms with particular phenotypes or genotypes.

genetic diagram A diagram depicting the results of a genetic cross by showing the types of gametes produced and the results of fertilisation.

genetic modification Modification of the genetic material of an organism (also called genetic engineering).

genetic screening Testing cells such as embryo cells for faulty alleles.

genotype The genetic make-up (alleles present) of an organism.

germinate To start growth of a seed or spore.

glands Small organs controlling bodily functions by chemical means.

glucose A simple sugar produced by photosynthesis and from starch by digestion.

glycerol A substance formed when fats and oils are digested.

glycogen A form of carbohydrate that is stored in muscle tissue and in the liver.

GM crops Genetically modified crops. These are crop plants that have had new genes added from another species.

greenhouse effect Gases such as carbon dioxide help to trap heat energy in the atmosphere.

guard cells Sausage-shaped cells which control the opening and closing of stomata.

Haber process The industrial process used to make ammonia from nitrogen and hydrogen.

haemoglobin The substance in red blood cells which combines with oxygen to form oxyhaemoglobin.

HDLs High-density lipoproteins which carry cholesterol.

herbicides Chemicals used to kill weeds.

heterozygous An individual is heterozygous when the alleles in a pair are different, e.g. Bb.

high blood pressure A higher than average value of pressure in arteries; associated with an increased risk of heart attacks and strokes.

homeostasis Regulation of the environment inside humans and other mammals to keep conditions at a steady level.

homogenised Milk in which the fat is dispersed as tiny droplets and does not separate out as cream.

homozygous An individual is homozygous when the alleles in a pair are identical, e.g. BB (dominant), and bb (recessive).

hormones Chemicals produced by one part of an organism, that control a process in another part of the organism.

Huntington's disease An inherited disease caused by a faulty dominant allele which damages the brain and other nerve tissue.

hydrolysis Breaking down a compound by chemical reaction, using water.

hyphae Microscopic thread-like structures that make up the bodies of moulds.

immune Protected against disease by the production of antibodies.

immunisation Injecting or swallowing vaccines to develop immunity.

immunised Given a vaccine containing dead or inactive pathogens, which stimulates the immune system to produce antibodies and memory cells.

immunity Protected against disease by the production of antibodies.

impulses Form in which information is transmitted by nerve cells.

***in vitro* fertilisation (IVF)** Fertilisation of an egg outside the body of a female by the addition of sperm, as a means of producing an embryo.

independent variable The variable whose value is selected by the experimenter.

ingest Take food into the body.

inherited Characteristics that are passed from parents to their offspring by gametes (eggs and sperm).

inoculate Inject with microbes (living or dead).

insecticide Chemical used to kill insects.

insulin A hormone produced by the pancreas that controls the concentration of glucose in the blood.

kidneys Organs that control the content of blood by filtering out waste substances such as urea, and regulating ion and water content.

Kyoto Treaty A treaty established in 1997, under the United Nations, which requires its signatories to reduce emissions of all greenhouse gases.

lactic acid A substance produced when glucose is broken down during anaerobic respiration.

lactose The principal sugar found in milk.

lagoons Large ponds.

landfill site Place where solid rubbish is dumped, then covered by soil.

LDLs Low-density lipoproteins which carry cholesterol.

limiting factor A factor such as light intensity that limits the rate of a process.

lipase A type of digestive enzyme that breaks down fats and oils into fatty acids and glycerol.

lipoprotein Protein that is combined with fats or other lipids.

magnesium A mineral that plants absorb as ions from the soil and that is used to make chlorophyll.
malnourished Suffering from inadequate nutrition.
meiosis A form of cell division that reduces the number of chromosomes and results in variation. This type of cell division forms gametes.
menstrual cycle The monthly cycle of changes in a woman's reproductive system, controlled by hormones.
metabolic rate A measure of the energy used by an animal in a given time period.
microbe Another word for microorganism.
microorganism An organism so small it can only be seen through a microscope.
microvilli Microscopic folds of the cell membrane which increase the surface area for absorption.
minerals Animals and plants use certain minerals to make substances such as proteins. Plants absorb minerals from the soil as ions.
mitochondria Structures found in the cytoplasm in which aerobic respiration takes place.
mitosis A form of cell division that forms cells identical to the parent cell.
motor neurone A nerve cell that carries information to an effector organ, such as a skeletal muscle.
MRSA A type of bacteria resistant to antibiotics (methicillin-resistant *Staphylococcus aureus*).
mutations Natural changes in a gene that may result in different characteristics.
mycoprotein Protein produced by moulds.
nerves Fibres consisting of bundles of neurone.
neurones Cells specialised to transmit electrical nerve impulses and so carry information from one part of the body to another.
nitrates A type of mineral absorbed from the soil by plants which is used to make amino acids and proteins.
non-renewable Once used it cannot be replaced.
nucleus The part of a cell that contains genetic information.
obese Very fat or very overweight.
osmosis The movement of water molecules from a region of higher water concentration to a region of lower water concentration through a partially permeable membrane.
oxygen debt The amount of oxygen needed to oxidise the lactic acid produced during anaerobic respiration.
oxygenated Blood which contains a high concentration of oxygen and a low concentration of carbon dioxide.
painkillers Drugs that reduce pain.
palisade cells Cells under the upper epidermis of a leaf, where most photosynthesis occurs.
pandemic A disease that is spread rapidly across many countries.
partially permeable membrane Membrane which allows only certain types of molecules to pass through.
pathogens Microbes that cause disease.
penicillin Antibiotic produced by the mould Penicillium.
pesticides Chemicals that kill pests such as insects.
phenotype The physical characteristics of an organism controlled by genes.
phytoplankton Microscopic organisms which can photosynthesise.
plasma The liquid part of blood containing dissolved substances such as glucose and urea.
potometer An apparatus used to measure the rate of transpiration in a plant shoot.
precursor A chemical used early in a process in which another chemical is made.
primary consumers Organisms that eat plants.
probiotics Foods that contain bacteria which are helpful to the body.
producers Green plants that use light energy to make food.
protease A digestive enzyme that breaks down proteins into amino acids.
pyramid of biomass A diagram that shows the mass of living organisms at each stage in a food chain.
receptors Organs or cells that are sensitive to external stimuli.
recessive An allele which shows in the phenotype only when an organism is homozygous for it (see also dominant).
red blood cells Blood cells which contain haemoglobin to transport oxygen.
reflex action Rapid involuntary response to a particular stimulus.
relay neurone A neurone that carries information from a sensory nerve cell to a motor nerve cell.
reliable Evidence that can be reproduced by other experimenters.

resistance A bacterium that has antibiotic resistance can't be destroyed by the action of antibiotics.
respiration A series of reactions in which living organisms release energy from food.
ribosomes The parts of a cell where protein synthesis occurs.
root hairs Extensions of the outer cells growing near the tips of plant roots to increase surface area for absorption of water and mineral ions.
secondary consumers Animals that eat animals that feed on plants.
sense organs Organs of the body which are sensitive to external stimuli.
sensory neurone A nerve cell that carries information to the brain or spinal cord.
sexual reproduction Biological process of reproduction involving the combination of genetic material from two parents.
specialised Each type of cell is adapted or specialised to its function.
species A group of organisms having some characteristics in common that are able to breed together.
sperm The male sex cell (gamete).
spontaneous generation A theory that living organisms can be produced from non-living materials.
starch The form of carbohydrate stored by most plants. Starch is made from glucose molecules joined together.
statins Drugs which act to reduce levels of cholesterol in the blood.
stem cells Cells, such as embryo cells, which can divide to form other types of cell.
sterile Containing no microbes.
stomata Small pores, usually found on the lower surface of leaves, which allow gases, such as water vapour and carbon dioxide, to diffuse in and out of the leaf.
substrate The molecules upon which an enzyme acts.
sustainable A process or development that can carry on for ever because it makes no net demands on the environment.
symptoms Effects of a disease on the body, such as high temperature and runny nose.
synapse A junction of two nerve cells.
synthesise To build up a chemical compound.
tertiary consumer An animal which eats animals that feed on other animals.
thermoregulation Maintaining a steady body temperature.
tissue culture A cloning technique used to grow groups of cells into new plants.
toxins Poisons produced by microbes.
transpiration The loss of water from leaves by evaporation.
transplant Removing an organ from a dead or living person and using it to replace a diseased or damaged organ.
tubules Tiny tubes found in the kidneys which filter waste substances from the blood and then reabsorb useful substances.
urea A waste product formed in the liver from excess amino acids, and removed from the blood in the kidneys.
vaccination Injecting or swallowing vaccines to develop immunity (also called immunisation).
vaccine Preparation made from dead or inactive pathogens, which can be injected so that white blood cells make antibodies to destroy live pathogens of that type.
vacuole A fluid-filled sac found in most plant cells.
variegated A leaf which has green areas containing chlorophyll and yellow/white areas with no chlorophyll.
vasoconstriction Narrowing of small arteries so that blood flow is lowered.
vasodilation Widening of small arteries so that blood flow is increased.
veins Blood vessels which carry blood, under low pressure, back to the heart.
villi Folds of the surface of the small intestine which increase the surface area for absorption.
viruses Types of microbes that cause disease; examples of such diseases are measles and the common cold.
water tables Levels below which the ground is saturated with water.
withdrawal symptoms Symptoms in a person with drug dependence that occur when they stop taking the drug or reduce the dosage.

Index